Undergraduate Lecture Notes in Physics

Undergraduate Lecture Notes in Physics (ULNP) publishes authoritative texts covering topics throughout pure and applied physics. Each title in the series is suitable as a basis for undergraduate instruction, typically containing practice problems, worked examples, chapter summaries, and suggestions for further reading.

ULNP titles must provide at least one of the following:

- An exceptionally clear and concise treatment of a standard undergraduate subject.
- A solid undergraduate-level introduction to a graduate, advanced, or non-standard subject.
- A novel perspective or an unusual approach to teaching a subject.

ULNP especially encourages new, original, and idiosyncratic approaches to physics teaching at the undergraduate level.

The purpose of ULNP is to provide intriguing, absorbing books that will continue to be the reader's preferred reference throughout their academic career.

More information about this series at http://www.springer.com/series/8917

Andy Lawrence

Probability in Physics

An Introductory Guide

Springer

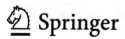

Andy Lawrence
Institute for Astronomy
University of Edinburgh
Edinburgh, UK

Royal Observatory
Edinburgh, UK

ISSN 2192-4791 ISSN 2192-4805 (electronic)
Undergraduate Lecture Notes in Physics
ISBN 978-3-030-04542-5 ISBN 978-3-030-04544-9 (eBook)
https://doi.org/10.1007/978-3-030-04544-9

This Springer imprint is published by the registered company Springer Nature Switzerland AG
The registered company address is: Gewerbestrasse 11, 6330 Cham, Switzerland

In Metaphysics, in Science, and in Conduct, most of the arguments, upon which we habitually base our rational beliefs, are admitted to be inconclusive in a greater or less degree. Thus for a philosophical treatment of these branches of knowledge, the study of probability is required.
(John Maynard Keynes, "*A Treatise on Probability*", 1921.)

Preface

The careful reader will note that the word 'statistics' does not appear in the title of this book. Althought Part III does cover the subject of Statistical Inference, this book is both more than and less than a textbook on Statistics. It is more than a Statistics textbook, in that it's purpose is to discuss how the concept of probability is central to modern physics, and how frequency distributions and probabilistic concepts arise in a wide range of physical problems. I discuss things you won't normally see in Statistics textbooks such as Information Theory and Quantum Measurement. On the other hand, it is less than a Statistics textbook, in that it is not a cookbook; it is quite brief on how to apply statistical techniques in model fitting and so on. You won't find a step-by-step explanation of how to do analysis of variance.

Aims of the Book

(a) I want the reader to understand the *basic principles* of how probabilistic concepts are used in physical problems. You won't find all the gory detail you need for applications in industry or research, but once you understand the principles, advanced technical works should be easier to follow—and to critically assess.
(b) I try to strike a sane balance between the "Bayesian" and "Frequentist" approaches to statistical reasoning. My experience is that to mathematicians, this division is a bit of a non-issue, but that Physicists and Astronomers still tend to treat it as kind of holy war. I take the attitude that you certainly can and should assign sliding-scale degrees-of-belief numbers to hypotheses, and adjust them according to the evidence; but also that it is often helpful to use the ideas and methods of significance testing. It's all fine as long as you are aware of what you are doing.
(c) As in my last book, *Astronomical Measurement*, I try to supply *the missing middle*, for my own benefit, but hopefully also for the benefit of the reader. For example, statistics books tend to be either simple cookbooks, which never really prove or explain anything, or rather terrifying full-blown mathematical

textbooks. There seems to be no bridge between the two. Likewise, some of the topics I cover, such as Quantum Entanglement, are covered either in hand-waving semi-popular articles, or in very detailed research papers, with the former really not preparing you for the latter. Where feasible, I try to give mathematical proofs, but really my aim is boiling down and de-mystifying.

An Overview of the Book

The book is in four parts, each with several chapters. In **Part I: The Basics** we start with a discussion of what probability is, where randomness comes from, and how to do probability algebra. We then take a look at probability distributions, moments, expectations, and how to deal with more than one random variable.

In **Part II: Frequency Distributions in the Physical World** we begin with the underlying ideas of combinatorics—how we count the different ways of doing things—and then move on to see how following that logic leads us to two important distributions that occur throughout the natural world—the Binomial and Poisson distributions. Next, we look at the single most important natural distribution—the Gaussian or "normal" distribution—and discuss how it arises as a quasi-universal distribution through the combination of many different random factors. We also look at how some distributions are Gaussians-in-disguise, for example, the Maxwell–Boltzmann distribution, and the log-Gaussian distribution. Finally, in Part II, we look at distributions that arise through random processes in time—random walks, shot noise, the Lorentzian, and power law distributions.

In **Part III: Probabilistic Inference: Reasoning in the Presence of Uncertainty**, we study the ideas and methods of statistical inference. Whereas in Part II, we learn how to calculate the expected observations from our knowledge of the physical situation, now we learn how to work backwards from observed data to the physical situation. First, we look at the basic ideas of hypothesis testing, emphasising the central role of likelihood, and using both Bayesian and significance-testing methods. Then we look at how to estimate a parameter, seeing it as a tuning knob that produces a continuous range of hypotheses, and how to calculate an uncertainty range on our estimate. Next, we look at how to assess the relationship between two variables, calculating our degree of certainty on whether they are connected at all, and learning how to produce "best fit" lines that model the relationship between them mathematically. Finally, we move on to more general model fitting, examining numerical methods of finding best fits in a multidimensional parameter space, and looking at the notoriously confusing issue of different types of multi-parameter uncertainty intervals. (Its not so confusing really!)

In **Part IV: Selected Topics**, we look at four selected areas of Physics, or the natural world more generally, where the role of probability is particularly important, interesting, or controversial. First, we look at Information Theory, stressing how it is at root just a different way to look at probability, but with very interesting implications and applications. These ideas are at the core of modern information technology. Next, we take another look at stochastic processes, and the problem of how to treat the erratic semi-random variations in time that we see throughout Nature, from quasars to the stockmarket. Following this, we look at Quantum Physics, the emergence of true randomness in the macroscopic world, and the rather strange way that probability works through complex amplitudes and partial correlations. We attempt to demystify Quantum Entanglement and Bell's Theorem, but make no claim to have solved these deep mysteries! Finally, we examine another area that seems to be enduringly controversial—the relationship between Entropy, Complexity, and the Arrow of Time. As with our look at Quantum Physics, our aim here is to simplify and demystify what other authors are talking about, without claiming to arrive at a final solution.

Exercises

Each chapter concludes with a set of exercises. These are mostly of the "plug-in" variety, rather than being factual or conceptual. The idea is that if you have followed the material you should be able to do these exercises, but if you haven't followed they will seem mysterious. Solutions are provided. If you are using the book as part of a structured course, then doing the exercises is highly recommended. Even for a general reader, it can be useful to take a look, as the solutions sometimes spell out things more slowly that may be a little condensed in the text.

Edinburgh, UK
June 2019

Andy Lawrence

Acknowledgements

This book grew out of a course on "Fourier Analysis and Statistics" given to Junior Honours Physics and Astronomy students at Edinburgh University. I am grateful to my co-conspirators on that course, John Peacock and Jorge Peñarrubia. The Statistics half of that course was in turn a longer but simpler version of a course that was given to the fourth-year Astronomers, "Astronomical Statistics", which was gradually evolved over many years by a stellar list of valued colleagues, including Lance Miller, Alan Heavens, Andy Taylor, John Peacock, and Peder Norberg.

Several colleagues have helped me with example datasets used in various places in the book. Nigel Hambly pointed me towards the stellar mass function datasets used in Fig. 5.8. Mikhail Bashkanov provided Δ^{++} cross-sectional data used in Fig. 6.3. Aaron Clauset clarified some issues with the city size and earthquake magnitude datasets used in Figs. 6.6 and 6.7. John Magorrian helped with the black hole mass versus bulge mass data of Fig. 9.1, and made some alternative suggestions which I didn't use in the end. Alastair Bruce reduced the data for the quasar light curve data shown in Fig. 10.2, and produced the plot in Fig. 10.4 as part of his Ph.D. thesis.

Finally, I am especially grateful to many smart Ph.D. students who have been involved in tutoring these courses, and who have helped to make the material sharper and clearer—especially David Edwards.

Contents

Part I
The Basics

In this first part of the book, we investigate what it means to be *random*, what *probability* means, and how to deal mathematically with random variables that may represent real quantities in nature. We meet the idea of a *probability distribution*; discuss how we can characterise the shapes of such distributions; and look at how we can capture mathematically the idea of "error". These basic techniques set us up for the later parts of the book.

Chapter 1
Randomness and Probability

1.1 Outline of Content

- The origin of unpredictability
- Probability as frequency
- Combining probabilities
- Probability as degree of belief
- A look ahead

The ideas of probability and statistics allow us to discuss situations in Nature that we cannot fully describe or predict. Ideally, this would not be necessary. Underlying classical physics (as opposed to quantum mechanics) is the belief that natural events are *deterministic*—that one event is strictly caused by another. If we know the starting condition of a system, and the physical laws that govern it, then we can "turn the handle" and exactly predict its condition at all future times. In fact we can look back in time and know exactly what the condition of the system was at all times in the past. Real life isn't normally like this however. If I plan to measure the velocity of a single molecule in a box of gas (perhaps using some particle counting time-of-flight device) I cannot tell you what value I will get. However, this does not mean that I can say nothing. If I know the temperature of the gas I can have a good idea of what values are likely and what values are unlikely. What do I mean by 'likely'? Can we sharpen that idea up? We will need to examine three closely related but not quite identical ideas—determinism, predictability, and randomness. Let's start by looking at where unpredictability comes from.

1.2 Where Does Unpredictability Come From?

The concepts of unpredictability, randomness, and probability are closely related but not quite the same. The idea of unpredictability is the least ambiguous, so lets look at that first. Unpredictability can arise in several different ways.

A. Lawrence, *Probability in Physics*, Undergraduate Lecture Notes in Physics,
https://doi.org/10.1007/978-3-030-04544-9_1

1.2.1 Incomplete Knowledge

Sometimes we simply don't know everything about the state of a system. If we roll a die,[1] in principle we could predict the side that will land face up. We would need to know the size and density of the die, the angle and velocity at which we throw it, and the friction of the surface it lands on. But in reality we don't know some of these things, so the number rolled is not in practice predictable.

1.2.2 Large Numbers

Any reasonable volume of gas contains an extremely large number of atoms. With the best will in the world we will never know the velocities of all the atoms. As far as any observer is concerned, the individual velocites are unpredictable. However, this doesn't mean we know nothing. What we see in Nature is that the relative numbers of atoms with different relative velocities follows a very precise law. In fact the exact same law applies to lots of other apparently unrelated things in Nature, like the distribution of heights of people. In Chap. 5 we will look at how to explain this strange universality.

1.2.3 Sensitivity to Initial Conditions

Suppose two systems are identical but with the initial conditions differing by a small amount. What happens as these systems evolve? In some systems the tracks converge, i.e. the difference gets smaller. But some diverge—the two systems get further apart. This makes the system predictable on short timescales but effectively unpredictable on long timescales, because we can never know the starting conditions *accurately* enough to keep our prediction on track. (This is the famous "butterfly effect" in climate modelling). Some systems end up bouncing erratically between values—this is the phenomenon of *dynamical chaos*.

1.2.4 Open Systems

Another practical problem is that we may know everything about the system we are trying to study, but it is not a perfect closed system—it may be subject to external influences. For example, the trajectory of a tennis ball may in principle be simple, even including the effects of air friction, but not if it's windy. Events starting a long

[1] The first of many controversial issues we shall face is whether the singular of *dice* is *die* or *dice*! I have taken a fairly arbitrary decision...

way outside the tennis court—for example, a depression over the North Atlantic—can end up changing the behaviour of the tennis ball.

1.2.5 Quantum Mechanics

The theory of quantum mechanics proposes that the Universe is *not* after all deterministic—events at the microscopic level are *intrinsically* unpredictable, different every time we look. As well as helping to explain the strange behaviour of the quantum world, this can have consequences in the macroscopic world—the times at which a lump of radioactive material emits an alpha particle cannot be predicted. Historically, some scientists have been uncomfortable with the intrinsic unpredictability of quantum mechanics, suggesting that there must be "hidden variables" that we don't yet know about, that would restore Nature to being deterministic. This is is a question which is in principle decidable by experiment. We will look in more detail at these issues in Chap. 13.

1.3 Randomness and Probability

We have seen that situations can be deterministic but still in practice unpredictable. What does it mean for an event to be *random*? You will encounter subtly different usages of this word. We will take it to mean "intrinsically unpredictable". Then situations where events are in-practice unpredictable can be treated as if they were random, and we can use the mathematics of probability theory to have rational discussions about uncertain events. If you are a determinist, you could take the attitude that the probability theory is a mathematical abstraction which we can use to discuss things which in reality are never truly random. Alternatively, you might care to believe that unpredictability is the true Nature of reality, with the apparent determinism of the macroscopic world being a kind of illusion. These are deep waters! Luckily, we can develop our theory and solve practical problems without having to come to a decision on these philosophical issues.

You will sometimes see the term "random" used to mean, that of all the possible values that an event might have, each is equally likely. This is not what we will mean by "random" in this book. If you roll one die, the numbers 1–6 are indeed all equally likely, but if you roll two dice, the various summed values are *not* equally likely. We want to use the term "random" for both these experiments. Below, we spell this out more carefully.

Putting quantum mechanics to one side for now, is it the case that randomness is *subjective*? Person A might in principle know everything necessary about some system, whereas Person B lacks some knowledge and has to treat it as behaving unpredictably. But this doesn't mean that randomness is woolly and not physical. Its a fact of life that sometimes we have incomplete knowledge, but we still want to have some way to have a rational discussion. "Subjective" here just means "observer dependent". Later on we will see how to adjust probabilities numerically when our knowledge changes. Furthermore, when considering for example what the velocity of a particle in a box full of gas will be, *every* plausible observer is in the same situation, and we can make very concrete and reliable statements about, for example, the distribution of velocities in a gas. However, the underlying subjectivity of randomness does mean that you have to specify the problem you are considering very carefully. Confusion in statistics often arises because there are some background assumptions that haven't been spelled out.

So what do we mean by "probability"? It's a concept that tries to answer questions about the future. Imagine standing in front of an experimental apparatus, about to make a measurement, and wondering whether you will get a particular value x. Based on what you have seen in the past, and know about your experiment, what is reasonable to expect? Historically, the word "probability" has been used for two different concepts related to this problem—either the *frequency* with which unpredictable events occur, or the *degree of belief* that some hypothesis is correct. These approaches are known as the "frequentist" and "Bayesian" approaches to probability theory. There have been long and heated debates between these two camps, but in fact both approaches are useful, and indeed as we shall see later, we can combine them mathematically. It is very useful to understand both approaches. The frequency approach gets us to the key mathematics more quickly, so lets look at that first.

1.4 Probability as Frequency

It is useful to start with the idea of a *random variable*. This is a quantity X whose value is different each time you enquire of it. (Or rather, could be different—sometimes you may happen to get the same value twice in a row.) This enquiry usually involves some kind of *event* which is the outcome of a measurement or experiment. For example, we roll a die and note the number which lands face up. It can sometimes be useful to distinguish between the event (seeing the face with a six land upwards), the value attached to the event ($X = 6$), and the list of possible outcomes ($X = 1$, $X = 6$ etc). Sometimes an event may be a *compound event* made up of *elemental events*. For example we might roll two dice and look at the total on the two dice, and could see this as equivalent to combining the results of rolling a single die twice. The variable can be *discrete*, such as the number rolled with a die, or the number of heads out of nine tosses of a coin; or *continuous*, such as the velocity of a gas particle or the height of a person. Usually we will use upper case for a discrete variable and lower case for a continuous variable.

1.4.1 Frequencies

Although we cannot predict the outcome of a single specific measurement, we can note the number of times various values come up. Suppose we perform an experiment that has n different possible *outcomes*, which we can label X_i, where $i = 1$ to n. The set of all possible outcomes $X_i = X_1, X_2, \ldots X_n$ is known as the *sample space* S—for example when rolling a six-sided die, the list of possible outcomes is $S = \{X_1 = 1, X_2 = 2, \ldots, X_6 = 6\}$. Suppose now we run the experiment a total of N times, giving a long series of *data values*, which we can label x_k, where $k = 1$ to N. Each run of the experiment gives a data value which is equal to one of the expected outcomes. Suppose now we count how often $x_k = X_i$, and find that ith outcome X_i is seen N_i times. Then the set of N_i values can be referred as the *observed frequencies* and the *normalised observed frequencies* are given by $o_i = N_i/N$. Note that

$$\sum_{i=1}^{n} o_i = 1.0.$$

If the total number of experiments N is large enough, we might hope that our observed frequencies o_i converge on a set of values f_i that characterise the experiment we are carrying out. These numbers, seen as a function of the X_i outcomes, define the *normalised expected frequency distribution* $f_i(X_i)$, or just $f(X)$ for short. In some cases we can predict what $f(X)$ should be. For example, for our rolled die, assuming that the die is unbiased, we can assume that each outcome $X_i = 1, 2, \ldots 6$ is equally likely—so we expect that $f_i = 1/6$ for all the possible outcomes. Of course, in real finite experiments, the observed o_i values won't be quite the same as the theoretically expected f_i values. In Chap. 2, we will take a closer look at the difference between observed and expected frequency distributions. Note that both o_i and f_i always sum to 1.0.

In many cases, the f_i for the various outcomes X_i won't all be the same. However, we can work out the expected f_i values by considering our events to be compound events made up of elemental events which are themselves equally likely. The idea is illustrated in Fig. 1.1. Suppose we roll two dice. The set of all elemental outcomes is

$$S = \{(X, Y) \mid X = 1, 2 \ldots 6,\ \ Y = 1, 2 \ldots 6\}$$

i.e. the set of all 36 pairs of values X, Y such that X can be any of 1–6 and likewise for Y. But suppose we are interested in how often we get a given total $T = X + Y$. The set of all ways we can achieve the event $T = 8$ is

$$S_8 = \{(2, 6), (3, 5), (4, 4), (5, 3), (6, 2)\}.$$

This is 5 out of the 36 members of the full set of elemental outcomes, so the frequency with which we expect the event $T = 8$ is $f(T = 8) = 5/36$. We have assumed that all the elemental outcomes occur with equal frequency; but the various possible events T do *not* occur with equal frequency. There is only one way of getting a 12, and no ways of getting 1.

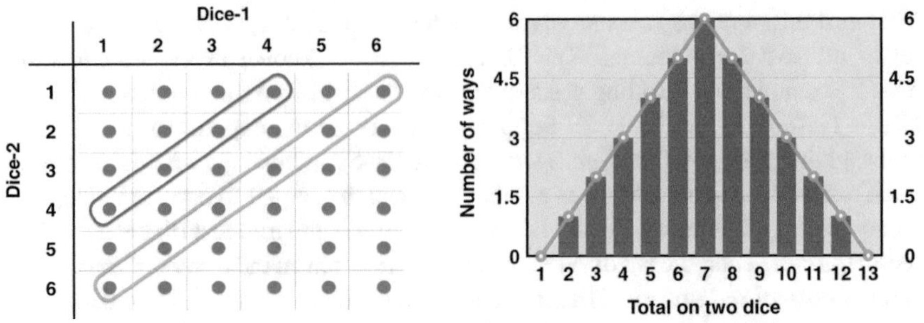

Fig. 1.1 **Left**: Sample space of elemental outcomes for rolling two dice. The green (lower) ring shows the set of all outcomes that give a total of $T = 7$, and the red (upper) ring shows those that give a total of $T = 5$. **Right**: Frequency distribution $n(T)$. To turn this into the normalised frequency distribution, we would divide by 36

1.4.2 Using the Frequencies as Probabilities

The frequency distribution $f(X)$ is then what you will find in the long run over many repeated measurements or experiments. From the viewpoint of a single measurement or experiment, the same function tells us the *probability* that we will get a particular value X. As you wait for a measurement to take place, you could imagine many potential realisations of your experiment. What fraction of the time does a particular value occur in those imagined realisations? We can then think of the expected normalised frequency distribution $f(X)$ as the *probability distribution* $P(X)$. Part II of the book is essentially about how we can work out what the probability distributions $P(X)$ should be in various different physical situations.

1.4.3 Continuous Random Variables: Frequency Densities

If we have a continuous random variable x, then the fraction of times we get exactly x is of course infinitesimal, so instead we have to define the *normalised frequency density* $f(x)$, defined such that the fraction of times we get values in the range x to $x + \mathrm{d}x$ is $f(x)\mathrm{d}x$. (See the illustration in Fig. 1.2.) Just like for discrete variables, the frequency density functions for continuous variables are properly normalised if $\int_{-\infty}^{+\infty} f(x)\mathrm{d}x = 1.0$. Likewise, analogously to discrete variables, we can use past observed or theoretically expected frequency densities as our estimate of probability density $p(x)$. The function $p(x)$ is known as the probability density function (PDF).

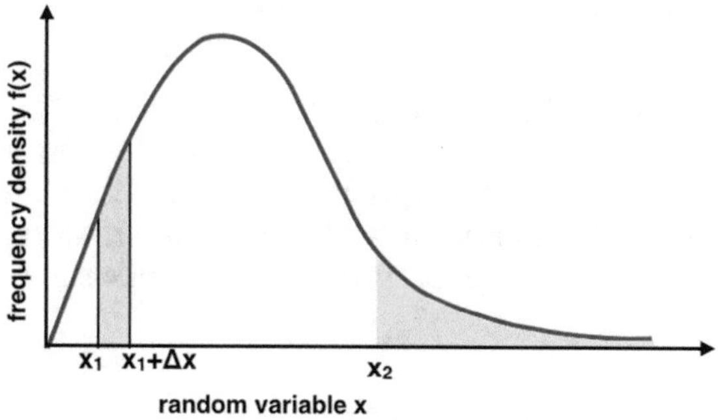

Fig. 1.2 The curve represents some frequency density $f(x)$. For small $\mathrm{d}x$, $f(x_1)\mathrm{d}x$ is the probability that x is in the range between x_1 and $x_1 + \mathrm{d}x$. The right-hand shaded area shows the probability that $x > x_2$

1.4.4 The Importance of Relative and Integrated Probabilities

The absolute values of probability or probability density are often not very useful, and what we need are relative values, or values integrated over a range.

For example, suppose you have an urn (statisticians love urns!) containing N balls. One ball is red, two balls are green, and the rest are white. The probability of picking a red ball is roughly 0.0833 if $N = 12$ and 0.00114 if $N = 876$, but in both cases you are twice as likely to pick a green ball as a red ball. So you can see that in this case the ratio of the probabilities is much more useful than the absolute values; this very often turns out to be the case.

For continuous random variables, the probability density depends on the units you choose. An example is the distribution of human heights. For example, a table given by the US Centers for Disease Control (CDC; see "Further reading" in Sect. 1.9) shows that the average female height in the USA is $h = 162.1$ cm, or if you like 1.621 m. The probability density function $p(h)$ has a typical "bell curve" shape with a characteristic spread of about ± 9.8 cm (we will be more precise about this shape in Chap. 5). The *density* of probability near the average depends on the units you choose—its about 0.041 per cm, or 0.0041 per m, or even 0.016 per inch if you prefer. However, if you ask "what fraction of women have a height $h > 169$ cm?" you will find

$$F = \int_h^\infty p(h) = 0.15,$$

and the answer is the same whether you measure h in metres, millimetres, inches, or light-years. When we come to studying statistical inference, this lesson that either the ratio of probabilities, or the integral of probability over a range, is more useful than point density values, is particularly crucial.

The integral of the probability density function $p(x)$ is sometimes known as the *cumulative distribution function (CDF)* $F(x)$.

1.5 Combining Probabilities

If we know the probabilities of two events A and B, what is the probability of getting *both* A and B? Or of getting *either* A or B? To answer such questions, we have to be quite careful about specifying the experiment we are conducting. We also have to decide whether the two events are *dependent*, and whether or not they are mutually *exclusive*. This is easiest to see if we step through an example.

1.5.1 *Probability of* A *and* B

To get the probability of A and B, we need to know whether the two events are *dependent* or not. Suppose we have a pack of 52 playing cards with four Aces, and we pick a card at random, twice. A is the event of getting an Ace in the first pick; B is the event of getting an Ace in the second pick. However, we could do this two different ways:

Experiment-1. After the first pick we put the chosen card back into the pack and shuffle it, before making the second pick. In this case, the two events A and B are *independent*—to know the probability of picking an Ace in the second pick, we don't need to know what happened in the first pick. If we imagine doing the first pick many many times, we will get an Ace a fraction 4/52 = 1/13 of the time. In the second pick, we will get an Ace again for a fraction 1/13 out of the cases where we got an Ace the first time. So, fairly obviously we have

$$P(A \text{ and } B) = P(A, B) = P(A) \times P(B).$$

Experiment-2. After we have picked the first card, we put it to one side before making the second pick. Then the probability of getting an Ace in the second pick clearly *does* depend on what happened in the first pick. If the first pick was not an Ace, then $P(B) = 4/51$. If the first pick was an Ace, then $P(B) = 3/51$. To express the probability of B given that A already happened, we write $P(B|A)$. This is known as a *conditional probability* and can be read as "probability of B given A". In this more general case we have

$$P(A \text{ and } B) = P(A, B) = P(A) \times P(B|A). \tag{1.1}$$

Note that if A and B are independent, then $P(B|A) = P(B)$ and (1.1) becomes the simple $P(A) \times P(B)$ formula.

We have stepped through this simple example rather laboriously, but its good practice; the most common source of error in thinking about probability problems is rushing things. The key lesson is that you have to define your problem very carefully.

1.5.2 Combining Continuous Variables

Note that the same logic works for probability densities. If we have two continuous variables x and y we can define a *joint probability density* $p(x, y)$ such that the probability that x is in the range x to $x + \mathrm{d}x$ *and* y is in the range y to $y + \mathrm{d}y$ is $p(x, y)\mathrm{d}x\mathrm{d}y$. In general $p(x, y)$ might be some complicated 2D function, but if x and y are independent, then it decouples into separate functions for x and y so that

$$p(x, y) = f(x)g(y),$$

where $f(x)$ is the PDF for x alone etc.

1.5.3 Bayes's Theorem

The conditionality works both ways, so we can write either

$$P(A, B) = P(A)P(B|A) \quad \text{or} \quad P(A, B) = P(B)P(A|B).$$

Equating the two versions and re-arranging, we get *Bayes's Theorem*

$$P(B|A) = \frac{P(A|B)P(B)}{P(A)}. \tag{1.2}$$

As it stands, this is a fairly innocuous re-statement of the laws of probability, and is sometimes useful for practical calculations. However, it re-appears in a more interesting way when we come to talk about statistical inference later in the book.

1.5.4 *Probability of* A *or* B

In this case, the probability of getting either A or B depends on whether A and B are *mutually exclusive*, i.e. whether or not it is possible for *both* events to occur. For example, suppose A is the event that the Scottish football team reaches the final of the football World Cup, and B is the event that Wales reaches the final of the football World Cup. It is perfectly possible for both those things to happen (even if each of them is somewhat unlikely!). On the other hand, if event A is Scotland *winning* the World Cup in some specific year, and event B Wales winning the World Cup the same year, then clearly it is impossible for both these things to happen—not just very unlikely, but logically exclusive—only one team can win.

It is helpful to think of such "either/or" problems graphically as Venn diagrams, as illustrated in Fig. 1.3. The box represents the sample space S, i.e. the set of all

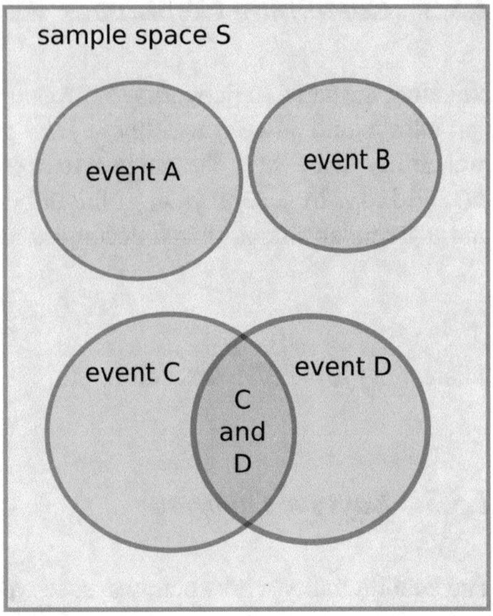

Fig. 1.3 The difference between exclusive and non-exclusive events. The light grey box represents the set of all elemental outcomes, and our events are composed of subsets of these elemental outcomes. Events A and B are mutually exclusive, whereas events C and D have elemental outcomes in common

elemental outcomes, and the circle A as the area representing the set of all outcomes giving the event A. The probability of A is the area of A, normalised to the area of S. For *mutually exclusive* events A and B, the sets do not overlap, and $P(A \text{ or } B) = P(A) + P(B)$. However, if some elemental outcomes could give either A or B, then the areas overlap. In this case

$$P(A \text{ or } B) = P(A) + P(B) - P(A, B) \tag{1.3}$$

i.e. to get the probability of either event occurring, we add the probabilities of each, but correct for the times where both occur. Lets look at an example. Suppose we roll two dice, and consider two different questions:

Question-1. Suppose event A is *getting a total of* $T = 9$ and event B is *getting a total of* $T = 5$. What is the probability of A or B? Clearly A and B cannot both occur—the two events are *mutually exclusive*. There are 4 ways to get event A ($T = 9$), and 5 ways to get event B ($T = 5$), so that there are $4 + 5 = 9$ ways to get either A or B, and so $P(A \text{ or } B) = 9/36 = 1/4$.

Question-2. Now suppose event C is *getting a total of* $T = 12$ and event D is getting a double, where both dice are the same. What is the probability of C or D? There is only one way to get 12, so $P(C) = 1/36$. There are six ways to get a double, so $P(D) = 6/36$. However C and D are not exclusive—rolling a double six is an example of both. The double six is already included in the calculation of $P(D)$. So $P(C \text{ or } D) = 6/36 + 1/36 - 1/36 = 6/36$.

1.5.5 Notation

If we see events as sets of elemental outcomes, then the event "C and D" is the intersection of set C and set D, and so we could write $P(C \text{ and } D)$ as $P(C \cap D)$, and likewise the event "C or D" is the union of sets C and D, and so we could write $P(C \text{ or } D)$ as $P(C \cup D)$. However, in this book we will normally use $P(C, D)$ for the probability of C and D, and $P(C \text{ or } D)$ for the probability of C or D.

1.6 Probability as Degree of Belief

Historically, the word "probability" has been attached to two different concepts—either the frequency of particular outcomes of an imagined ensemble of experiments, or the degree of belief in a hypothesis. We have already looked at frequency—now lets look at the other concept.

We need to start with the idea of a *hypothesis*. A hypothesis H is just the assertion that some statement is true—"it will rain today", or "the velocity of this particle is less than 3 m/s", or "the ball selected will be red". Some people argue that the sensible approach to hypotheses is to consider that they are always strictly either true or false. We could attach *truth values* of 1 or 0 to a hypothesis. If we had complete knowledge, we would know which statements are true and which are false. The problem of course is that we don't have complete knowledge, and so cannot make definite and completely reliable statements about truth and falsehood. However that doesn't mean we can say nothing.

In real world reasoning we instinctively find ourselves expressing degrees of belief in hypotheses, depending on the evidence we have. If we see dark clouds above, we have a stronger belief in the statement "it will rain today" than we do if the sky is blue. Perhaps rather than just 1 or 0, we could attach a numerical value to a hypothesis on a sliding scale between 1 (definitely true) and 0 (definitely untrue). We could call such a number the *plausibility* of a hypothesis. The trouble is, there are endless different ways to do this. You could imagine stretching the sliding scale in many different ways. However, it has been shown by Cox and Jaynes and others that if you require your plausibility numbers to follow the same rules as frequencies of events, you get a unique well defined scale, which we can call the *credibility* C of a hypothesis. To spell it out, this means

- If we have a stronger belief in H_1 than H_2 we must have

$$C(H_1) > C(H_2)$$

- If we can list all the possible hypotheses H_i that can apply in a given situation, then we must have

$$\sum_i C_i(H_i) = 1.0$$

- If we have two hypotheses H_1 and H_2, then the credibility we attach to the hypothesis that *both* H_1 and H_2 apply should be

$$C(H_1, H_2) = C(H_1)C(H_2|H_1)$$

- Likewise, the credibility attached to the hypothesis that *either* H_1 or H_2 apply should be

$$C(H_1 \text{ or } H_2) = C(H_1) + C(H_2) - C(H_1, H_2)$$

You will note that the way we have defined our credibilities, they behave just the same way as the concept of probability that we derived from the idea of frequencies, and so we might decide simply to think of credibility as the "probability of a hypothesis". In Part III of the book we will look at how we can use the observations we make to *adjust* our credibilities/hypothesis probabilities. When we do this, we apply Bayes's theorem. For this reason, such techniques are traditionally known as Bayesian methods, and the whole approach of using a sliding scale for the probabilities of hypotheses is known as the Bayesian approach.

1.6.1 What Is the "Correct" Definition of Probability?

To some extent this is a non-question. The ideas of "frequency of an imagined ensemble" and "degree of belief" are both valid and useful concepts, and it is a historical accident that the same word has been attached to both. A strict "frequentist" would deny that credibilities have any meaning. A strict Bayesian would say that the probability of an event should actually be interpreted as your degree of belief in the hypothesis that the event will occur, and that experimental frequencies are just a way we can estimate such hypothesis probabilities. However, mathematicians can put both approaches into a common framework, and scientists can and do use both frequentist and Bayesian concepts to solve problems, and given that event frequencies and hypothesis credibilities follow the same algebra of combination, we can mix and match them in calculations. This will become clearer when we look at examples in Part III.

For the rest of this book, we will mostly use "probability" for both events and hypotheses. However sometimes we will revive the term "credibility" to remind ourselves of the conceptual distinction between frequencies and degrees-of-belief.

1.7 Look Ahead

We have established the basic ideas of random variables and probabilities, and how they combine. In the second chapter of Part I we will look more in more detail at the properties of probability distributions, and how we analyse errors, to give us some mathematical tools for the rest of the book. The remainder of the book divides into three main chunks.

In Part II, we "turn the handle forward", that is, given a variety of different physical circumstances, we work out what frequency distributions we expect to see. So for example, if we have a hot gas, and we know its temperature, and then make some particle velocity measurements, how often should we expect to see various different velocity values?

Next, in Part III, we look at how to "turn the handle backwards"—how we can deduce what the physical circumstances are, from the observed data. This is known as the problem of *statistical inference*. So for example, if we make some measurements of a few particle velocities, can we now work out what the temperature must be? The answer will be "yes, but not with any certainty", so we also need to understand how we get uncertainties on our estimates.

Finally, in Part IV, we look at how probability theory is crucial in several key areas of physics, often arising in the most historically controversial areas, such as quantum mechanics, entropy, and the arrow of time.

1.8 Key Concepts

Some of the key concepts from this chapter are:

- Unpredictability comes from incomplete knowledge
- We can characterise the values attached to events as random variables—either discrete or continuous
- Often, what we want is the frequency distribution of events/ random variable values
- To equate frequencies to probabilities, they need to be normalised, so that

$$\sum_i P_i = 1.0 \quad \text{or} \quad \int_{-\infty}^{+\infty} p(x)\mathrm{d}x = 1.0$$

- To combine the probabilities of events, we need to know whether they are dependent, and whether they are mutually exclusive
- We can attach a number to the degree of belief in a hypothesis, and think of this as its probability or credibility
- Point values of probability density can be a bit unhelpful. Its better to look at relative values, or probability integrated over a range.

Make sure you can understand the key formulae: how to calculate the probability of getting both of two events (1.1), the closely related Bayes formula (1.2), and how to calculate the probability of getting either of two events (1.3)

1.9 Further Reading

There are course many good textbooks on the basics of probability and statistics, so this is a small personal choice. For those who have had little exposure to statistics before, a good choice is Clarke and Cooke (2011). This starts at high school level, but covers a lot of what we need by the end, although not always with mathematical proof. A beautifully clear and quite short book which covers much of our material at the right level is Bulmer (2003). For a thorough and rigorous mathematical treatment, there is a series of excellent books by John Freund and various collaborators and acolytes. A recent incarnation is Miller and Miller (2013), which is really a reworking of Freund and Walpole (1986). Lupton (1993) is also useful.

Most of the above books are centred on the traditional "frequentist" approach—the Bayesian approach tends to be covered in research literature. A prominent exception is Sivia and Skilling (2006), which is very simple and readable. If you are interested in where these ideas came from, I strongly recommend Keynes (2010; originally 1921), Jeffreys (2000; originally 1939), and Jaynes (2003; originally 1967). The research paper credited with first establishing that seeing probability as "reasonable expectation" leads to a unique algebra is Cox (1946).

From the mathematician's point of view, modern probability theory sees no essential difference between the frequentist and Bayesian points of view. The mathematician defines a "probability space" as a set S of elements which can be combined using the rules of a Boolean algebra $B(S)$, and which have an associated probability measure P, of total mass 1. However this rather sweeps the question under the carpet, because, if the elements of our set are hypotheses, it avoids the question of whether it is philosophically legitimate to associate a probability measure with those elements. A good clear introduction to the mathematical approach to probability is given in the early chapters of Applebaum (2008), who also provides a good set of references. The origin of much of twentieth century probability theory is in the classic short book first published by Kolmogorov in 1933. (The references list a 2018 reprint).

There are also course many websites providing useful statistical resources. I won't attempt to detail all these. A comprehensive listing of resources is provided at the statpages.info website. Two resources I find particularly useful are the random.org website which provides a variety of random number generators, and geogebra.org, a very general online mathematics tool, which in particular provides a variety of probability calculators.

1.10 Exercises

1.1 A deck of cards has 52 cards, 4 of which are aces. You draw two cards at random. What is the probability of drawing two aces if (a) the first card is replaced, and (b) the first card is not replaced?

1.2 You roll a die. What is the probability of the result being divisible by three? You then roll two dice and examine the total. What is the probability of the result being divisible by three?

1.3 A farmer leaves a will saying that they wish their first child to get half of their property, the second child to get a third, and the third child to get a ninth. As they have left seventeen horses, the children are distressed because they don't want to cut any horses up. However a local statistician lends them a horse so that they have eighteen. The children then take nine, six, and two horses respectively. This adds up to seventeen, so they give the statistician her horse back and everybody is happy. What is wrong with this story?

1.4 From the records of a dental practice, it is found that when a patient visits, the probability s/he will have their teeth cleaned is 0.44; the probability of having a cavity filled is 0.24; the probability of an extraction is 0.21. Furthermore, the probability of having cleaning and a filling is 0.08; the probability of cleaning and an extraction is 0.11; and the probability of a filling and an extraction is 0.07. Finally, the probability of having all three of cleaning, filling and an extraction is 0.03. What is the probability of having **at least one** of cleaning, filling or extraction?

1.5 A manufacturer of airplane parts knows from past experience that the probability is 0.80 that an order will be ready for shipment on time, and it is 0.72 that an order will be ready for shipment on time and will also be delivered on time. What is the probability that such an order will be delivered on time given that it was ready for shipment on time?

1.6 A test for cancer is known to be 90% accurate either in detecting cancer if present or in giving an all-clear if cancer is absent. The prevalence of cancer in the population is 1%. How worried should you be if you test positive? Try answering this question with and without Bayes' theorem.

1.7 A sock is selected at random and removed from a drawer containing five brown socks and three green socks. A second random sock is then removed. What is the probability that two different colours are selected?

1.8 A game show host shows you three doors, and tells you that behind two of them is a cuddly toy, and behind one of them is a car. You pick a door. The host opens one of the other doors, revealing a cuddly toy, and asks whether you want to switch your choice to the other unopened door. Does switching improve your chance of winning the car? (Hint: consider all the possible permutations of what is behind each door.)

1.9 In Russian roulette, a single bullet is loaded into a six chambered gun. The chamber is spun, and the first player fires at their head. If they survive, the gun is handed to the second player, who spins the chamber again, and then fires at their own head. If the second player survives, they hand the gun back to the first player, and so on. If you go first, what is the probability you will lose?

References

Applebaum, D.: Probability and Information: An Integrated Approach, 2nd edn. Cambridge University Press, Cambridge (2008)
Bulmer, M.: Principles of Statistics. Dover Books, Mineola (2003)
Clarke, G.M., Cooke, D.: A Basic Course in Statistics, 5th edn. Wiley, New York (2011)
Cox, R.T.: Probability, Frequency, and Reasonable Expectation. Am. J. Phys. **14**, 1 (1946)
Freund, J.E., Walpole, R.E.: Mathematical Statistics with Applications. Prentice-Hall, Upper Saddle River (1986)
Jaynes, E.T.: Probability Theory: The Logic of Science: Principles and Elementary Applications, vol. 1. Cambridge University Press, Cambridge (2003)
Jeffreys, H.: Theory of Probability, 3rd edn. Oxford University Press, Oxford (2000)
Keynes, J.M.: A Treatise on Probability. Wildside Press, Rockville (2010)
Kolmogorov, A.N.: Foundations of the Theory of Probability, 2nd edn. Dover Publications Inc, Mineola (2018)
Lupton, R.: Statistics in Theory and Practice. Princeton University Press, Princeton (1993)
Miller, I., Miller, M.: John E. Freund's Mathematical Statistics with Applications, 8th edn. Pearson, London (2013)
Sivia, D.S., Skilling, J.: Data Analysis: A Bayesian Tutorial, 2nd edn. Oxford University Press, Oxford (2006)

Websites (all accessed March 2019):

Stat Pages Listings: http://statpages.info
Geogebra probability tool: https://www.geogebra.org/classic/probability
Random.Org random number service: https://www.random.org

Chapter 2
Distributions, Moments, and Errors

2.1 Outline of Content

- Sample versus population distributions
- Multi-variate distributions
- Summarising quantities for distributions
- Expectations and moments
- Transformation of probability distributions
- Error analysis

In the first chapter we began to look at how to reason in the presence of uncertainty. This led us to the idea of a probability distribution $P(X)$ or $p(x)$ for a random variable. This might for example represent the distribution of molecular velocities in a hot gas. In order to perform reasoning based on such distributions, we need to know how to characterise and manipulate them. What is the typical value? What is the spread? How do we deal with multiple variables? If we know the distribution for a variable, can we deduce the distribution for a related variable? In this chapter we look at some mathematical techniques to answer these questions. This will set us up for dealing with real world distributions, and understanding how to perform statistical inference. First however, we need to carefully distinguish between theoretically expected and empirically observed distributions.

2.2 Sample Versus Population Distributions

Suppose we toss six coins and count how many heads we get. There is a theoretical prediction for this, which we will derive in Chap. 4:

$$P_n(X) = \frac{n!}{n!(n-X)!}\left(\frac{1}{2}\right)^n .$$

A. Lawrence, *Probability in Physics*, Undergraduate Lecture Notes in Physics,
https://doi.org/10.1007/978-3-030-04544-9_2

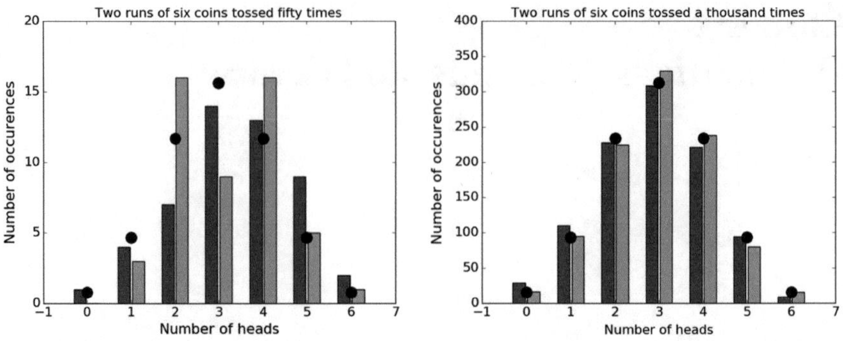

Fig. 2.1 Illustration of the difference between population and sample distributions. The theoretical population (parent) distribution is shown by the filled circles, and shows the expected number of $n = 0, 1, 2\ldots$ heads out of six coin tosses. Each histogram is the result of a simulated experiment using a random number generator. **Left**: The simulated coins are tossed fifty times. The two histograms represent two separate runs of this numerical experiment. **Right**: Same, but the simulated coins are tossed a thousand times

Here, n is the number of coins we toss, X is the number of heads, and $P_n(X)$ is the probability that any individual n-coin toss will give us X heads. For $n = 6$ and say $X = 2$ we get the predicted probability $P = 0.234375$. If we toss six coins an infinite number of times, that is the fraction of times that we will get two heads. Suppose however we do a real experiment in which we toss our six coins ten times. The prediction is that out of those ten experiments, we will typically get two heads $10 \times 0.24375 \sim 2.4$ times. Of course any one experiment can only give us an integer number of heads. In reality we might get two heads perhaps two times out of the ten. Then if we do another run of ten we might get two heads three times, and so on.

In any such situation, the theoretically expected perfect distribution is known as the *population distribution*, or sometimes the *parent distribution*.[1] A specific real experiment produces what is known as a *sample distribution*, because the observed values are sampled from the underlying true distribution. Figure 2.1 shows several sample distributions generated from the above formula, using a random number generator. First, we take six coins and toss them fifty times, noting the number of times we get no heads, one head, two heads etc. We then do a second run of fifty tosses of six coins. The two sample distributions roughly follow the theoretical prediction, but do not precisely agree with it, or with each other. Next we repeat the experiment, but toss the six coins a thousand times. The agreement is still not perfect, but is closer.

One of the key problems of physical science is that often we don't know what the underlying process is, but would like to *find out*. All we have is the sample distribution, and we wish to use this to try to understand what is going on. This is the problem of *statistical inference*, which will occupy us in Part III of the book.

[1]Personally I prefer the term "parent distribution' because it feels more explanatory, but 'population distribution" is more common in the literature.

2.2.1 Sampling a Discrete Population Distribution

This follows the discussion in Chap. 1, Sect. 1.4.1, which we recap briefly. A discrete random variable X can be seen as having a set of n possible outcomes X_i. Although the outcomes are distinct, it is quite possible to have an infinite set of them—for example the possible outcomes might be any positive integer. The different outcomes occur with a well determined set of probabilities P_i, which constitute the population distribution. The population distribution will be usually be defined by some mathematical expression $P(X)$. By contrast the sample is simply a list of observed data values x_k. (Note that i indexes the possible outcomes, whereas k indexes the list of data values). There can be many repeat values—five 2s, eight 3s, etc. If there are N_i repeats of a given X_i, the *sample frequency distribution* is then the set of these values $N_i(X_i)$. To turn this into a normalised sample probability distribution, we need the total number of samples $N = \sum_i N_i$. Then we have $f_i(X_i) = N_i(X_i)/N$. The f_i values are then *estimates* of the true population values P_i.

2.2.2 Sampling a Continuous Population Distribution: Binning

For a continuous random variable, the population PDF will normally be given by some smooth mathematical function $p(x)$. Note that we might not necessarily know what this function is, or be able to write it down, but we assume that it nonetheless exists. Meanwhile, the measurements are still just a list of distinct values x_k. We need to somehow relate these to the density of probability $p(x)$ in the local region. We can do this by *binning* the values. We choose a range Δx centred on each of a set of values x_b, where b can be seen as the "bin number", and count how many times the individual measurements x_k fall inside the range $x_b \pm \Delta x/2$. This gives us a binned sample frequency distribution $N_b(x_b)$. At each position, we can get the sample PDF $p_b(x_b) = N_b(x_b)/N\Delta x$, where $N = \sum_b N_b$ is the total number of measurements. The p_b values are then estimates of the population PDF, $p(x)$.

For these to be good estimates, we need Δx to be large enough that we get a reasonable number N_b in each bin, but not so large that p is changing a lot across the bin. This is not always an obvious choice, as you can see in Fig. 2.2. Note that although we have assumed that x_b is the value at the centre of the bin, sometimes people define x_b as the lower edge of the bin. Likewise, we assumed that Δx is fixed, but in principle it could be different for each bin—for example you might use wider bins towards the wings of the PDF in order to get larger counts. Watch out for what different authors are doing!

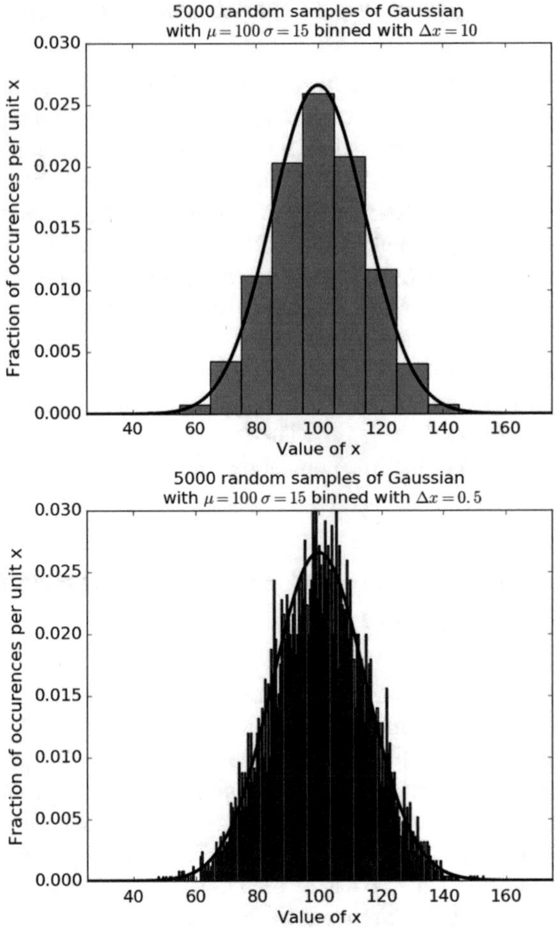

Fig. 2.2 Comparing a binned sample distribution to its population (parent) distribution. The smooth population distribution is the same Gaussian distribution in each case (see Chap. 5 for the definition of the Gaussian). A random number generator was used to generate 5000 random samplings from that Gaussian PDF, which was then binned into ranges of x. The upper and lower figures show different bin widths Δx

2.3 Multi-variate Distributions

Quite often we want to deal with two or more variables. The methods we have used above generalise quite simply, but we now have the additional question of how the variables relate to each other.

Suppose X and Y are discrete random variables, with n and m possible outcomes respectively, which we can label X_i and Y_j. Then we can define the joint probability for any i, j pair, to give the *joint probability distribution* P_{ij}, where i runs from 1 to n and j from 1 to m. The sample is a list of individual x_k, y_k data value pairs. We can count how many times N_{ij} the various possible outcome pairs X_i, Y_j occur, normalise by the total number of experiments $N = \sum_{ij} N_{ij}$, and so get a normalised sample distribution which will be an estimate of the true P_{ij} values.

If x and y are continuous random variables, then the population probability density will correspond to some smooth function $p(x, y)$, such that $p(x, y)\mathrm{d}x\mathrm{d}y$ is the probability that in a specific trial the two variables will fall in the range x to $x + \mathrm{d}x$ and y to $y + \mathrm{d}y$. We can extend these definitions to any number of variables using $p(x, y, x)\mathrm{d}x\mathrm{d}y\mathrm{d}z$ and so on. The sample data values are once again a list of x_k, y_k data value pairs. Just as in the 1D case, we could count occurrences within a chosen set of bins centred at x_b, y_c, with binwidths $\Delta x, \Delta y$, divide the count by the bin area and the total number of experiments, and so estimate $p(x, y)$ at each bin centre.

Figure 2.3 shows two examples of bivariate distributions. The upper panel shows a smooth bivariate probability density function, simulated using the Gaussian function that we will discuss in Chap. 5. The lower panel shows an example of a bivariate sample distribution, represented as a table of numbers—the observed binned frequency distribution of the measured breadth and length of human heads, in a sample of 3000 people. This sample of head sizes is assumed to be drawn from some true underlying smooth population distribution of probability density. We do not know what that true population distribution is, and certainly don't know how to write it down mathematically, but we assume that it exists. Note that the cell sizes are 0.5 cm in each direction. To estimate the population probability density distribution, we would start by taking the numbers in each cell, dividing by $N = 3000$, and then dividing by the bin area $0.5 \times 0.5 = 0.25$, to give estimated probability densities in units of probability per cm^2.

Suppose we are interested in the distribution of just y, rather than the joint distribution of x, y. How do we extract that from the bivariate distribution? As often in statistics, it depends on what you mean exactly. We might mean (i) at this particular value of x, what is the distribution of y? Or we might mean (ii) what is the overall distribution of y, if we don't care what the value of x is? Lets examine these two questions.

2.3.1 Conditional Distributions

To answer question (i), we choose a particular value of x and freeze it. Then the frequency distribution of y is given by taking a vertical slice along y at the chosen x value. This is called the *conditional distribution* of y. Note however that these values will not sum to 1.0; so we need to re-normalise by summing all the values in the slice and then dividing each value by the sum. For the discrete and continuous cases this gives

$$g_i(j) = g(Y|X_i) = \frac{P_{ij}}{\sum_{j=1}^{m} P_{ij}} \quad \text{or} \quad g_x(y) = g(y|x) = \frac{p(x, y)}{\int p(x, y)\mathrm{d}y}. \tag{2.1}$$

We write i or x as a subscript and j or y in the brackets to indicate that we are seeing the result as a function varying over j or y, at a chosen value of i or x. In a similar

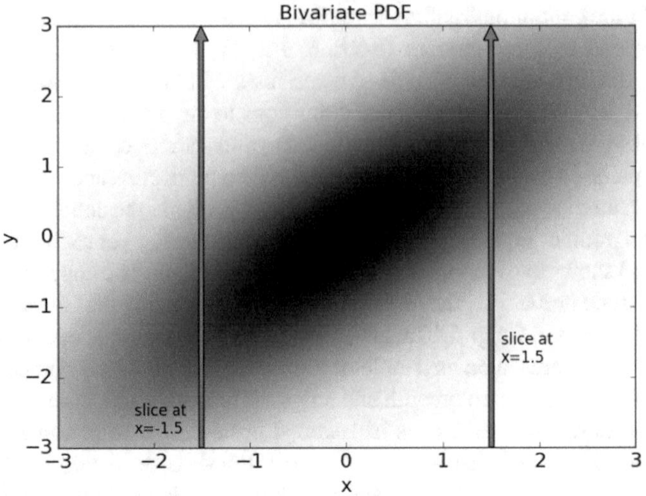

Head Length(cm)	**13–**	**13.5–**	**14–**	**14.5–**	**15–**	**15.5–**	**16–**	**16.5–**	Total
	Head breadth (cm)								
16–	0	0	0	0	1	0	0	0	1
16.5–	0	0	1	0	1	0	0	0	2
17–	0	5	4	4	1	0	0	0	14
17.5–	1	8	17	15	11	2	0	0	54
18–	0	6	55	119	74	14	1	0	269
18.5–	0	5	108	264	234	75	6	1	693
19–	0	10	72	360	400	156	26	2	1026
19.5–	0	1	28	174	239	160	36	7	645
20–	0	2	4	31	86	100	24	2	249
20.5–	0	0	1	4	17	17	5	0	44
21–	0	0	1	0	0	1	0	1	3
Total	1	37	291	971	1064	525	98	13	3000

Fig. 2.3 Illustrations of bivariate probability distributions. **Upper**: This shows a smooth continuous probability density distribution, with the value of probability density represented as a greyscale as a function of x and y. The mathematical formula used was a bivariate Gaussian, which we will define in Chap. 5. The "slices" are explained in the text. **Lower**: This shows a binned sample distribution. The dataset is a real one, representing the measured breadths and lengths of the heads of a sample of 3000 criminals, published by MacDonell (1902). The numbers show the number of people with breadth/length in the binned ranges shown. The entries in the rightmost column shows the sum of the values in the row to the left. Likewise, the entries in the bottom row shows the sums of the columns above

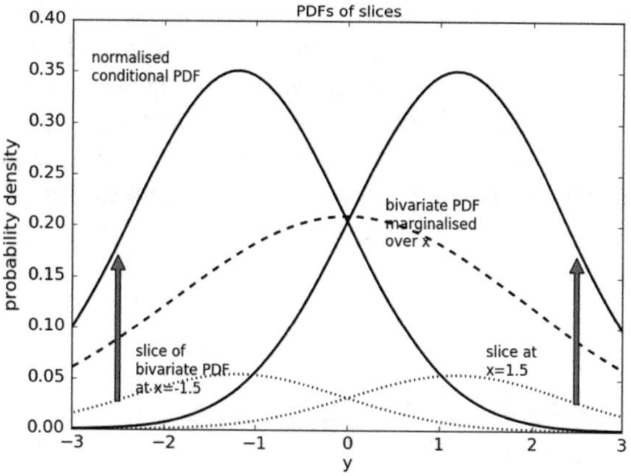

Fig. 2.4 Conditional versus marginal distributions. The dotted curves show the result of taking a vertical slice through the bivariate PDF shown in the upper part of Fig. 2.3, at the two x-values indicated. The solid curves show the adjustment when these curves are normalised as explained in the text. The dashed curve shows the marginal distribution of y obtained by integrating over x

manner, we could take a slice at a given y to construct the conditional distribution of x, $g_y(x)$. Note that the integration is over the full range of y or x as appropriate, which may be 0 to ∞, or $-\infty$ to ∞, or potentially some other range.

The upper panel in Fig. 2.3 shows two different slice positions through the continuous PDF, and the result of constructing these slices is shown in Fig. 2.4, before and after re-normalisation. For the binned sample distribution in the lower panel of Fig. 2.3, the conditional distribution corresponds to taking the values in a specific column, $N_{ij}(Y_j|X_i)$. To turn these frequencies into a normalised probability distribution, we have to divide by the total in that column, $\sum_j N_{ij}$, which is shown at the end of the columns—for example taking the numbers in the column with head breadths in the range 14.5–15 cm, we divide by 971.

2.3.2 *Marginal Distributions*

To answer question (ii), the overall distribution of y regardless of x, lets start by looking at the binned sample version. We start by choosing a value of y, say $y = 18.75$. We count *all* the occurrences in cells consistent with that value. This corresponds to adding the values in the row with $y = 18.5$–19.0, giving 693. We then repeat for different y values, giving the final column on the right of the table in Fig. 2.3. This process is known as "marginalising over x", giving the *marginal distribution* of y, because the result is written in the margin of the table. For a continuous PDF, we can apply exactly the same logic, by dividing the x, y plane into differential $\mathrm{d}x$, $\mathrm{d}y$ cells. For the discrete and continuous cases, we can then write the marginal distribution for y as

$$g(j) = \sum_{i=1}^{n} P_{ij} \quad \text{or} \quad g(y) = \int p(x, y)\mathrm{d}x. \tag{2.2}$$

Likewise, we can construct the marginal distribution of x by integrating over y. Note that as long as $p(x, y)$ is a properly normalised PDF, there is no need for any further normalisation. If we marginalise over an observed frequency distribution such as the table in Fig. 2.3, then we normalise over the sum of values over the whole of the x, y plane. Figure 2.4 contrasts conditional and marginal distributions, showing the results we get for the bivariate PDF example of Fig. 2.3.

2.3.3 Dependence

Note that the grey-scale ellipse in the bivariate example of Fig. 2.3 is tilted diagonally. As a result, when we take the two vertical slices, we get a different answer for each slice, as shown in Fig. 2.4. This shape-changing means that the variables are *dependent*. This is another way of looking at the result we got in Chap. 1. If the variables x and y are independent, then $g(y|x)$ is the same as $g(y)$, and we can write $p(x, y)$ in the form $p(x, y) = f(x)g(y)$, where $f(x)$ and $g(y)$ are the marginal distributions of x and y. Likewise for the discrete case, we can write P_{ij} as $f_i g_j$. In this case $g(y)$ is also the same as the conditional distribution of y, for all values of x. If the variables are dependent, this will not be the case; the conditional distribution $g(y)$ will be a function of x. Note that we can always form the marginal distributions $f(x)$ and $g(y)$, but if the variables are dependent, then we can't write $p(x, y)$ in the separated form.

2.4 Summarising Quantities for Distributions

For a population distribution we will in principle have a mathematical expression that fully defines the distribution. For a sample distribution, we don't have this, and would like an objective way to characterise the observed distribution. What is the "typical" value, what is the spread of values, is it a flat topped or peaky distribution? Even for population distributions with mathematical expressions, these simple numbers are useful for comparing one distribution to another. Furthermore, if we calculate these properties for an observed sample distribution, we can take them as estimates of the equivalent quantities in the underlying population distribution.

How do we define such summarising quantities carefully? In this section, we will start by doing this in an intuitive way, looking at the concepts of mean, variance, and covariance. Then in Sect. 2.5 we will look at a more formal general method for generating summarising quantities—calculating the moments of a distribution. Bear in mind that a sample distribution is essentially a list of x_i's, although we might typically present the data as a binned histogram.

2.4.1 Measures of Location

First we might ask "what is the *typical* value?" There are three common ways of translating this qualitative concept into something rigorous. The first is to use the histogram and estimate the *most probable value* or *mode*, where the local density of x_i values is highest. This is intuitive but sensitive to how you bin the histogram. Another method is the *median*—the value for which half the values are above and half below. This has the great virtue of being completely robust against transformations of x, but it is hard to work with mathematically. The commonest estimate is the *arithmetic mean* or *average*:

$$\text{sample mean} \quad \bar{x} = \frac{\sum x_i}{N} \qquad \text{population mean} \quad \mu = \int x\, p(x)\mathrm{d}x \tag{2.3}$$

where once again the integration is over the full range of x, whatever that may be. Note that convention is to carefully distinguish the sample mean and the population mean by using two different symbols, $\bar{x}$ and μ. For any real sample, they won't quite be the same, but in the limit of large N, $\bar{x} \to \mu$. For symmetrical population distributions, the mean, mode and median will all be the same on average. For an asymmetric PDF, they will not be the same, as illustrated in Fig. 2.5.

2.4.2 Measures of Dispersion

Next we want to know the *spread* of values. We could start by calculating the deviation of each point from the mean, $x_i - \bar{x}$. The mean value of this will typically be zero for many population distributions, which is not helpful. We could find the average

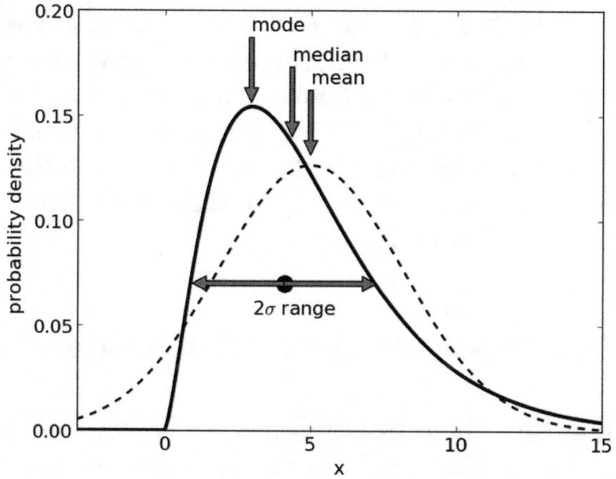

Fig. 2.5 Measures of location and dispersion for an asymmetric PDF, with a symmetric PDF shown for comparison

of the absolute value, but this is hard to work with mathematically. Instead it is usual to define the *variance*

$$\text{sample} : V = s^2 = \frac{\sum (x_i - \bar{x})^2}{N} \quad \text{population} : \sigma^2 = \int (x - \mu)^2 \, p(x) \, \mathrm{d}x \tag{2.4}$$

Once again we carefully use separate symbols for the sample and population versions, $V = s^2$ and σ^2, and expect that $V \rightarrow \sigma^2$ in the limit of large N. The square root of the variance, $s = V^{1/2}$ or σ, is then the *standard deviation*. The ratio $s/\bar{x}$ is sometimes called the *coefficient of variation*. It gives us a dimensionless quantification of spread that we can compare from one distribution to another.

2.4.3 Percentiles of a Distribution

Sometimes we want to know what range of x values will encompass "most" of the outcomes of an experiment, where by "most" we mean, say, 90% or perhaps 95% or 99%, depending on how cautious we want to be. What we need to do then is to integrate $p(x)$ and find two values x_1 and x_2 such that

$$P_{\text{int}} = \int_{x_1}^{x_2} p(x) \, \mathrm{d}x$$

where $P_{\text{int}} = 0.9$ or 0.95, etc. There are two problems with implementing this idea. The first is that most distributions are not analytically integrable. Of course, all the well known distributions have been numerically integrated many times, so the custom is to use tables of integrals, or computer routines. The second problem is that there are many possible pairs of values x_1 and x_2 which would give the same value of P_{int}, so one needs a further specification. One solution is to specify a symmetrical distance x either side of the mean, and look for the integrated probability in the range $\mu - x$ to $\mu + x$. Another might be to look at the integral from x from ∞ (or the maximum of the x range), so that one is asking "what value of x would I need so that most outcomes are at least that big?". The general lesson, as often in statistics, is to be careful to frame your question precisely. This question—ambiguity in integrated probability—comes up often in traditional hypothesis testing problems—see Chap. 7, Sect. 7.6.2, and in particular Fig. 7.2.

2.4.4 Summarising Quantities for Multivariate Distributions

If we have a bivariate distribution $p(x, y)$ then we can find quantities characterising x and y by first forming the marginal distributions $f(x)$ and $g(y)$ and then using these to calculate μ_x, μ_y and σ_x^2 and σ_y^2. However, we would also like quantities which summarise how the variables are related together. Potentially, there are many different ways we could do this, but the most common question is to ask "when y is bigger, does x also then tend to be bigger, and vice versa?". The simplest quantity which captures the sense of this question is the *covariance*:

$$\text{sample} \quad s_{xy} = \left[\frac{1}{N}\sum(x_i - \bar{x})(y_i - \bar{y})\right] \qquad \text{popn.} \quad \sigma_{xy} = \lim_{N\to\infty} s_{xy}. \tag{2.5}$$

If there is no tendency for x and y to vary together, then the terms inside the sum are just as likely to be negative as positive, and so the sum tends to zero. On the other hand, the "tilted" bivariate distribution illustrated in Fig. 2.3 will clearly give a positive value for σ_{xy}. A distribution tilted from top left down to bottom right would give a negative value. This begins to look like a test for whether two variables are dependent or independent, but in fact the issue is a little more subtle, and we will return to it in Chap. 9.

We can extend these ideas to any number of random variables $x_1, x_2, x_3 \ldots x_i, \ldots$ for a general multi-variate distribution.. We can form the marginal distribution for any variable x_i by marginalising over all the others, and then find μ_i and σ_i^2. We can think of the list of each of these as a vector characterising the distribution. Likewise, we can look at the covariance of each possible pair of variables in turn, and find all the covariances σ_{ij}. The collection of covariance values σ_{ij} is known as the *covariance matrix*, with the diagonal values $\sigma_{ii} = \sigma_i^2$ being the variances of the individual variables.

2.5 Expectation Values and Moments

As discussed at the start of Sect. 2.4, we would like a standardised way of generating a small set of numbers which can summarise the features of a distribution. The usual method is to calculate the *moments* of the distribution. Before we explain what this means, we first need a new mathematical concept—the expectation value of a random variable.

2.5.1 Expectation Values

The *expected value* of a random variable, alternatively called the *expectation value*, is simply the average of all its possible values, weighted by the probability of occurrence of each value. We can define this concept for either discrete or continuous random variables:

$$E[X] = \sum_X X P(X) \qquad \text{or} \qquad E[x] = \int_{-\infty}^{\infty} x p(x) \mathrm{d}x \tag{2.6}$$

Here as usual $P(X)$ is the set of probabilities of the discrete random variable X, and $p(x)$ is the PDF for the continuous random variable x. (You will often see expected values written as $\langle x \rangle$ rather than $E[x]$. Personally I find this less readable, but it is very common, especially in the mathematical literature.) Because a function of a random variable is also a random variable, we can calculate the expected value of any function of x

$$E[f(x)] = \int_{-\infty}^{\infty} f(x) p(x) \mathrm{d}x.$$

2.5.2 The Algebra of Expectations

With a little manipulation, you can see how expectation values combine:

$$\begin{aligned} E[X+Y] &= E[X] + E[Y] \\ E[X-Y] &= E[X] - E[Y] \\ E[aX+b] &= aE[X] + b \\ E[aX+bY] &= aE[X] + bE[Y] \end{aligned}$$

Note also that if b is a constant $E[b] = b$; and because an expected value is itself a constant, having integrated over x, then $E[\ E[x]\ ]$ is just $E[x]$.

2.5.3 Moments of a Distribution

The *moments* of a random variable x are just the expected values of x^n for various n:

$$m_n = E[x^n]. \tag{2.7}$$

The moments calculated for successively higher powers n are just the quantities we want to represent the features of a distribution, starting with the simplest features and moving up towards more subtle ones. Stepping through these:

The zeroth moment $m_0 = \int_{-\infty}^{+\infty} p(x) \mathrm{d}x = 1$ as long as $p(x)$ is properly normalised.

The first moment $m_1 = E[x] = \mu$ is the **mean**—it is the expected value of x.

The standard second moment is $m_2 = E[x^2]$, but from here on it is more useful to define the *centred moments* obtained by shifting the origin of x to the mean:

$$\mu_n \equiv E\left[(x - E(x))^n\right] = E\left[(x - \mu)^n\right].$$

Then $\mu_0 = 1$ and $\mu_1 = 0$. Note that μ, the mean value, is not the same as μ_1. It is annoying that the same letter is used, but unfortunately this is the convention. The second centred moment is then

$$\mu_2 \equiv E\left[(x - \mu)^2\right] = \mathrm{Var}(x) = \sigma^2.$$

In an analogous manner, for a bivariate distribution,

$$E[(x - \mu_x)(y - \mu_y)] = \mathrm{Cov}(x, y) = \sigma_{xy}.$$

A useful result which can be obtained (see exercises) is that

$$\sigma^2 = E\left[(x - E[x])^2\right] = E\left[x^2 - 2xE[x] + E[x]^2\right] = E\left[x^2\right] - E[x]^2.$$

In other words, the variance can be obtained from the mean of the square minus the square of the mean.

2.5.4 Sample Moments

The moments for the sample distribution are defined using the same idea—the average value of powers of x. They can be defined with respect to any point x_a, i.e. by calculating the average of $(x_i - x_a)^n$, but the most commonly used values are the central moments, i.e. with respect to the position of the mean. Then we have

$$\mu_n = \frac{1}{N}\sum_{i=1}^{N}(x_i - \bar{x})^n.$$

2.5.5 Higher Moments: Skewness and Kurtosis

For either population or sample distributions, we can calculate higher moments such as μ_3, μ_4 and so on. We can use these to characterise the degree of asymmetry of a distribution (skewness), or its degree of peakiness (kurtosis). These moments carry the units of x of course, so to compare one distribution with another, it's normal to define dimensionless coefficients by normalising to an appropriate power of the second moment:

$$\text{skewness } \alpha_3 = \frac{\mu_3}{\sigma^3} \qquad \text{kurtosis } \alpha_4 = \frac{\mu_4}{\sigma^4}.$$

Skewness can be positive or negative, depending on the direction of skewness, with a symmetric distribution having $\alpha_3 = 0$. A moderately skewed distribution might have $\alpha_3 \sim \frac{1}{2}$ to 1. The exponential distribution which we shall meet later has an

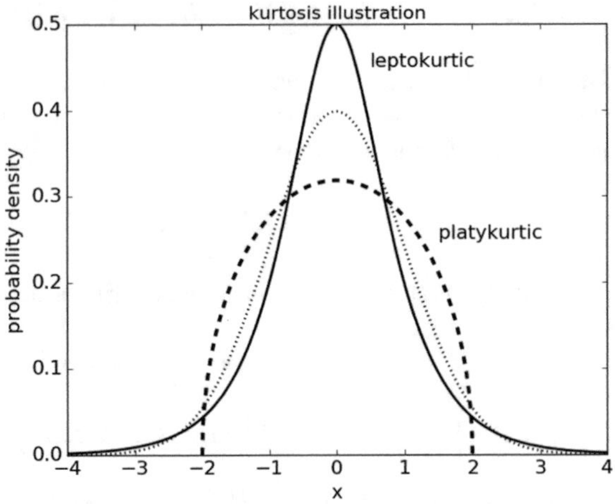

Fig. 2.6 PDFs with the same variance ($\sigma = 1$) but different values of excess kurtosis. The middle (dotted) curve is a Gaussian, and by definition has zero excess kurtosis. The upper (solid) curve is a hyperbolic secant and has excess kurtosis 1.2. The lower (dashed) curve is a stretched semi-circle and has excess kurtosis of −1.2. This figure is based on that originally made by Mark Sweep for the wikipedia article on kurtosis

extreme skewness of $\alpha_3 = 2$. Kurtosis can in principle be anywhere between 1 and ∞. The normal or Gaussian distribution has $\alpha_4 = 3$, so it is common to compare to this by defining the *excess kurtosis* as $\alpha - 3$. A distribution with positive excess kurtosis (peaky) is known as *leptokurtic*. A distribution with negative excess kurtosis (squat) is known as *platykurtic*. Examples are shown in Fig. 2.6.

2.6 Transformation of Probability Distributions

If we have a random variable x then any function of that variable, $z = z(x)$, is also a random variable. If we have a mathematical expression for the PDF of x, say $f(x)$, then we can derive an expression for the PDF of z, $g(z)$ using the usual rules of calculus:

$$g(z) = f(x)\frac{\mathrm{d}x}{\mathrm{d}z}.$$

Likewise for multivariate probability distributions we can transform to other variables, but of course the algebra will be more cumbersome, involving Jacobians and so on. We could then proceed to calculate the moments of the new distribution e.g. $\mu_z = \int z g(z)\mathrm{d}z$. In practice by far the most common question is to ask what happens to the variance in the transformation—if we know σ_x^2, what is σ_z^2?

2.6.1 Transformation of Variance: Univariate Case

We can write down the variance of z as

$$\sigma_z^2 = \lim_{N\to\infty}\left[\frac{1}{N}\sum(z_i - \bar{z})^2\right].$$

Now consider deviations in x. To a first order approximation we have

$$z_i - \bar{z} = (x_i - \bar{x}) \left(\frac{\mathrm{d}z}{\mathrm{d}x} \right),$$

so then we get

$$\sigma_z^2 = \lim_{N \to \infty} \frac{1}{N} \sum (x_i - \bar{x})^2 \left(\frac{\mathrm{d}z}{\mathrm{d}x} \right)^2.$$

The first part of that expression is just the variance of x. So finally we can see that

$$\sigma_z = \sigma_x \left(\frac{\mathrm{d}z}{\mathrm{d}x} \right). \tag{2.8}$$

2.6.2 *Transformation of Variance: Bivariate Case*

Things get more interesting if we have a bivariate PDF which is a function of the two variables x and y, with the new variable z being a function of both, $z = f(x, y)$. Now we need to consider the partial derivatives of z with respect to x and y, so that the deviations in z, to first approximation, are

$$z_i - \bar{z} = (x_i - \bar{x}) \left(\frac{\partial z}{\partial x} \right) + (y_i - \bar{y}) \left(\frac{\partial z}{\partial y} \right),$$

and the variance in z will be

$$\sigma_z^2 = \lim_{N \to \infty} \frac{1}{N} \sum \left[(x_i - \bar{x}) \left(\frac{\partial z}{\partial x} \right) + (y_i - \bar{y}) \left(\frac{\partial z}{\partial y} \right) \right]^2,$$

which gives

$$\sigma_z^2 = \lim_{N \to \infty} \frac{1}{N} \sum \left[(x_i - \bar{x})^2 \left(\frac{\partial z}{\partial x} \right)^2 + (y_i - \bar{y})^2 \left(\frac{\partial z}{\partial y} \right)^2 + 2(x_i - \bar{x})(y_i - \bar{y}) \left(\frac{\partial z}{\partial x} \right) \left(\frac{\partial z}{\partial y} \right) \right].$$

The first two terms inside the sum are the variances of x and y, scaled by the dependence of z. The third term includes the covariance which we defined in (2.5). So we have

$$\sigma_z^2 = \sigma_x^2 \left(\frac{\partial z}{\partial x} \right)^2 + \sigma_y^2 \left(\frac{\partial z}{\partial y} \right)^2 + 2\sigma_{xy} \left(\frac{\partial z}{\partial x} \right) \left(\frac{\partial z}{\partial y} \right). \tag{2.9}$$

Much of the time we assume that x and y are independent variables, so that the covariance term vanishes. A simple example is the *sum of two random variables*. If we have $z = x + y$, and assume x and y are independent, then by calculating $\partial z/\partial x$ and $\partial z/\partial y$ we find that

$$\sigma_z^2 = \sigma_x^2 + \sigma_y^2. \tag{2.10}$$

So we see that uncertainties "combine in quadrature". It is the variances that add, not the standard deviations. For this simple example, if x and y are dependent, then we have $\sigma_z^2 = \sigma_x^2 + \sigma_y^2 + 2\sigma_{xy}$. With a little algebra, and assuming independence, we can derive some further handy examples of combining variances:

$$\begin{array}{ll}
f = ax \pm by & \sigma_f^2 = a^2\sigma_x^2 + b^2\sigma_y^2 \\
f = xy \text{ or } x/y & \left(\sigma_f/f\right)^2 = (\sigma_x/x)^2 + \left(\sigma_y/y\right)^2 \\
f = ax^{\pm b} & \left(\sigma_f/f\right) = b\,(\sigma_x/x) \\
f = a\ln(\pm bx) & \sigma_f = a\,(\sigma_x/x) \\
f = ae^{\pm bx} & \left(\sigma_f/f\right) = b\sigma_x \\
f = a^{\pm bx} & \left(\sigma_f/f\right) = b\ln(a\sigma_x)
\end{array}$$

2.7 Error Analysis

An important practical application of the formulae for combining variances is in understanding the errors on quantities derived from measurement. Before we spell that out, let us take a more careful look at the concept of error, and how it relates to probability distributions.

2.7.1 *Types of Error*

From high school physics onwards, we are taught that we should associate uncertainties or errors with our measurements. In fact the term "error" in science is a rather subtle one, with several different meanings that are worth carefully distinguishing.

In normal English, an "error' is some kind of actual *mistake*. This can certainly apply to physical experiments. If we believe we are measuring a lump of Sodium, and it's actually a lump of Potassium, we will certainly get an incorrect value. There might also be some kind of fixed *bias*, i.e. an offset from the true value. We measure some effect with a voltmeter, but a problem with our circuitry means that all our measurements have 0.03 V added to them.

Mistakes and biases can in principle be corrected. However, central to the idea of error analysis is that a range of measured values is unavoidable, because the act of measurement is itself a random process. We assume that some property has a true value x_t, but the measured value x is a random variable. Each time we run the experiment and produce a value of x, it is drawn from some probability distribution $e(x)$, known as the error distribution, centred on the true value x_t. For the most careful analysis, we would in principle like to know the full $e(x)$, but if we are to pick a single number to represent the random error, it would be the standard deviation σ of the distribution $e(x)$. In many cases it is reasonable to assume that the error distribution is a Gaussian distribution, which we discuss in Chap. 5. The Gaussian distribution is fully specified by its mean and variance, so in fact we can calculate the probability of any other value.

We should carefully distinguish the idea of randomness in the measurement process from the idea of randomness in actual physical properties. People have a variety of heights, and so the height h of a person selected at random from a large population has a spread. If we measure the height of one specific person, we should in principle get a single definite value. However, when we measure the height of that specific person repeatedly, we actually get a small spread of results, because of randomness in the measurement process.

Quite often measurements involve both truly random errors, and *systematic errors*, a term which includes both fixed biases and unknown uncertainties of various kinds. The key question is whether making more measurements helps or not. If a sample of measurements is drawn from an assumed underlying random error distribution, then we can use the sample mean as an estimate of the true value. The more measurements we make, the more *precise* will be our estimate of the population mean. The error on 30 measurements is smaller than the error on 3 measurements. (We will sharpen that statement in Chaps. 3 and 4.) However, if we are dominated by *systematic* errors, the precision does not improve—the error stays just as big, however many measurements we make. If both a random error and a systematic error apply, it is healthy if we quote each separately, rather than a combined error.

2.7.2 Evaluating Errors

Sometimes we know what is causing the randomness in the measurement process. For example, when an experiment involves counting things, such as the number of photons detected from a star, the number counted is subject to *Poisson statistics*, which we will look at in Chap. 4. Often however, we do not know the source of error; we then need to estimate the error empirically by making a sequence of measurements and calculating the sample variance. Of course, as Donald Rumsfeld might have said, our problem is not just the known errors, or the known unknown errors, but the unknown unknown errors.

Often the quantity that we measure is not the same as the quantity of interest. We might for example have an experiment where we measure a voltage and a time delay, but then we use some formula to calculate the value of electron mass that these measurements imply. We might perhaps perform the experiment several times, and use the spread of measured values of voltage and time delay—in other words their standard deviations—as a measure of the error on each of those quantities. Given those errors on voltage and time delay, what then is the error on our calculated value of electron mass? To find this, we can *propagate* the errors using formulae such as those derived in Sect. 2.6, if necessary allowing for the possibility that our measurements of voltage and time delay are not independent, so that we need to calculate the covariance as well the variances.

There is however a danger in this process of propagating errors using the standard formulae. Our tendency is to assume that measurement errors follow a Gaussian or normal distribution. This is generally a fairly good assumption, for reasons we will discuss in Chap. 5. One of the advantages of assuming a normal distribution is that if we know the value of σ, we have a pretty good idea of how common it is to find points within $\pm 2\sigma$ or $\pm 3\sigma$ etc. However, when we have transformed to some other variable z, although the formula will give the correct value for the dispersion σ_z, the PDF of z will in general *not* be Gaussian, so our instinct for how likely aberrant points are is not reliable.

As an example, consider how we might plot error bars for data points when our data processing has involved taking logs. Suppose we have measured a value $x = 256$ and have estimated an error $\sigma_x = 27$, so that the 1σ range of x is from 229 to 283. We then decide to transform our data to base 10 logs before plotting. Our x measurement corresponds to $z = \log_{10}(x) = 2.408$. If we take the $\pm 1\sigma_x$ limits of x and also transform these to logs, we get the range $z = 2.360$–2.452. However, the propagation formula tells us that $\sigma_z = 0.110$. Taking $\pm 1\sigma_z$ limits, we get the range $z = 2.298$–2.518. As often in statistics, there is no unique correct method—you just need to be explicit about what you have done.

2.8 Key Concepts

Some of the key concepts from this chapter are:

- The distinction between population (parent) and sample distributions
- How you can make 1D distributions from a 2D (bivariate) distribution two different ways: either by taking a slice (conditional distribution) or by integrating over the variable you don't care about (marginal distribution)
- The difference between mode, median, and mean
- The idea of variance to characterise the spread of a distribution
- The idea of the expectation value of a random variable
- The idea of the moments of a distribution as the expectation values of x^n
- How variances transform when you make a function of one or more random variables
- The importance of transmission of variances for error analysis, and especially how errors add in quadrature.

There were lots of equations in this chapter, but make sure you can understand the key formulae: how to calculate conditional and marginal distributions ((2.1) and

(2.2)); the formulae for mean and variance of a distribution ((2.3) and (2.4)), and the covariance of a bivariate distribution (2.5); the definitions of expectation value (2.6) and moments of a distribution (2.7); the method for transforming variance for a function of random variables in the univariate case (2.8) and the bivariate case (2.9); and derived from this, how errors add in quadrature (2.10).

2.9 Further Reading

Most of the material in this chapter is covered by the same basic textbook references as for Chap. 1. There are however some textbooks which specifically discuss the analysis of errors, such as Taylor (1996) and Hughes (2010).

The distribution of head lengths and breadths, used as an example of a sample bivariate distribution, has a curious history. I came across it in the textbook of Bulmer (2003), but the sample is from a research paper by MacDonell (1902). This was a study which was part of the now discredited science of phrenology—the idea that the shapes of heads tells you something about the moral character of a person.

2.10 Exercises

2.1 From the data in the table shown in Fig. 2.3, estimate the mean, median, and mode for the marginalised distributions for head length and head breadth. In each of those two cases, are mean, median and mode the same?

2.2 A random variable x has the PDF $f(x) = ke^{-3x}$ for $x > 0$ and $f(x) = 0$ elsewhere. Find the probability that x lies between 0.5 and 1.0.

2.3 Two random variables x and y follow the joint PDF $f(x, y) = \frac{2}{3}(x + 2y)$ for x and y between 0 and 1, and 0 elsewhere. Find the marginal PDFs for x and y.

2.4 If x is the number scored with one roll of a dice, what is the expected value of the random variable $g(x) = 2x^2 + 1$?

2.5 Prove the result given in Sect. 2.5.2, that $\sigma^2 = E[x^2] - E[x]^2$.

2.6 In the notes, we defined the n'th moment as $E[x^n]$, and then considered the centred moment as $\mu_n = E[(x - \mu)^n]$. More generally we could define the moment with respect to the arbitrary point x_a as $E[(x - x_a)^n]$. Derive expressions for the zeroth, first and second moments of a sample distribution in terms of the standard sample mean and variance, with respect to the point $x = x_a$.

2.7 When tossing a coin, what is the average number of tosses before getting heads? If rolling a die, what is the average number of rolls before getting a 6?

2.8 Two particle physics labs are attempting to measure the mass of the zarquon particle. Lab A gets $m_A = 1326 \pm 150\,\text{MeV}$; Lab B gets $m_B = 1560 \pm 125\,\text{MeV}$. There is a suspicion that Lab B always has some offset in its measurements. How does the difference between the mass measurements compare to the expected error on that quantity? Does it look like evidence for the suspected offset? What about the ratio of the two mass estimates?

2.9 The apparent brightness of a star can be measured using the flux F of light from the star—i.e. the amount of light energy per second per square metre due to that star arriving at Earth. However, astronomers often quote the apparent brightness of a star as its *magnitude*, defined using

$$m = k \times \log_{10}\left(\frac{F}{F_0}\right)$$

where F is the flux of the star; F_0 is a constant, a standard "zeropoint" flux that the star is compared to; and k is a constant that is used to match the magnitude scale onto the values used historically. Astronomers use this strange method partly because star fluxes cover an enormous range, so using a logarithmic scale makes some sense; and partly because getting absolute measurements is very hard, so using a relative scale makes sense—you can compare one star to another.

If the error on the flux is σ_F, what is the error on the magnitude, σ_m? To match historical magnitude values, astronomers use $k = -2.5$. (This means that "sixth magnitude" is a hundred times fainter than "first magnitude".) Show that this gives magnitude errors that are numerically close to the fractional flux error, σ_F/F. What value of k would make this exactly correct?

References

Bulmer, M.: Principles of Statistics. Dover Books, New York (2003)

Hughes, I.: Measurements and Their Uncertainties: A Practical Guide to Modern Error Analysis. Oxford University Press, Oxford (2010)

MacDonell, W.R.: On criminal anthropometry and the identification of criminals. Biometrika **1**, 177 (1902)

Taylor, J.R.: An Introduction to Error Analysis: The Study of Uncertainties in Physical Measurements, 2nd edn. University Science Books, California (1996)

Websites (accessed March 2019):

Wikipedia page on kurtosis, including figure by Mark Sweep: https://en.wikipedia.org/wiki/Kurtosis

Part II
Frequency Distributions in the Physical World

In Part I, we met the idea of a probability distribution, and in particular, the interpretation of this as the relative frequency with which unpredictable events occur. In Part II, we look at how such frequency distributions occur naturally in various physical circumstances. At the root of how to treat such issues mathematically is the subject of *combinatorics*—how to count the number of ways different permutations and combinations of events can occur.

In Chap. 3, we first look at the basics of how to perform such calculations, and then examine the ideas of partitions, macrostates, and microstates, and how such calculations lie behind some of the core ideas of statistical physics—for example, the Second Law of Thermodynamics, and the Boltzmann distribution of particle energies.

In Chap. 4, we use the ideas of combinatorics to look in more detail at situations that involve any kind of either/or, or hit/miss question, and see how this leads to one of the most common frequency distributions—the Binomial distribution. Narrowing down further to cases where the "hit" probability is small, we arrive at another important distribution of widespread physical application—the Poisson distribution. Together, the Binomial and Poisson distributions are sometimes known as *counting statistics*.

In Chap. 5, we look at the single most famous distribution in statistics—the Gaussian or "normal" distribution, which is extremely widespread in Nature, in a huge range of apparently completely unrelated circumstances. This mysterious physical universality actually turns to be a kind of (almost) inevitable mathematical convergence that happens when we combine many different random events. To see how this works, we will need to understand how adding two random variables produces a *convolution* of their probability distributions. We also see how some other well-known physical distributions—for example, the Maxwell–Boltzmann distribution of velocities—are really Gaussians in disguise.

Finally, in Chap. 6, we look at frequency distributions that arise from *stochastic processes*—i.e. sequences of random events in time. This leads to the concept of a *random walk*, and two more frequency distributions of great importance in Physics—the Lorentzian distribution, and the power law distribution.

Chapter 3
Counting the Ways: Arrangements and Subsets

3.1 Outline of Content

- Balls, slots, boxes and labels
- Multi-step operations
- Arrangements or permutations
- Subsets or Combinations
- Macrostates and Microstates
- Finding the most probable macrostate
- Examples in Statistical Physics

This chapter is all about counting the ways we can arrange things, or create combinations or subsets of objects. Mathematically, the topic of *combinatorics* is at the heart of understanding a variety of naturally occurring distributions, and also underlies much of statistical inference, but it is also of direct physical interest. For example, we learn that the entropy of a closed system never decreases—but why is this? It is essentially because some configurations of molecules in a box are more common than others. But how much more common? As we shall see, the answer is *hugely* more common. In a similar fashion, the remorseless logic of combinations shows that particles must be distributed exponentially in energy. A little simple combinatorics goes a long way.

3.2 Balls, Slots, Boxes, and Labels

Combinatorics can get very confusing. I find it helpful to think in terms of a simple uniform physical picture, that any other situation can be mapped on to. Some problems boil down to starting with a collection of balls, and thinking about putting the balls into a set of slots, where each slot holds one ball. (There could be some empty slots however, or there might be more balls than slots.) Other problems boil down

A. Lawrence, *Probability in Physics*, Undergraduate Lecture Notes in Physics,
https://doi.org/10.1007/978-3-030-04544-9_3

to putting the balls into a set of boxes, where each box could hold several balls. The question is always "how many different ways can I carry out this operation?"

What counts as a "different way" depends however on which elements are *distinguishable* and which are not. You can always think of this as depending on whether the balls or slots or boxes have *labels* on—if they don't, you can't tell them apart. You might have various ways of arranging some labelled balls; but if you take the labels off, you can't tell which arrangement is which, so those don't count as different ways. Sometimes you will see a distinction made concerning whether the *order* of selected elements matters, but this is essentially the same issue as labelling. Imagine putting balls into some order—then position-1, position-2 etc is just like labelling slot-1, slot-2 and so on. If you take the labels off the position-slots, you can then shuffle them round and not know the difference. When a problem in combinatorics is getting confusing, it can often be helpful to stop and ask yourself "could I tell the difference?"

You may worry that the question of distinguishability is subjective. For example, most people can tell red and green balls apart, but some colour-blind people cannot. But as we discussed in Chap. 1, "subjective" doesn't mean "vague and woolly". It just means 'observer dependent"; and in many physical circumstances, such as knowing the properties of molecules in a container of gas, every plausible observer is in an equal position of ignorance.

3.3 Multi-step Operations

First, the most basic rule. If an operation is made up of multiple consecutive steps, with step-1 having n_1 choices, step-2 having n_2 choices and so on, the total number of ways I could make these choices is

$$W_{\text{choice}} = n_1 \times n_2 \times n_3 ... \tag{3.1}$$

If a menu has 5 starter choices, 4 main course choices, and 6 dessert choices, we could pick 120 possible different dinners. Sometimes it can be helpful to illustrate the situation as a tree diagram.

3.4 Arrangements or Permutations

How many different ways are there to arrange n distinguishable items? Suppose I have an apple, and orange and a banana. In how many different orders can I eat them? For my first item I can pick one of three; for my second, there are only two left; and for the last item, there is only one choice—so the answer is clearly $W = 3 \times 2 \times 1 = 6$.

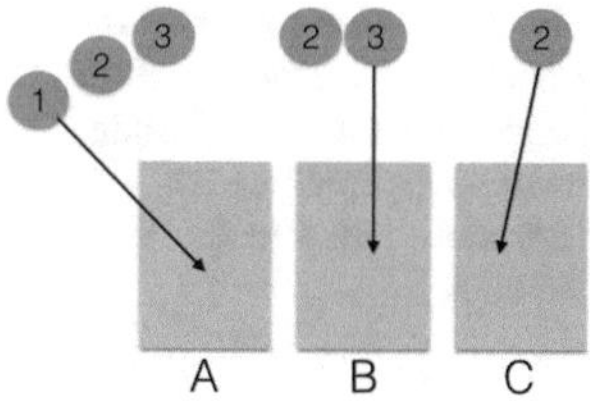

Fig. 3.1 Arranging labelled balls in labelled slots

In general, we can imagine putting n labelled balls into n labelled slots. Its important that *both* the balls and the slots are labelled. There are n ways to fill the first slot, and then $n - 1$ choices for the second slot, and so on, as illustrated in Fig. 3.1. Then we apply the multi-step multiplication rule, and the number of possible arrangements is $W = n!$.

Now suppose the labels are taken off either the balls or the slots so that they are *indistinguishable*. Then the number of *distinguishable arrangements* is just 1. You could shuffle the balls around—or shuffle the slots around—and not know the difference.

3.4.1 *Arranging* r *Things Out of* n

How many ways are there to arrange n things if they are taken r at a time? This is like having r labelled slots and n labelled balls, where $r < n$. For example suppose I have an apple, an orange, a banana, and an apricot, but I decide only to eat two things. The number of ways is now $W = 5 \times 4 = 20$. In general, we have the same slot-filling process, but it stops earlier, so that $W = n \times (n - 1) \times (n - 2) ... \times (n - r + 1)$. This can be written as

$$W_{\text{perm}} = \frac{n!}{(n-r)!} = {}_nP_r = P(n, r). \tag{3.2}$$

The notation ${}_nP_r$ or $P(n, r)$ is used because mathematicians refer to such arrangements as *permutations*. It can be read as "perm r out of n". With a little thought you can see that putting r things into n slots—i.e. where we leave some of the slots empty, as opposed to some of the balls unused—has the same number of permutations as putting some of n things into r slots.

Another example might be picking two letters out of the list a, b, c, d. There are twelve ways—ab, ac, ad; ba, bc, bd; ca, cb, cd; and da, db, dc. This is indeed $4!/(4 - 2)!$. In setting out these choices, we have cared about the order or slot labelling—ab is not the same as ba. But suppose we don't care about the order? That is the problem of picking subsets, which we look at next.

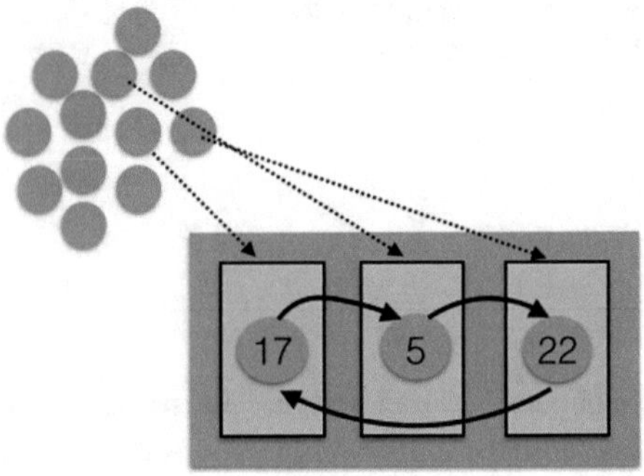

Fig. 3.2 Picking a subset of r balls from a larger collection of n balls. We can imagine first picking balls for r slots, and then shuffling the balls

3.5 Subsets or Combinations

How many ways are there to pick r objects out of n, if we don't care how the r balls are arranged? This is like having a box that can hold r loose balls. We could start by temporarily imagining the box to have r slots inside it, with these slots labelled. We can put balls into these slots ${}_nP_r = n!/(n-r)!$ different ways, as above. Next, imagine taking any one of those arrangements and shuffling the r balls, as illustrated in Fig. 3.2—there are $r!$ different ways we could put the r balls in the r slots. If we then imagine taking the labels off the slots, all those arrangements look the same, so in our earlier total we have overcounted by that factor. Finally then, the number of distinguishable subsets is

$$W_{\text{comb}} = \frac{n!}{r!(n-r)!} = {}_nC_r = \binom{n}{r} = C(n,r). \tag{3.3}$$

The notation ${}_nC_r$ can be read as "choose r from n". Subsets are usually known by mathematicians as "combinations" which is why the symbol C is used. The other notation $\binom{n}{r}$ is also popular, and could be read as "n over r" or again as "choose r from n".

Stirling's approximation. Choosing two out of four pieces of fruit is simple enough, but what if we are choosing 27 students out of a class of a 150? Or considering millions out of billions of molecules? Calculating large factorials is extremely laborious. A convenient approximation is *Stirling's formula*:

$$n! \sim \sqrt{2\pi n}\left(\frac{n}{e}\right)^n.$$

This is proved in many maths textbooks. It is fairly accurate for any n larger than a few. For example, in the game of bridge, you select a hand of 13 cards from a deck of 52 cards. The order doesn't matter, so the number of ways of doing this is 52!/13!39!. Calculating those factorials by hand would take you a long time, but with Stirling's formula you can quickly confirm that the answer is about 635 billion. For larger n, it is often convenient to work with the natural log of $n!$, in which case

$$\ln n! \sim n \ln n - n + \frac{1}{2}\ln n + \frac{1}{2}\ln 2\pi .$$

For very large n, this approximates further to

$$\ln n! \sim n \ln n - n, \tag{3.4}$$

which is the form you will see most often in Physics textbooks. However, be careful to use this only for $n > 100$ or so. For $n = 10$ the true value is $\ln n! = 15.014$; Stirling's formula gives 15.096; but the $n \ln n - n$ version gives 13.026. This is a 13% error in $\ln n!$, and a factor of 8 error in $n!$ itself. By $n = 100$ we have only a 1% error in $\ln n!$, although of course still a considerable error in $n!$.

3.6 Partitioning: Macrostates and Microstates

What if, rather than choosing a single subset of a pool of objects, we want to divide up our pool of objects into a number of boxes? For example, we might have 40 children we want to divide into 8 different 5-a-side football teams in order to hold a tournament. We could be said to be *partitioning* our collection of objects. Much of Statistical Physics is based on the logic of partitioning—for example a large number of atoms can each be in one of a number of different energy states. How many ways can this happen, and which energy states are more likely to be populated? Lets start by taking the easiest problem—dividing into two states or boxes.

3.6.1 Two-Box Problems

We have already solved this problem. In the subset-choosing problem above, when we have picked the r balls, we have $n - r$ left over. We have therefore partitioned our n balls into a group of r and a group of $n - r$. This is therefore the same as dividing our n balls amongst two boxes. In standard Physics terminology, if we specify *how many balls* are in box A and how many in box B that tells us the *macrostate*. If we specify exactly which balls are in box A and which are in box B, that tells us the *microstate*. Older books refer to "configurations and complexions" rather than "macrostates and microstates". I rather like this older terminology because it is more general, but many authors use "configuration" rather loosely to mean any of these things, whereas "macrostate and microstate" are fairly unambiguous.

Some systems will continuously explore all the possible microstates. For example, if our boxes are imaginary regions of a container of gas, the motions of the gas molecules will cause the populations of those regions to be constantly changing. A system that, over time, spends equal amounts of time in each microstate is known as *ergodic*. However, such a system will not spend equal amounts of time in a given macrostate. The key question is—how many microstates correspond to the same macrostate? We can refer to this as the *multiplicity* W of that macrostate. If the i'th macrostate corresponds to W_i different microstates, then the total number of microstates is $N = \sum W_i$, and $P_i = W_i/N$ is the probability of finding our system in macrostate i.

3.6.2 Example: The Two-State Particle Problem

Suppose we have a number of particles, each of which has a quantum spin which can be in one of two states, up or down. How likely is it that you will find all of the particles in the up-state? Suppose we have just 4 particles and we distribute them at random between the two states. If we take the macrostate (4, 0), that is, 4 particles in the up state, and none in the down state, we can see that the number of microstates is $W = 1$—there is only one way of doing this. On the other hand, for the macrostate (2,2), we have $W = 4!/(2! \times 2!) = 6$ possible microstates. So having the particles evenly spread is six times as likely as them being all up. If the particles flip between states at random, they will *sometimes* all be up; they should spend $1/2^n = 1/16$th of the time in that macrostate. However many particles there are, there is only one way of having them all in the up-state. For $n = 10$ the even macrostate (5,5) has $W = 252$ microstates. For $n = 100$ we get $W = 1.01 \times 10^{29}$; for larger numbers W becomes enormously large very quickly. You can see that for realistic numbers of particles it basically never happens that they are all in the up-state at the same time.

3.6.3 A First Look at the Second Law of Thermodynamics

The same logic applies to the classic gas-molecules all-at-one-end problem. Suppose we divide a container of gas into two virtual boxes, by imagining an invisible dividing line half-way across. If molecules move around the container at random, they will sometimes by chance all gather in just one of the boxes—just very, very, very, very rarely. More generally, if a system is in a state with some value of W below the maximum, then the next state it finds itself in will almost always have a larger value of W, simply because there are many more microstates to choose from. Roughly speaking, this is the microscopic cause of the second law of thermodynamics; but we will look at this issue more carefully in Chap. 14.

3.6.4 Most Probable Macrostate

It is instinctively obvious that the macrostate with an equal population in each box will have the most microstates, but lets prove this. Suppose we have n balls in total, and put r in box A and $n-r$ in box B. Let us suppose that r is less than $n/2$, and so $r < (n-r)$. Now suppose we move one ball from box A to box B, so that box A now contains $r-1$ balls. Lets calculate the multiplicity for the old macrostate and the new macrostate.

$$W_r = \frac{n!}{r!(n-r)!}, \qquad W_{r-1} = \frac{n!}{(r-1)!(n-r+1)!}.$$

Now we take the ratio of the two multiplicities, term by term

$$\begin{aligned}\frac{W_r}{W_{r-1}} &= \tfrac{n!}{n!}\tfrac{(r-1)!}{r!}\tfrac{(n-r+1)!}{(n-r)!} \\ &= 1.\tfrac{1}{r}.(n-r+1) \\ &= \tfrac{n}{r} - 1 + \tfrac{1}{r} \\ &\sim \tfrac{n}{r} - 1\end{aligned}$$

where in the last step we have assumed that both n and r are large. Because we started by assuming that $r < n/2$ then the ratio of multiplicities is > 1. In other words, if a box has less than half the balls, and we transfer a ball out, we will *reduce* the multiplicity. At the maximum multiplicity, $W(r)$ will be flat, so transferring a ball will make no difference. We then find

$$\frac{W_r}{W_{r-1}} = 1 \quad \Longrightarrow 1 = \frac{n}{r} - 1 \quad \Longrightarrow r = n/2.$$

3.7 Multi-box Partitioning

Now suppose that rather than partitioning into two groups, we want to partition into many groups. This is like putting our n balls into k boxes. The boxes are labelled (distinguishable), and we have n_1 balls in box-1, n_2 balls in box-2, etc. The values of n_1, n_2 etc are the *populations* of each box. The whole set of numbers $n_1, n_2, n_3, \ldots, n_k$ is the *macrostate*, or the *configuration*. If the balls are also labelled, we could exchange balls between boxes, making many different distinguishable arrangements that correspond to the same macrostates. These arrangements are the *microstates* or *complexions*. Moving the balls around within a box makes no distinguishable difference. These concepts are illustrated in Fig. 3.3

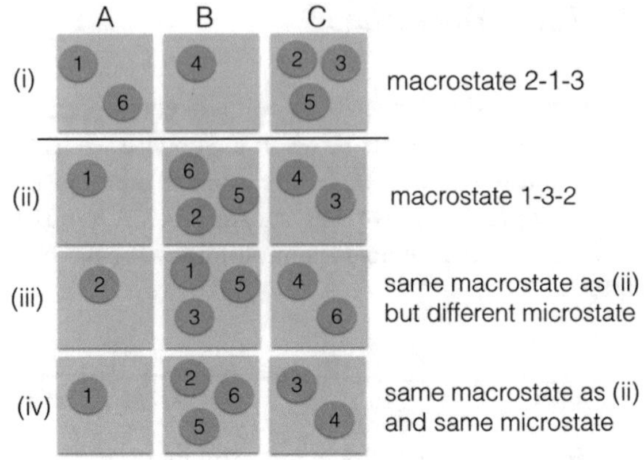

Fig. 3.3 Partitioning 6 balls into 3 boxes. The upper two rows show two different macrostates. The third row is the same macrostate as row (ii), as the boxes have the same populations—1, 3, 2—but its a different microstate, because different balls are in each box. The fourth row is the same macrostate as row (ii), and is also the same microstate—the numbered balls in each box are the same. Although the balls are drawn for the purposes of illustration in apparently different positions from our "God's eye" point of view, there are actually no distinguishable positions within the boxes

So how many microstates are there for a given macrostate? We do this calculation in a similar fashion to the two-box problem above. You could temporarily imagine the boxes to have many little slots in, and once again initially consider these slots to be labelled. There are n slots in total, across all the boxes, so we have $n!$ arrangements as usual. Now take the slot-labels off. Within any one box we can shuffle the balls around and not know the difference; but we can't shuffle balls between boxes because we know which box is which. Considering box-1, there are $n_1!$ arrangements which we have overcounted, and likewise for all the other boxes. So finally, the number of distinguishable ways to achieve that macrostate is

$$W = \frac{n!}{n_1!n_2!...n_k!} = \binom{n}{n_1, n_2, \ldots, n_k}. \tag{3.5}$$

3.7.1 *Maximising Multiplicity for a Multi-box System*

Just as with the two-box problem, the macrostate with the largest multiplicity is the one with a uniform distribution. You can see this as follows. Consider two specific boxes, and vary the number in each of those two boxes, while keeping the populations of all the other boxes fixed. From our two-box analysis, we know that we will maximise W if we make the population of those two boxes the same. However, we can apply this logic to *any* two boxes. So all the boxes must have the same population as all the other boxes.

3.7.2 *Continuous Systems*

Suppose we want to consider the spatial distribution of molecules in a container. Putting quantum physics to one side, this seems a problem because there are no discrete boxes. However, we can arbitrarily divide space into cells, and use the above logic to conclude that the most likely division is where there are equal numbers of molecules in each cell. As long as we only care about the *relative* multiplicity, we can apply this analysis for any cell size we like—the answer is always that a uniform distribution is very strongly favoured. Quantum physics of course does give us a kind of natural gridding of space, although the calculation is a bit subtle.

3.8 Systems with Extra Constraints

Suppose we are handing out prizes to a small class of 4 students. Each student either gets a prize or no prize. How many different ways can we do that? Think of the students as the balls, allocated at random to a no-prize box and a prize-box. The number of ways is given by the usual two-box formula, and the macrostate with the largest multiplicity is the one with two students in each prize category, with $W = 6$. However, suppose we have only prize to give out. We can't put two students in each category. We have to put one student in the prize-box, and the other three in the no-prize-box. There are $W = 4$ ways to do this.

Now, suppose we decide instead we have three prizes to hand out (perhaps for different categories of performance). We now have four boxes—no prize, one prize, two prizes, three prizes. Most of the allocations of students to prize-boxes are not allowed because we wouldn't have enough prizes. There are just three allowed macrostates, where the box populations are 3-0-0-1, 2-1-1-0, 1-3-0-0. These have multiplicities of $W = 4$, 12 and 4. (Fairly obvious in for the first and last cases, but needs a little bit more working out in the middle case.) It is in general much easier to populate the "lower" levels, and the populations steadily decline. Averaging across all the microstates, the four prize-boxes have average populations of, 2.0, 1.2, 0.6, and 0.2. If we increase the number of balls and boxes, the decline becomes more and more obvious. As we shall see below, for large numbers there is a simple analytic solution to the box-population distribution.

3.8.1 *Particle Energy Distributions*

The most important physics example of a system with an extra constraint is the distribution of particles amongst their allowed energy states, under the condition that the total summed energy of all the particles is constant. Suppose we have a series of possible states i, with an energy jump between each of size ε. For simplicity we will label the lowest energy state $i = 0$. There are n_i particles in state i. Then we have two constraints. The first is that the total number of particles is fixed:

$$n = n_0 + n_1 ... = \sum_i n_i,$$

and the second is that the total energy is fixed:

$$E = 0 \times n_0 + \varepsilon n_1 + 2\varepsilon n_1 ... = \sum_i i n_i \varepsilon.$$

The sum over i runs to a large enough number to account for all the particles. Under these constraints, what set of n_i values, i.e. the population distribution, gives the maximum W? We proceed in a similar manner to Sect. 3.6.1. We look at neighbouring energy states, move particles between them, and argue that at the maximum of W this will have no effect. Take the i'th energy state. We cannot simply move one particle to its neighbour, because this will change the total energy. However we can remove *two* particles and move one up and one down. Then the before and after values of W are:

$$W = \frac{N!}{n_0!n_1!n_2!...}, \qquad W_* = \frac{N!}{n_0!n_1!...(n_{i-1}+1)!(n_i-2)!(n_{i+1}+1)!...}.$$

Taking the ratio of these we have

$$\frac{W_*}{W} = \frac{n_i(n_i-1)}{(n_{i-1}+1)(n_{i+1}+1)} \sim \frac{n_i^2}{n_{i-1}n_{i+1}},$$

where the last step assumes n_i is large. If we then require the ratio to be 1, we find

$$\frac{n_{i-1}}{n_i} = \frac{n_i}{n_{i+1}}, \quad \text{and so} \quad \frac{n_0}{n_1} = \frac{n_1}{n_2} ... = \frac{n_i}{n_{i+1}} = \text{const.}$$

A constant ratio is of course satisfied by an exponential function, i.e. we have shown that the most likely population distribution is given by

$$n_i = n_0 e^{-Ai}. \qquad \text{The Boltzmann distribution} \tag{3.6}$$

The value of the constant A can be found from the total energy E, and is of course related to the *temperature* of the system of particles. We won't fill out those details here, which you can find in any textbook on Statistical Mechanics. The main point here is to show that the exponential distribution comes from relatively simple combinatoric considerations, and will also apply to any similar "sum-constrained" situation.

3.8.2 Lagrange Multiplier Method

Finally, a brief introduction to a more general method for solving maximisation problems with constraints. Consider a function of two variables, $f(x, y)$. With standard calculus techniques, we can find the x, y location where f is a maximum by setting $\partial f/\partial x = 0$ and $\partial f/\partial y = 0$. This gives us two simultaneous equations which we can solve for x and y. Now however, suppose that allowed values of x and y are constrained to lie on the line $y = ax + b$. You can imagine tracking along that locus, watching how f changes, and seeing where f comes to a maximum. This will not in general be the same place as the unconstrained maximum of f. Any such constraint can be expressed in the form $g(x, y) = 0$—in our straight line case, $g(x, y) = y - ax - b$. The general way to solve such a problem is to define a new function called the *Lagrangian*:

$$L(x, y, \lambda) = f(x, y) + \lambda g(x, y),$$

where λ is a new variable called the "Lagrange multiplier". We then set

$$\frac{\partial L}{\partial x} = 0 \quad \text{and} \quad \frac{\partial L}{\partial y} = 0 \quad \text{and} \quad \frac{\partial L}{\partial \lambda} = 0,$$

which gives us three simultaneous equations to solve for x, y, λ. Usually we don't care about the value of λ, but we get the location x, y of the constrained maximum.

This technique can be extended to any number of variables $x_1, x_2, \ldots, x_k$, and any number of constraints $g_1, g_2, \ldots, g_m$, and a Lagrangian $L(x_1, x_2, \ldots, x_k, \lambda_1, \lambda_2, \ldots, \lambda_m)$. We then set each $\partial L/\partial x_i = 0$ and each $\partial L/\partial \lambda_j = 0$, which gives us $k + m$ simultaneous equations to solve. We can apply this general technique to our multi-box partitioning problem. The k variables are the population values $n_1, n_2, \ldots$, with the function we want to maximise being $W(n_1, n_2..)$ as in equation (3.5). The constraint is that $\sum n_i = n$. We won't wade through the solution of the k equations, but you can then show that W is maximised if for all i, $n_i = n/k$. In a similar fashion, we can then add a second constraint, keeping $E = \sum i n_i \varepsilon$ constant. This also leads to the same solution we had before, an exponentially declining distribution in n_i.

3.9 Key Concepts

Some of the key concepts from this chapter are:

- The distinction between distinguishable and indistinguishable elements
- How to calculate the number of permutations (ordered arrangements) of objects, and the number of combinations (subsets)
- The difference between a microstate and a macrostate
- How to calculate the multiplicity of a macrostate
- The idea of maximising the probability of a macrostate.

Key formulae in this chapter include: how to calculate the number of choices (3.1) in a multi-step operation; the number of permutations (3.2) and the number of combinations (3.3) when picking r objects from n; Stirling's approximation for calculating large factorials (3.4); the formula for the number of ways to do a "multi-bucket" partitioning (3.5); and population distribution resulting from a sum-constrained situation, i.e. the Boltzmann distribution (3.6).

3.10 Further Reading

The ideas and techniques of combinatorics are covered in all basic statistics textbooks, but there are also textbooks dedicated specifically to combinatorics, with an example being Cameron (1994). A very short, fun, and readable approach is in Wilson (2016).

The application of these ideas in statistical physics is likewise covered in many textbooks at a variety of levels. Two excellent books that we recommend here in Edinburgh are Baierlein (2010) and Ford (2013). A much older book which I still like very much, as it is so clear on the statistical underpinnings of the physics, is Brown (1968).

3.11 Exercises

3.1 How many seating arrangements are there for a dinner party of 5? (Think it through ... its not quite the same as a normal permutation..)

3.2 The e-reader known as "kobo" is clearly an anagram of "book". How many distinct anagrams could the designers have considered? How many would there have been if they had been using "books" instead of "book"?

3.3 A departmental committee of senior professors must contain 3 men and 3 women. There 11 eligible male professors and 5 eligible female professors. How many different committees can be formed?

3.4 How many different ways can you pick two five a side football teams from 12 students?

3.5 A cricket team takes a squad of 16 players on tour. For any one game, how many ways could you pick a team of 11 players? Calculate exact and approximate answers and compare.

3.6 Six people are playing Dungeons and Dragons. Every ten minutes on average the game involves all players except the Dungeon Master rolling a D20, that is, a twenty-sided die. How long might we expect before everybody rolls a score of 20?

References

Baierlein, R.: Thermal Physics. Cambridge University Press, Cambridge (2010)
Brown, A.: Statistical Physics. Edinburgh University Press, Edinburgh (1968)
Cameron, P.J.: Combinatorics: Topics, Techniques, Algorithms. Cambridge University Press, Cambridge (1994)
Ford, I.: Statistical Physics: An Entropic Approach. Wiley-Blackwell, Hoboken (2013)
Wilson, R.W.: Combinatorics: A Very Short Introduction. Oxford University Press, Oxford (2016)

Chapter 4
Counting Statistics: Binomial and Poisson Distributions

4.1 Outline of Content

- The binomial distribution and its properties
- Variations on the binomial distribution
- Counting rare events
- Deriving the Poisson distribution
- Properties of the Poisson distribution
- Applications of the Poisson distribution

A large number of physical situations are at root "either/or" questions, or if you like, "hit/miss" questions. We toss a coin—is it heads or tails? During the previous one second of time, did our radiation detector detect a beta particle or not? A simple yes/no experiment like this is known as a *Bernoulli trial*, and the sequence of outcomes is a *Bernoulli process*.

In such a Bernoulli process, we can count the hits, and ask questions such as—after twelve tosses, whats the probability of getting four heads? Or nine heads? After running our detector for a minute, whats the typical number of beta particle events? If I find an average of say 3.7 events per minute, whats the probability I will actually get 6 events? Or no events? The solution to these "counting statistics" problems follows smoothly on from our analysis of two-box partitions in the previous chapter. The key extra feature we need is to allow the hits for each trial to have a different probability from the misses. However, we will start with the simplest version, assuming that hits and misses are equally likely.

4.2 The Binomial Distribution

4.2.1 The Simplest Hit-and-Miss Problem

Let's proceed using our standard boxes-and-balls metaphor. The general idea is illustrated in Fig. 4.1. Box-A can be seen as holding the hits (getting heads, rolling a six, seeing a particle event, etc). If we have n trials, the number of ways to put r balls in

A. Lawrence, *Probability in Physics*, Undergraduate Lecture Notes in Physics,
https://doi.org/10.1007/978-3-030-04544-9_4

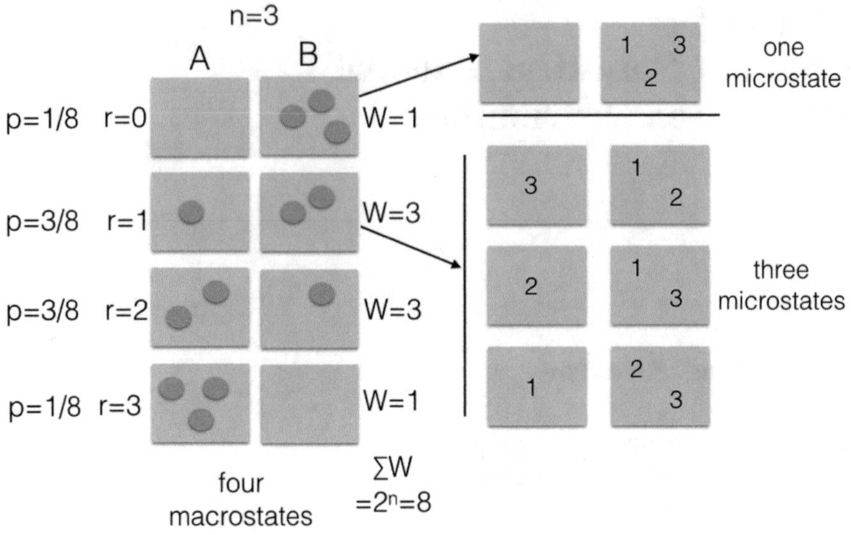

Fig. 4.1 How to distribute 3 balls amongst 2 boxes. On the left we show the four different macrostates, with $r = 0, 1, 2, 3$. For the first two, we illustrate the different possible microstates. To get the probability of each macrostate, we need to normalise its multiplicity W by the sum of W for all the possible macrostates

Box-A will just be $W(r) = {}_nC_r$. If the two boxes are equally likely, then to turn $W(r)$ into a probability we just need to divide by the *total* number of microstates, across all the possible macrostates, so that $f(r) = W(r)/\sum_r W(r)$. The total number of microstates is just $\sum W = 2^n$. Imagine taking each ball one at a time, and deciding whether to place it in Box-A or Box-B. This is just the menu-choice problem, so the total number of choices is $2 \times 2 \times 2 \cdots = 2^n$. The net result then is that the probability of getting r heads out of n coin tosses is

$$f_n(r) = \binom{n}{r}\left(\frac{1}{2}\right)^n = \frac{n!}{r!(n-r)!}\left(\frac{1}{2}\right)^n.$$

We have written this as $f_n(r)$ to emphasise that we are interested in how this probability varies with r, for a given value of n.

4.2.2 *Varying the Hit-Probability*

What if the two boxes do not have equal probability? Suppose Box-A, the hit-box, has probability p. For example, when rolling a six-sided die, if we are counting sixes as the hits, then $p = 1/6$, whereas if we are tossing a coin and counting heads as the hits, then $p = 1/2$. Box-B of course has probability $1 - p$. Consider the macrostate with r balls in Box-A, and $n - r$ balls in Box-B. Lets think about a specific microstate corresponding to that macrostate, i.e. with r specific labelled balls in the box. Imagine taking all the n balls one at a time and choosing where to put them; there is a probability p of landing in Box-A and $1 - p$ of landing in Box-B. The probability of getting all of the correct balls in box A is therefore p^r. Likewise the probability of getting a specific set of $n - r$ balls in box B is $(1 - p)^{n-r}$. The net probability of *that microstate* is $p^r(1 - p)^{n-r}$. To get the probability of the *parent macrostate* we multiply by the multiplicity W of that macrostate. Finally then the probability of r successes is

$$f_{n,p}(r) = \binom{n}{r} p^r(1-p)^{n-r} = \frac{n!}{r!(n-r)!} p^r(1-p)^{n-r}. \tag{4.1}$$

This is known as the *binomial distribution*. The term "binomial" occurs in another important place in mathematics, and the use of the same term is not a coincidence. Consider expanding the expression $(x + y)^n$ for various values of n. For example $(x + y)^3 = x^3 + 3x^2y + 3xy^2 + y^3$. In general you can show that

$$(x+y)^n = \sum a_r x^{n-r} y^r \qquad \text{where} \qquad a_r = \binom{n}{r}.$$

This is essentially the same result as our binomial distribution, with x and y playing the parts of our two probabilities p and $1 - p$. This happens because the coefficients of the expansion are given by the number of different ways you can get r lots of y and $n - r$ lots of x.

4.3 Properties of the Binomial Distribution

Lets explore the binomial distribution and see how it behaves.

4.3.1 Mean Value of the Binomial Distribution

It is intuitively obvious that the mean (expected) value of r is just $\mu = np$. If we toss a coin 200 times, we expect 100 heads; if we roll a die 36 times we expect to get 6 sixes. Lets prove this however. Recall that the expected value of r for a discrete probability distribution $p(r)$ is $\sum r \cdot p(r)$ and so in this case

$$\mu = \sum_{r=0}^{n} \left[r \frac{n!}{r!(n-r)!} p^r(1-p)^{n-r} \right].$$

The first term in the sum, with $r = 0$, is 0, so we can take this out. Next, note that $r! = r(r - 1)!$, so the r there cancels with the initial r. So we now have

$$\mu = \sum_{r=1}^{n} \left[\frac{n!}{(r-1)!(n-r)!} p^r (1-p)^{n-r} \right].$$

Next, anticipating the answer, we can take out one factor of n and one factor of p and get

$$\mu = np \sum_{r=1}^{n} \binom{n-1}{r-1} p^{r-1} (1-p)^{n-r}.$$

Now we define $q = r - 1$ and $m = n - 1$ so that this becomes

$$\mu = np \sum_{q=0}^{m} \binom{m}{q} p^q (1-p)^{m-q}.$$

However, the terms inside the sum are just the binomial distribution terms for our new variables q and m. Because it is a normalised probability distribution, the sum of these terms must add up to 1. So finally we have

$$\mu_{\text{bin}} = np. \tag{4.2}$$

Note that although the possible values of r can only be integers, their mean value μ will generally be a real number, because the probability p is a real number.

4.3.2 *Variance of the Binomial Distribution*

The variance is given by $\sigma^2 = E[(r - \mu)^2]$. The sum can be solved by a very similar technique to that shown above but we won't wade through it. The answer is that

$$\sigma^2_{\text{bin}} = np(1-p) = \mu(1-p). \tag{4.3}$$

For the simplest case with $p = 1/2$ we get $\sigma^2 = \mu/2 = n/4$. It is instructive to look at the coefficient of variation: $\sigma/\mu = 1/n^{1/2}$. For large numbers of trials, the distribution gets narrower in relative terms. For large numbers, this has a dramatic effect. For $n = 10$, the uniform macrostate with $r = 5$ is 252 times more likely than the macrostate with $r = 0$. For $n = 1000$ the macrostate with $r = 500$ is $\sim 3 \times 10^{299}$ times more likely than the state with $r = 0$; but it is even $\sim 5 \times 10^8$ times more likely than the state with $r = 400$. By the time we reach the values appropriate to the number of molecules in a box full of gas, the probability of being anything other than a miniscule distance from the uniform macrostate is vanishingly small.

4.3.3 Shape of the Binomial Distribution

Figures 4.2 and 4.3 plot the binomial distribution for different values of n and p, comparing two values of n (5 and 20) and two values of p (0.5 and 0.2).

For $p = 0.5$ the shape is always symmetrical. At low n the distribution appears quite broad, but appears narrower as n increases, as shown by the algebra above.

For small p the shape is strongly asymmetric. As you increase n, the asymmetry decreases, and is hardly noticeable even for a quite modest value like $n = 20$.

Notice also that the absolute value of the probability at the peak of the distribution gets smaller with larger n. This is an example of the point made in Chap. 1, that the absolute value of probability is not always what you want. The integrated probability over a range is often more useful.

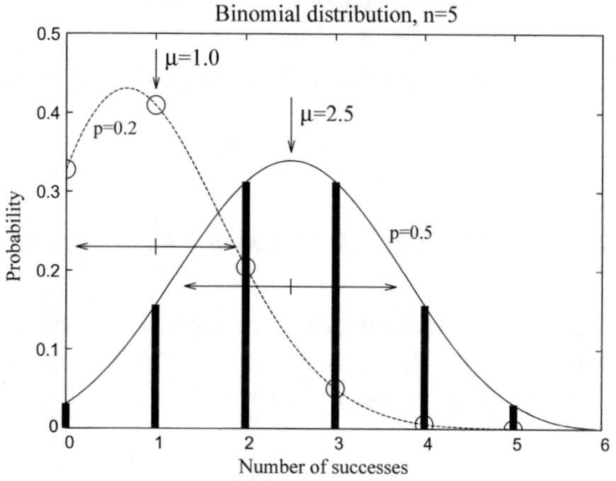

Fig. 4.2 Illustrating the binomial distribution for $n = 5$ and two different values of p. The horizontal lines indicate $\pm 1\sigma$ from the mean; but note that the $p = 0.2$ curve is not symmetrical. Note that the distribution is only defined at discrete values of m; a smooth curve has been drawn however using the Γ function, as explained in the text

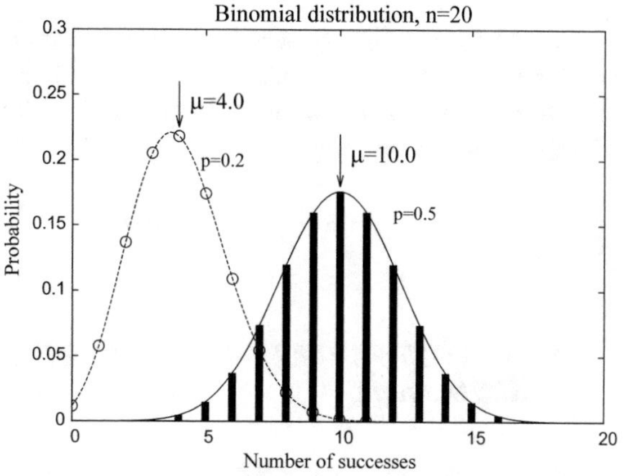

Fig. 4.3 Binomial distribution for the slightly larger value of $n = 20$. Even the smaller p version is now almost symmetrical

4.3.4 Continuous Approximation to the Binomial Distribution

The function representing the binomial distribution is only defined at integer values of r; however we can plot a smooth curve through these points using the gamma function, which is equal to the factorial for integer values:

$$\Gamma(n) = (n-1)! \quad \text{where} \quad \Gamma(z) = \int_0^\infty t^{z-1} e^{-t} \mathrm{d}t.$$

This can be useful for calculations, but also simply to make smooth plots.

4.4 Variations on the Binomial Distribution

There are some variations on the binomial that come in handy in various circumstances.

4.4.1 The Negative Binomial Distribution

Imagine a series of children each exposed to a virus, with a probability p that each child will catch the virus. What is the probability that the xth child will be the kth one to catch the virus? The answer is

$$b_{k,p}(x) = \binom{x-1}{k-1} p^k (1-p)^{x-k}.$$

This is known as the *negative binomial distribution* or sometimes as the *binomial waiting time distribution*. A useful special case is where $k = 1$, i.e. the probability that we have to wait until the xth trial to get the first hit. This is known as the *geometric distribution* and is given by

$$g_p(x) = p(1-p)^{x-1}.$$

4.4.2 The Hypergeometric Distribution: Binomial Without Relacement

Imagine an urn containing N balls of which W are white and $N - W$ are black. If we pick a ball at random it has probability $p = W/N$ of being white. If we pick n

balls in succession, what is the probability that r of the n will be white? The answer depends on whether we replace the balls after each pick. If we do replace the balls, then the probability $p = W/N$ is the same for each pick, and the situation is exactly like our standard binomial case. However if we do not replace the balls, we get the *Hypergeometric distribution*

$$P(r) = \frac{\binom{W}{r}\binom{N-W}{n-r}}{\binom{N}{n}}.$$

4.4.3 The Multinomial Distribution

In Chap. 3 we looked at how to partition N objects into k boxes, and derived the number of ways to achieve this partition. To complete the picture, if rather than having equal probabilities we have a situation where box i has probability p_i, defined of course so that all the p_i values add up to 1, then the joint probability of getting r_1 objects in box-1, r_2 objects in box-2 and so on is given by

$$f(x_1, x_2, \ldots, x_k) = \binom{N}{r_1, r_2, \ldots, r_k} p_1^{r_1} p_2^{r_2} \ldots p_k^{r_k}.$$

This has direct relevance to many statistical physics problems. Particles may for example be distributed at random amongst various quantum states with differing energy. If the states are equally likely, but with the total energy of the system constrained, then we will end up with the Boltzmann distribution equation (3.6). However, the energy-states may not be equally likely; there may be several quantum states with the same energy, and so the distribution is adjusted by the "degeneracy" of each energy state.

4.5 Counting Rare Things: The Poisson Distribution

The *Poisson distribution* is a special case of the Binomial distribution, where $p \to 0$ and $n \to \infty$, such that the mean $\mu = np$ remains a finite, middling sort of number. Why is this useful?

Imagine a radiation source emitting alpha particles which we detect with a Geiger counter and hear as a series of clicks. For a large piece of radioactive material, the rate of clicks is quite fast, and seems to be more or less constant. For a smaller piece of material perhaps we are down to something like two clicks per second, but now it seems to be quite erratic—sometimes there are four or five clicks in a second, and sometimes none. Why does this happen? Can we predict the distribution of clicks per second?

The trick is to consider each second of time for a single atom as an individual hit-or-miss trial, but the behaviour of all the atoms in the lump of material as the n repeats. The probability p of an individual atom emitting a particle in that second is extremely small; but the number of atoms n is extremely large. Then the overall number of events per second will follow a binomial distribution with very small p and very large n. As another example, imagine a large department store that sells expensive diamond rings. Thousands of shoppers glance at the rings, but there is only a small chance that any one person will buy one. However, over time, the shop perhaps finds that it consistently sells an average of 2.7 rings per week. What then is the chance that in a particular week none will be sold, or six?

In principle, if we know p and n, we could treat such problems with the binomial formula. However, in the limit of very small p and very large n we can show that the distribution *only* depends on the mean μ, so that we don't need to know p and n. Lets look at how to prove this.

4.6 Derivation of the Poisson Distribution

The Poisson distribution is a special case of the binomial distribution, where $p \to 0$ and $n \to \infty$, in such a way that the product $\mu = np$ remains finite. We can write the binomial distribution as

$$\underset{①}{\frac{1}{r!}} \qquad \underset{②}{\frac{n!}{(n-r)!}} \qquad \underset{③}{p^r} \qquad \underset{④}{(1-p)^{-r}} \qquad \underset{⑤}{(1-p)^n}$$

Lets look at the five terms one by one. Term ① we will just leave. Term ② is $n(n-1)(n-2)\cdots(n-(r-2))(n-(r-1))$. This is made of r factors, where each is $\simeq n$ because n is large; the second term therefore tends towards n^r. Taking the second and third terms together, we have $n^r p^r = (np)^r = \mu^r$.

Term ④, $(1-p)^{-r}$, is a multiplication of r factors each of which is very close to 1 because p is small; so this term tends towards 1. For the final term, $(1-p)^n$, we cannot assume all factors equal to one; even though each factor is arbitrarily close to 1, an arbitrarily large number of them are multiplied together. Instead, we expand as a power series. The McLaurin series approximation for $(1+y)^\alpha$ is $1 + \alpha y + \alpha(\alpha - 1)y^2/2! + \alpha(\alpha-1)(\alpha-2)y^3/3!\ldots$. This give us

$$1 - pn + \frac{p^2}{2!}n(n-1) - \frac{p^3}{3!}n(n-1)(n-2)\ldots$$

The various $n-1$, $n-2$ factors are all $\simeq n$ so this becomes

$$1 - pn + \frac{(pn)^2}{2!} - \frac{(np)^3}{3!}.$$

However, recall the series expansion for $e^x = 1 + x + x^2/2! + x^3/3! \ldots$ so our series is just $e^{-np} = e^{-\mu}$. Finally putting all the terms together we get the the result:

$$f_\mu(r) = \frac{\mu^r}{r!} e^{-\mu}. \tag{4.4}$$

The probability is written as $f_\mu(r)$ to emphasise that we are seeing this as a distribution over r for a given μ. Note that like the binomial, it is a discrete probability distribution. It is defined only at integer values of $r = 0, 1, 2 \ldots$. On the other hand, the mean μ is a continuous real number.

4.7 Properties of the Poisson Distribution

4.7.1 *Variance of the Poisson Distribution*

The Poisson distribution is defined by its mean μ, but what is its variance? Because the Poisson distribution is a special case of the binomial, we can consider an imaginary combination of n and p which gives the right mean $\mu = np$. The variance is then $np(1-p)$. However because p is small, this is just $\simeq np$. In the limit then we have the striking result that for the Poisson distribution

$$\sigma^2 = \mu, \tag{4.5}$$

or equivalently that the standard deviation is equal to the square root of the mean.

4.7.2 *Shape of the Poisson Distribution*

Figure 4.4 plots the Poisson distribution for various values of μ. The distribution is only defined at discrete values of r, but as before we have used the gamma function to plot smooth curves through the distribution. For small values of μ, the distribution is highly asymmetric; with increasing μ it becomes increasingly symmetrical. Figure 4.5 compares the Poisson and Binomial distributions, picking a fixed value of $\mu = 1.5$ and using different n, p pairings all of which have the same binomial mean $\mu = np$. It can be seen that the binomial distribution converges on the Poisson distribution quite rapidly.

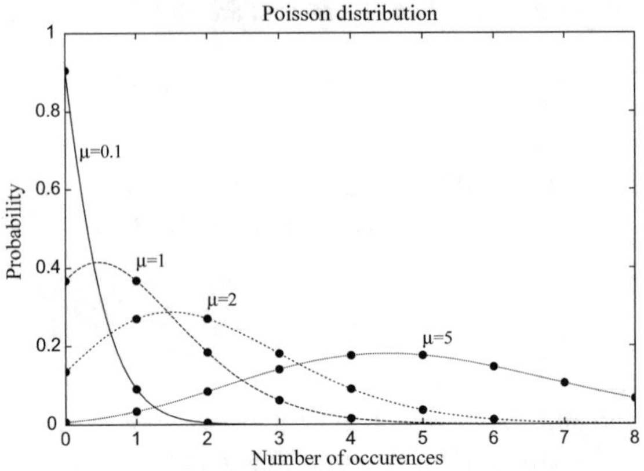

Fig. 4.4 The Poisson distribution for a range of means, indicating how it is very asymmetric for low means, and becomes more symmetric at larger means. As with the binomial distribution, the Poisson distribution is defined only at discrete values, but we have drawn a smooth curve using the Γ function

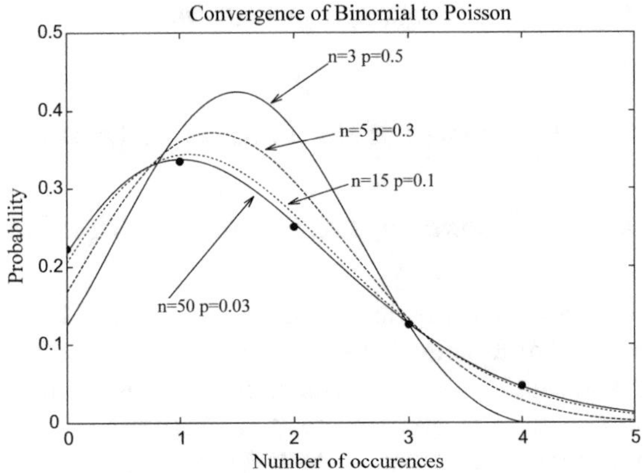

Fig. 4.5 Illustrating how the Poisson distribution is the limiting form of the Binomial for large n, but at Small probability p. The circles show the Poisson distribution for $\mu = 1.5$ at discrete values of r. The various curves show the Γ function based smooth curve fitting the binomial distribution for various combinations of p and n, but in each case such that $\mu = np = 1.5$

4.8 Applications of Poisson: Counting Statistics

The term "counting statistics" refers to any situation where the binomial distribution applies, but it is often used in reference specifically to the Poisson distribution limit, because it so often works for situations where we count the number of objects or events—for example the counts per second caused in a Geiger counter by a radiation source, or the number of people replying to an online advert. In these cases, the uncertainty is completely determined by the mean; if in a single run of the experiment we get N counts, that gives our estimate of the true mean, $\mu = N$, and the error on our estimate is just $\pm\sqrt{N}$.

4.8.1 Why Large Samples Are Better

In experimental design, it is important to realise that the counting statistics error is *unavoidable*. We cannot design a perfect experiment that has no error, as long as our measurement involves some random counting of events. However, note that an implication of the Poisson distribution is that measurements producing large numbers are more precise than those with small numbers—the coefficient of variation is $\sigma/\mu = \sqrt{N}/N = N^{-1/2}$.

Suppose for example we wish to estimate the count rate R, photons per second, of the light from a star. Lets say that we integrate for 10 s and get 13 counts. The estimated rate is $R = 13/10 = 1.3$ and the uncertainty is $13^{0.5}/10 = 0.36$, which is 28% of our estimate. If we integrate for 1000 s we might get say 12,176 counts, giving $R = 0.122$ and an uncertainty of $12{,}176^{0.5}/1000 = 0.11$, which is now only 9% of our estimate. So although we cannot completely avoid counting statistics uncertainty, we can minimise its effect by designing an experiment to have large count rates.

4.8.2 The Something-or-Nothing Problem

The number of people who reply to a online advert should follow Poisson statistics—presumably large numbers of people glance at my advert, but nearly all of them are uninterested. In a given week, what is the probability somebody replies to my advert? I would be equally happy if I get one, two, three or more replies—just unhappy if I get none. There is of course no need to add up $P(1) + P(2) + \cdots$. The probability of *at least one success* is

$$p(>0) = 1 - p(0) = 1 - e^{-\mu}.$$

In the something-or-nothing problem, a count rate of $\mu \sim 1$ is a qualitative dividing line. For example, for a mean count rate of $\mu = 0.2$ we get $p = 0.18$. We usually get nothing. For $\mu = 1.0$ we get $p = 0.63$, close to an even chance of something or nothing. For $\mu = 5$ we get $p = 0.993$—almost always something.

4.8.3 Cautions

There are two typical circumstances where it is common to apply Poisson statistics incorrectly. First, the **small number caution**. Note that when dealing with small numbers, the Poisson distribution is very asymmetric, so the rule-of-thumb $\sqrt{N}$ error shouldn't be seen as a symmetric "$\pm\sqrt{N}$" guide. Consider the $\mu = 2$ curve in Fig. 4.4. Although the $r = 1$ and $r = 3$ values are equally distant from the mean, $r = 3$ is much more probable than $r = 1$.

The second caution is about the tendency to assume that Poisson statistics applies whenever we have an experiment that involves counting events. Suppose people buying ice-cream in a cafe can choose between strawberry and vanilla. It is noted that on an average day $N = 25$ people choose vanilla. Does this mean that day-to-day this has a spread given by $\sigma = 5$? It does not. In general, the number of people choosing vanilla will follow the binomial distribution. Suppose people choose randomly, so that $p = 1/2$. Then, for our $N = 25$ vanilla-choosers, there must have been $n = 2N$ trials in total, and $\sigma = \sqrt{N/2}$, not $\sqrt{N}$. In general for the Binomial, as we saw in Sect. 4.3.2, $\sigma^2 = \mu(1 - p)$. The Poisson result of course follows directly from this and $p \to 0$. The number of vanilla-choosers will only follow Poisson statistics if vanilla is extremely unpopular.

4.9 Key Concepts

Some of the key concepts from this chapter are:

- The binomial distribution and how it changes shape with p and n
- The idea of a limiting case of the binomial distribution where we select rare events from a large pool
- The fact that the Poisson distribution is determined only by its mean, as this automatically fixes the variance
- The way that the shape of the Poisson distribution changes with mean μ
- The use of the Poisson distribution in a wide variety of "counting statistics" problems
- The idea that relative error improves with sample size.

Key formulae in this chapter include: the binomial probability distribution (4.1) and the formulae for its mean (4.2) and variance (4.3); the Poisson distribution formula (4.4), and the formula for the variance of a Poisson distribution (4.5).

4.10 Further Reading

The Binomial and Poisson distributions are of course covered in all standard statistics textbooks—see the Chap. 1 references. A short, and lively introductory e-book specifically about the binomial distribution is Hartshorn (2017).

The Binomial and Poisson distributions were amongst the first probability distributions to be understood, because they are at the heart of the gambling problems which so fascinated early mathematicians. If you find these early developments, and the stories attached to them, interesting, a standard book on the history of statistics is Stigler (1990).

A particularly fascinating story is how one of the pioneers of modern statistics, Ronald Fisher, used a binomial distribution analysis to claim that the famous results of Mendel were possibly fudged, because they were more perfect than they ought to have been (Fisher 1936). This claim has been the source of argument ever since. A definitive study of this controversy is the book by Franklin et al. (2008).

Calculating binomial coefficients can be very cumbersome. In the next chapter, we will look at how to use the Gaussian distribution to approximate the binomial. Meanwhile, an excellent way to calculate binomial coefficients is using the Geogebra online calculator.

4.11 Exercises

4.1 Ten patients have a form of cancer from which the recovery rate is 80%. What is the probability that exactly 7 of them will recover? What is the probability that exactly 3 of them will recover? What is the probability that 7 or fewer will recover?

4.2 Consider rolling a die n times and noting how many sixes are rolled. How large does n have to be for the coefficient of variation to be less than 0.1?

4.3 It has been suggested that the probability of passing a driving test on any one attempt is 0.75. What then is the probability that an applicant will pass only on the fourth attempt? Comment on whether the implied assumption in the question is likely to be correct.

4.4 A government inspector checks whether the lorries run by haulage companies have exhaust emissions compliant with the law. She turns up at a company and picks 6 lorries at random to inspect. One particular company has 24 lorries, and the truth is that 4 of them are defective. What is the probability that the inspector will find none of the defective lorries?

4.5 A shop sells fabric on a roll that is 1 m wide. They find that typically there is a defect in the fabric once every 5 m length. They don't like to sell a length of fabric if it has four or more defects. A customer wants a 30 m length. What is the probability that this will have four or more defects?

4.6 In Sect. 4.6 we derived the Poisson distribution as the limit of the Binomial distribution, and then took it as obvious that the mean was μ because $\mu = np$ for the Binomial. Working from the formula for the Poisson distribution itself, can you show that the mean is equal to the μ in the formula?

4.7 You are trying to spot a small bird through a distant clump of trees. In this clump, there are on average 0.033 trees per square metre, but the trees can be regarded as being distributed randomly. Each tree is 0.7 m wide, and the clump is 57 m thick. At any one moment in time, what is the probability that the bird is hidden from your view?

References

Fisher, R.A.: Has Mendel's work been rediscovered? Ann. Sci. **1**, 115–126 (1936)

Franklin, A., Edwards, A.W.F., Fairbanks, D., Hartl, D.: Ending the Mendel-Fisher Controversy. University of Pittsburgh Press, Pittsburgh (2008)

Hartshorn, S.: Probability with the binomial distribution and Pascal's triangle: a key idea in statistics. Kindle e-book (2017)
Stigler, S.M.: The History of Statistics: The Measurement of Uncertainty Before 1900. Belknap Press, Cambridge (1990)

Websites (all accessed March 2019):

Geogebra probability tool: https://www.geogebra.org/classic/probability

Chapter 5
Combining Many Factors: The Gaussian Distribution

5.1 Outline of Content

- A common distribution in diverse circumstances
- Physical origin of Gaussian distributions
- The Gaussian as the limiting form of the binomial
- Properties of the Gaussian
- How random variables combine
- The Gaussian form as a fixed point for convolutions
- Multi-variate Gaussians
- Gaussians in disguise: particle velocities and stellar masses
- Error distributions

Figure 5.1 shows three diverse examples of naturally occurring distributions—cluster star velocities, human heights, and the distribution of the sum of the last four digits of phone numbers in my address book. They all show a very similar shape—symmetrical, with a "bell curve" shape. They can all be reasonably well fitted by the same mathematical expression, which we will come to shortly. This mathematical form is named the "Gaussian Distribution", after Carl Friedrich Gauss, but it is often referred to as the "Normal Distribution", because it occurs so frequently in Nature. An important Physics example is the distribution of molecular velocities. Although the Maxwell–Boltzmann distribution for molecular velocities looks somewhat different, it is really a Gaussian in disguise, as we will see later in this chapter. How can such completely different physical circumstances produce the same result? The answer lies in what happens mathematically when many different factors combine. Before we start to examine how this works, lets take a closer look at the distributions in Fig. 5.1.

A. Lawrence, *Probability in Physics*, Undergraduate Lecture Notes in Physics,
https://doi.org/10.1007/978-3-030-04544-9_5

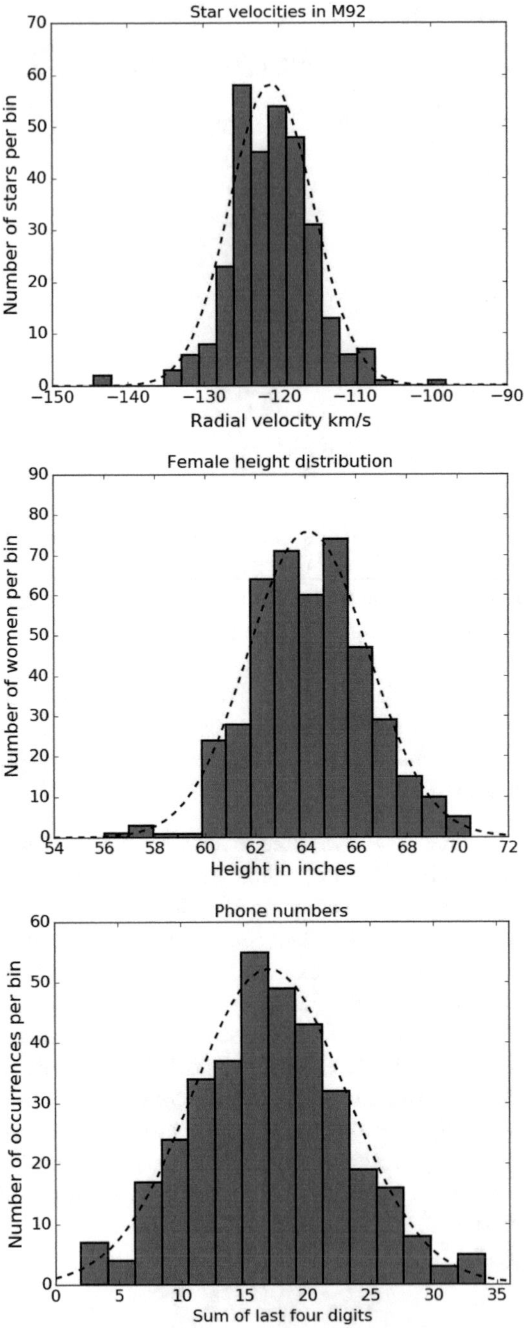

Fig. 5.1 Examples of naturally occurring distributions of observed/measured quantities, all showing an approximation to the classic "bell curve" shape. In all three cases the dashed curve shows a Gaussian curve which fits the data. **Top**: The line of sight velocities of stars in the globular cluster M92 (Drukier et al. 2007); **Middle**: Heights of female subjects in the classic study by Galton (1886); **Bottom**: The sum of the last four digits of numbers in my own address book

5.2 Rescaling to Reveal a Universal Distribution

The sample distributions in Fig. 5.1 are more than just qualitatively similar; they are essentially identical. We can see this with a little simple re-scaling. Of course the distributions all have different units, different means, and different widths. We can remove these particularities by calculating the sample mean $\bar{x}$ and sample variance s^2 for each dataset. The next step is to convert each x value to its deviation from the mean value, normalised to the standard deviation, $z = (x - \bar{x})/s$. The sample distributions look symmetrical; lets assume this, so really we just want the size of the deviation. We could consider $|z|$ but instead we shall use z^2. In Fig. 5.2 we look at the binned distribution of this normalised square deviation, z^2, for the star velocity and phone datasets. They look very similar, and also look as if they decline exponentially. To test this idea, we take each histogram point and normalise to the peak value; and we then plot the log of the normalised number versus z^2, putting the two datasets on the same plot. As you can see in the lower frame of Fig. 5.2, the rescaled datasets agree well with each other. The shape of the various distributions is not just similar, its *identical*.

Empirically then, we have suggestive evidence that the population distributions underlying each of these sample distributions has the form

$$f(z) = A\, e^{-kz^2} \qquad \text{with} \qquad z = (x - \mu)/\sigma.$$

All the distributions have the same values of A and k, but they have differing values of μ and σ. We can estimate k by seeing where f drops by a factor of $1/e$, i.e. a factor of -1.0 in $\ln f$. To a good approximation, this suggests $k = 1/2$. To fix the value of A, we take $f(z)$ to represent a probability density function, and so require that

$$\int_{-\infty}^{\infty} f(z) \mathrm{d}z = 1.0.$$

Now, the integral of e^{-z^2} is equal to $\pi^{1/2}$ (not proved here, but it is in standard maths textbooks), and so we find $A = 1/\sqrt{2\pi}$. Note that this value of A gives us the correct normalisation if $f(z)$ is the probability density per unit z. Alternatively if we express the density per unit x, this brings in a factor $1/\sigma$. In summary, we have

$$f(x) = \frac{1}{\sigma\sqrt{2\pi}} \exp\left[-\frac{1}{2}\left(\frac{x-\mu}{\sigma}\right)^2\right] \text{ or } f(z) = \frac{1}{\sqrt{2\pi}} \exp\left(-\frac{z^2}{2}\right). \tag{5.1}$$

This mathematical form is the *Gaussian distribution*, with the version on the right being the *standard form*.

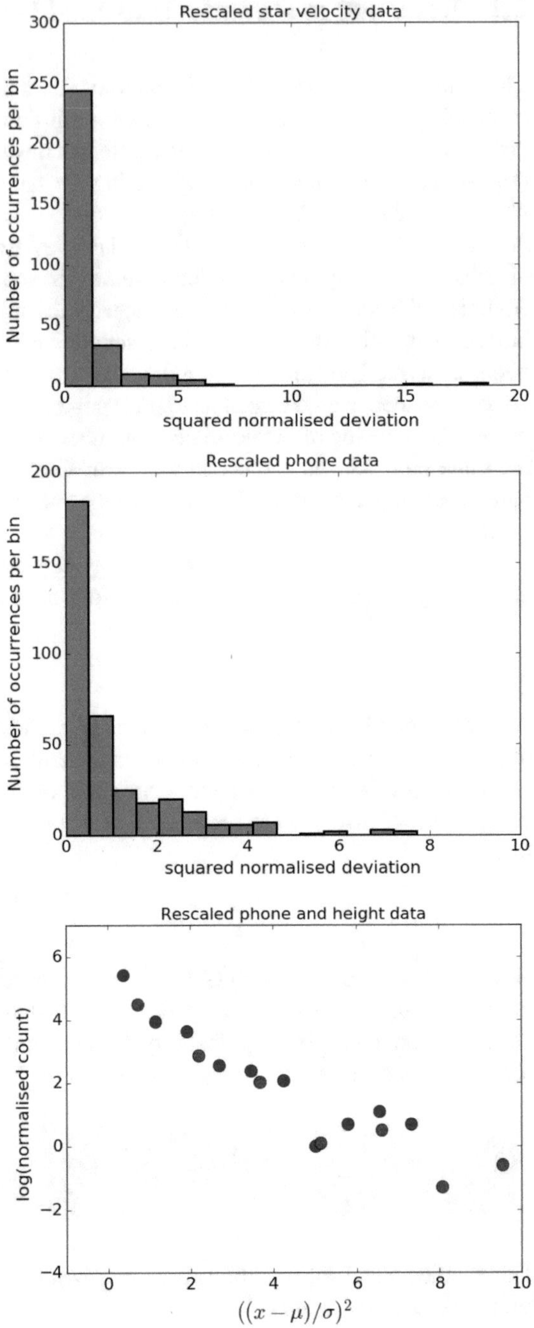

Fig. 5.2 Rescaling tests for datasets from Fig. 5.1, as described in the text. **Top**: Re-scaled star velocity data **Middle**: Re-scaled phonebook data. **Bottom**: Comparing the normalised distribution for the phonebook data and the height data

5.3 Properties of the Gaussian

Above, we have arrived at a guess for the mathematical form, based on observed data. In the next section, we will look at how this mathematical form arises. But first, let us look at some of the properties of the Gaussian distribution.

5.3.1 Moments of the Gaussian

Figure 5.3 illustrates the Gaussian function, showing the well known "bell curve" shape. It is positive at all values of x, for all positive values of σ. Let us consider the *moments* of this function, $m_n = E[x^n f(x)]$, as defined in Chap. 2, Sect. 2.5.2. The zeroth moment, $m_0 = \int f(x)\mathrm{d}x = 1.0$—in the previous section we in fact found the normalising factor $1/\sqrt{2\pi}$ which makes this true. The first moment, m_1 is the mean of the function $f(x)$. The fact that this is the same as the μ as used in the definition of $f(x)$ is obvious from symmetry. For higher moments, we use the central moments $\mu_n = E[(x-\mu)^n f(x)]$.

The second central moment is the **variance**, $\mu_2 = V(x)$. It might seem obvious that $\mu_2 = \sigma$, as we derived our educated guess for $f(x)$ based on the normalised deviations for our sample data. However, if we take $f(x)$ as our definition, we can check that $\int_{-\infty}^{\infty}(x-\mu)^2 f(x)\mathrm{d}x$ does in fact return σ. In order to do this, the trick is to manipulate the integral until it looks like the gamma function, $\Gamma(z) = \int_0^{\infty} t^{z-1}e^{-t}\mathrm{d}t$. We will then have reduced our problem to a previously solved one. This suggests that we use the transformation

$$t = \frac{1}{2}\frac{(x-\mu)^2}{\sigma^2} \quad \text{and so} \quad \frac{\mathrm{d}t}{\mathrm{d}x} = \frac{(x-\mu)}{\sigma^2} = \frac{\sqrt{2t}}{\sigma}.$$

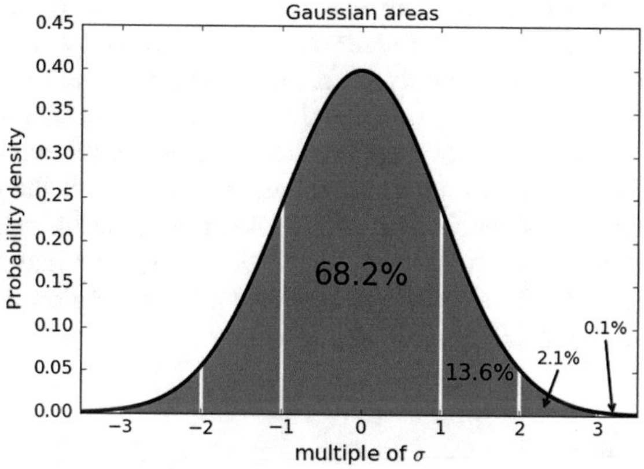

Fig. 5.3 Gaussian distribution, illustrating the amount of probability within different regions of the curve, with shaded areas at $\pm 1\sigma$, 2σ, 3σ and above

Next, because $f(x)$ is symmetrical, the integral between plus and minus infinity can be replaced by twice the integral between zero and infinity. Then we have

$$\begin{aligned} V(x) &= 2\frac{1}{\sigma\sqrt{2\pi}}(x-\mu)^2\int_0^{\infty}\exp -\frac{1}{2}\frac{(x-\mu)^2}{\sigma^2}\mathrm{d}x \\ &= \frac{2}{\sigma\sqrt{2\pi}}\, t\sigma^2 \int_0^{\infty} e^{-}t\mathrm{d}t.\frac{\sigma}{\sqrt{2t}} \\ &= \frac{2\sigma^2}{\sqrt{2\pi}}\frac{1}{\sqrt{2}}\int_0^{\infty} t^{1/2}e^{-t}\mathrm{d}t \\ &= \frac{2\sigma^2}{\sqrt{\pi}}\Gamma(3/2). \end{aligned}$$

Finally, standard mathematics texts will show us that $\Gamma(3/2) = \sqrt{\pi}/2$, and so we confirm that indeed $V(x) = \sigma^2$.

The third central moment μ_3 gives us the **skewness** $\alpha_3 = \mu_3/\sigma^3$, as described in Chap. 2, Sect. 2.5.4. As $f(x)$ is symmetrical, the skewness is zero. The fourth central moment μ_4 gives us the **kurtosis** $\alpha_4 = \mu_4/\sigma^4$. It can be derived in a similar fashion to the variance. The result is that $\mu_4 = 3\sigma^4$, so that $\alpha_4 = 3$. Because the Gaussian distribution is so common, it is common practice to compare other distributions to the Gaussian, and so define the *excess kurtosis* as $\alpha_4 - 3$—see the examples in Fig. 2.6.

5.3.2 Integrated Probability Within Standard Regions

A common practical question is to ask how likely a measurement is to be within some range. You can see that this integrated probability depends on distance from the mean in multiples of σ—in other words, the z value of the standard form. Unfortunately, although the definite integral over all z has a well known value, the indefinite integral does not have an analytic solution. However we can integrate approximately, either by making a power series approximation, or by summing trapezoids in the usual fashion. Furthermore, this is such a common question that of course the problem has been solved many times, and so standard tables and/or programming routines exist.

Figure 5.3 shows the fractional area under the curve for various ranges of z. The fraction of measurements falling within $\pm 1\sigma$ of the mean is 68.3%; within 2σ is 95.4%, and within 3σ its 99.7%. Events outside this range are therefore very unlikely; if we see too many such events we may suspect that there is something anomalous (and therefore important) happening. We will follow that logical trail when we study statistical inference in part III.

5.4 Physical Origin of Gaussian Distributions

We have seen that somehow Nature produces an almost universal frequency distribution in a diverse range of circumstances. How does this come about? The common element in the situations we have looked at is that the distribution arises from the combination of many elements or factors.

Sometimes this happens through the simple **addition of quantities**. Lets look at the phone book example. If we take any one of the digits, we should get a random number uniformly distributed between 0 and 9. In practice, we will indeed get a uniform distribution for the last digits, but not for the early digits—many of my numbers come from the same area code for example. The distribution of last digits will have a mean of 4.5, but will have a flat "top-hat" distribution. The sum of four such digits has a mean of $4 \times 4.5 = 18$ but now has the bell-curve distribution rather than a flat one, as we see in Fig. 5.1. This seems qualitatively reasonable—there is for example only one way to get a score of 36, i.e. four 9s, but a number of different ways to get a score of 22.

Sometimes the combination of factors seems to be more complicated. The height of a person is likely to be the result of many factors conspiring together—a number of different genes, some inherited from each parent, the person's diet, their medical history, and so on. Any one of these factors will have its own probability distribution. If the combination of these effects is *additive*, or equivalently some kind of *averaging* effect, then the situation is not so different from our digit-adding case, and the net effect is a normal or Gaussian distribution. Very often, as in the human heights example, we won't have a full quantitative understanding of the factors involved, and sometimes we won't even know what the component effects are—but Nature is still doing the averaging, and a Gaussian distribution results.

Finally, rather than being the effect of adding physically separate quantities, a quantity may be the result of **repeated drawings** from the same distribution, either in real practice or conceptually. For example, although the binomial distribution is mathematically different from the Gaussian, for a large number of trials n, the shape becomes very close to Gaussian. This is because n trials can be seen as equivalent to combining n single trials (Bernoulli trials) each of which has an even chance of scoring 1 or 0. In Sect. 5.7 we will look more closely at the relation between the binomial and Gaussian distributions.

The key issue seems to be the effect of adding random variables. How does that work mathematically?

5.5 Mathematical Origin of the Gaussian: Adding Random Variables

It seems that to understand the Gaussian we need to understand the addition of random variables. Let us start with just two variables. Suppose we have two random variables x and y, and form a third random variable as $z = x + y$. We will assume for

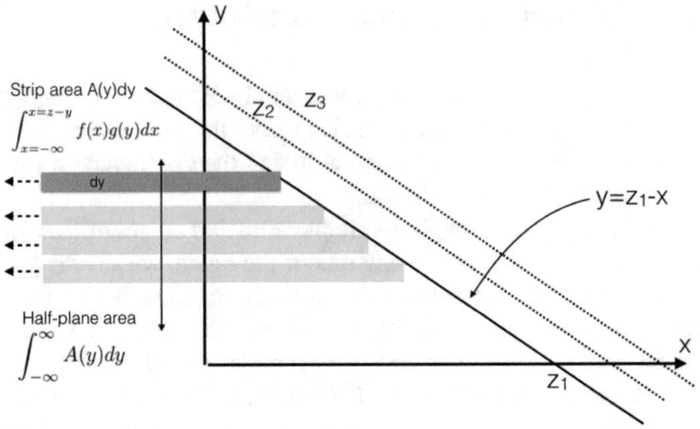

Fig. 5.4 Illustrating integration over the half-plane, as described in the text

simplicity that we are dealing with independent continuous variables. If the variable x has PDF $f(x)$ and the variable y has PDF $g(y)$, what will be the resulting PDF of z, $p(z)$? The way to solve this problem is to work in terms of cumulative distributions or CDFs, and then differentiate at the end. We will denote the equivalent CDFs with capital letters, i.e. $F(x)$ etc.

5.5.1 *Integration over the Half-Plane*

The joint PDF at x, y is $f(x)g(y)$. Suppose we pick a value of $z = z_1$ and ask, what is the cumulative CDF, $P(< z_1)$? We need to sum up all the $f(x)g(y)$ values over the x, y plane, but only at those points that satisfy $x + y \leq z_1$. In Fig. 5.4 we show the line $y = z_1 - x$; the points we need to add up are all those to the left of this diagonal line. Other possible values of z are indicated by the dotted lines. To do this half-plane integration, start with a strip at y with width dy, and sum all the values from $x = -\infty$ to $x = z - y$, as illustrated in the dark grey strip in the figure. Having integrated over that strip, we repeat for all values of y, as shown in the lighter grey strips. Finally we can see that the CDF must be

$$P(z) = \int_{y=-\infty}^{y=+\infty} \int_{x=-\infty}^{x=z-y} f(x)g(y)\mathrm{d}x\mathrm{d}y.$$

Of course we could just as well integrate over all x, and y to $z - x$. The integral of $f(x)$ up to $z - y$ is just the value of the CDF $F(x)$ for $x = z - y$, i.e. $F(z - y)$, so we can write

$$P(z) = \int_{y=-\infty}^{y=+\infty} F(z - y)g(y)\mathrm{d}y.$$

Now we can differentiate to get $p(z) = \mathrm{d}P/\mathrm{d}z$. Because $z = x + y$ we have $\mathrm{d}F/\mathrm{d}z = \mathrm{d}F/\mathrm{d}x$ so in differentiating F we just get back the density f. So finally we get

$$\text{for } z = x + y: \qquad p(z) = \int_{y=-\infty}^{y=+\infty} f(z-y)g(y)\mathrm{d}y. \tag{5.2}$$

The mathematical form we have just arrived at is known as the **convolution** of two functions, and comes up in many different areas of physics. It is sometimes written

$$p = f \star g.$$

The conclusion then is that *the PDF of the sum of two random variables is the convolution of the PDFs of the two original variables.*

5.5.2 *Understanding Convolutions*

The equation above may seem a little opaque. What is happening? In words, we can see the convolution as made of four operations—flip, shift, multiply, add.

- flip: $f(y) \to f(-y)$
- shift by z: $f(-y) \to f(-y+z)$
- multiply: for a given shift z, at each y, calculate $m(z, y) = g(y) \times f(-y+z)$
- add: for a given z, repeat and sum over all y, to give $p(z) = \int m(z, y)$.

Note that if $f(y)$ is symmetric, the flip makes no difference, and the remaining three operations are known as "correlation". The idea is probably best seen pictorially. In Fig. 5.5, we show graphically how to convolve/correlate a simple "top hat" function with itself. Here $f(y) = g(y)$ and each is defined so that $g(y) = 1$ over some range,

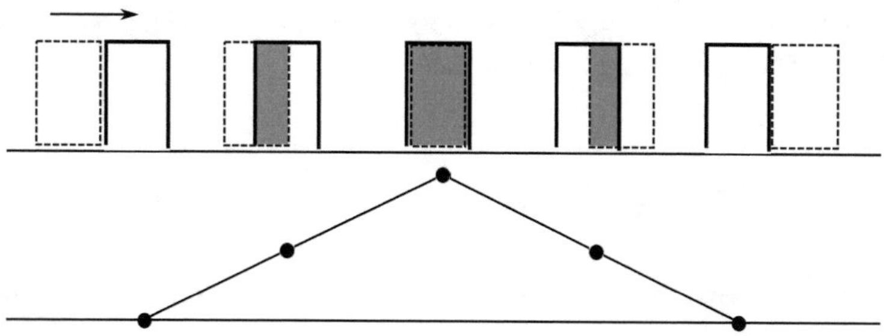

Fig. 5.5 Illustrating the convolution of a top-hat function with another top-hat. The varying values of z correspond to sliding one of the functions past the other, as shown in the upper panel. The product of the two functions in this case is just the overlapping area, indicated by the shading. The change in resulting values of $p(z)$ are shown in the lower panel

and is zero outside that range. The function $g(y)$ is drawn as a top hat with a solid line, and $f(y)$ as a similar top hat with a dotted line, but shifted by an amount z changing from frame to frame. The product $f \times g$ is 1 where the top-hats overlap, and zero outside this; so the integral is just the shaded area in each case. You can see that as you try different values of z, the output $p(z)$ is a triangle. In algebraic terms, if we have

$$
\begin{aligned}
f(x) = g(y) &= 1 \text{ when} \quad 0 < x < 1 \\
&= 0 \text{ otherwise} \\
z &= x + y \\
p(z) &= f \star g.
\end{aligned}
$$

Then you can easily see that

$$
\begin{aligned}
&\text{for } z < 0 && p(z) = 0 \\
&\text{for } 0 < z < 1 && p(z) = z \\
&\text{for } 1 < z < 2 && p(z) = 2 - z \\
&\text{for } z > 2 && p(z) = 0.
\end{aligned}
$$

5.6 Repeated Convolutions Leading to the Gaussian

We have seen that adding two random variables leads to a convolution of their PDFs. If we combine more random variables then, we will get repeated convolutions. Figure 5.6 shows the effect of adding a third top-hat variable; the result is already a curved shape that is starting to look a little like the Gaussian. Numerical experimentation shows that more convolutions leads quite rapidly to a shape close to the Gaussian. But what is going on mathematically? The proposal that repeated convolution leads inexorably to the Gaussian form is known as the *Central Limit Theorem*. It is possible, within certain constraints, to prove this rigorously, but the proof is somewhat laborious, and involves some mathematics (Fourier transforms) that I have been trying not to assume in this book. However, we can show something simpler, and nearly as good. The Gaussian has a very special property—if you convolve it with itself, you get another Gaussian.

5.6.1 Convolving the Gaussian

To simplify the algebra we will assume two identical Gaussians in standard form,

$$f(x) = f(y) = \frac{1}{\sqrt{2\pi}} e^{-y^2/2},$$

and then form $z = x + y$ so that $p(z) = f \star f$

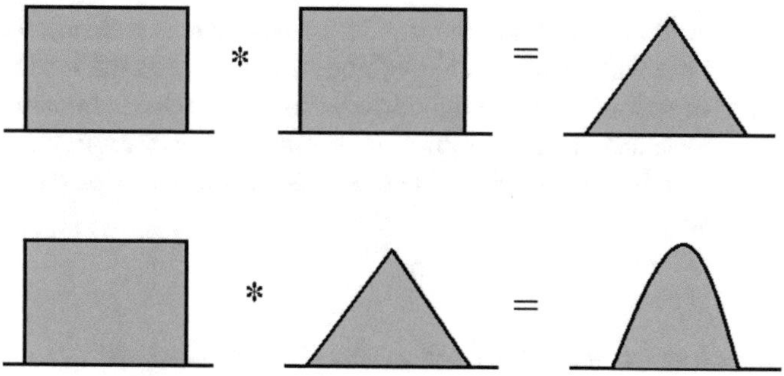

Fig. 5.6 Repeated convolution gradually moves us toward the normal curve

$$p(z) = \int_{-\infty}^{+\infty} f(z-y)f(y)\mathrm{d}y = \frac{1}{2\pi}\int_{-\infty}^{+\infty} e^{-\frac{(z-y)^2}{2}} e^{-\frac{y^2}{2}}.$$

Taking the terms inside the exponential, these add to give a combined exponent

$$-\frac{z^2}{2} + zy - \frac{y^2}{2} - \frac{y^2}{2} = -\frac{z^2}{2} + zy - y^2.$$

Anticipating the answer we want, we can re-write this as

$$-\frac{z^2}{4} + \left[-\frac{z^2}{4} + zy - y^2\right] = -\frac{z^2}{4} - \left[\left(y - \frac{z}{2}\right)^2\right].$$

So we get

$$p(z) = \frac{1}{2\pi}e^{-\frac{z^2}{4}} \int e^{-(y-z/2)^2}\mathrm{d}y \quad = \quad \frac{1}{2\pi}e^{-\frac{z^2}{4}}\sqrt{\pi}\left[\frac{1}{\sqrt{\pi}}\int e^{-(y-z/2)^2}\mathrm{d}y\right].$$

The expression being integrated inside the square brackets is a Gaussian with $\mu = z/2$ and $\sigma = 1/\sqrt{2}$, and so it integrates to 1. So finally we find that the convolved function is

$$p(z) = \frac{1}{\sqrt{4\pi}}e^{-z^2/4}.$$

This is a Gaussian with mean zero and variance $\sigma^2 = 2$. So by convolving two Gaussians we get another Gaussian with twice the variance. More generally if we convolve two Gaussians with means μ_1, μ_2 and variances σ_1, σ_2, we get a new Gaussian with

$$\mu = \mu_1 + \mu_2 \text{ and } \sigma^2 = \sigma_1^2 + \sigma_2^2. \quad (5.3)$$

Note that the variances add, not the standard deviations. The standard deviations are said to "add in quadrature". This is of course the same as the result we found in equation (2.10) when discussing the transmission of variances. Equation (2.10) is more general—variances always combine in quadrature, regardless of the underlying distributions—with this result for the Gaussian distribution being an example of the more general result.

5.6.2 *The Gaussian as a Fixed Point for the Convolution Process*

What we have shown is that the Gaussian is a *fixed point* for the convolution process—that is, once you have arrive at a Gaussian form, you won't go anywhere else. It doesn't strictly prove that from any given starting point you must end up there, but it does make it seem reasonable. In fact, a large variety of distributions head towards the Gaussian on repeated convolution, but it doesn't always happen. An important exception is the Cauchy/Lorentz distribution, which we will meet in the next chapter. If you convolve a Lorentzian, you get another Lorentzian, and it is therefore a distinct fixed point for the convolution process.

5.7 The Gaussian as a Limiting Form for Other Distributions

The averaging process means that many distributions have the Gaussian as a limiting form. For example, the binomial tends towards the Gaussian for large numbers of trials n; the Poisson distribution tends towards the Gaussian for large values of the mean μ; and the χ^2 distribution, which we will meet in part III, tends towards the Gaussian for large numbers of data points N. Let us look at the binomial case carefully.

5.7.1 *Limiting Form of the Binomial Distribution*

Suppose we toss a coin n times and ask how often we get heads. The probability of getting r heads will be given by the binomial distribution with $p = 1/2$. The mean value of r will be $\mu = n/2$, and the variance will be $\sigma^2 = np(1-p) = n/4 = \mu/2$. Suppose we consider a small deviation from the mean value, with $r = \mu + q$. Then the probability of getting a given value of q is

$$P(q) = \left(\frac{1}{2}\right)^n \frac{n!}{(\mu + q)!(\mu - q)!}.$$

Next, consider $P(q + 1)$.

$$P(q + 1) = \left(\frac{1}{2}\right)^n \frac{n!}{(\mu + q + 1)!(\mu - q - 1)!}.$$

Taking the ratio $P(q)/P(q + 1)$ we find that

$$P(q + 1) = P(q)\frac{(\mu - q)}{(\mu + q + 1)}.$$

From this, and noting that $\Delta q = 1$ we find that

$$\frac{\Delta P}{\Delta q} = P\left(\frac{\mu - q}{\mu + q + 1} - 1\right) = P\left(\frac{\mu - q - \mu - q - 1}{\mu + q + 1}\right) \sim P\,\frac{(-2q)}{\mu},$$

with the last step following if $\mu >> q$. From this we can write

$$\frac{\mathrm{d}P}{P} = \frac{-2q}{\mu},$$

which has the solution

$$\ln P = -q^2/\mu \quad \text{i.e.} \quad P = P_0 e^{-q^2/\mu} \quad = \quad P_0 e^{-q^2/2\sigma^2},$$

and this is just the Gaussian mathematical form in the deviation from the mean, as required.

5.7.2 *Limiting Form of the Poisson Distribution*

The Poisson distribution also converges on the Gaussian. This is illustrated numerically in Fig. 5.7. You can see that by $\mu = 12$ the correspondence is already pretty accurate. Of course, this is not too surprising, as the Poisson process is itself a limiting form of the binomial distribution, with small p value, so it is fairly obvious that for large μ the Poisson distribution will look normal, with $\sigma^2 = \mu$. Another way to look at this is that within a small region of x, the probability density is effectively the average of many local samples.

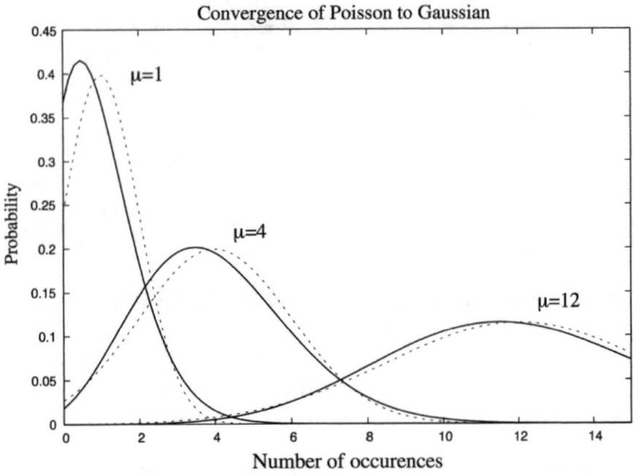

Fig. 5.7 Convergence of Poisson distribution to Gaussian form. The solids curves are a smooth version of the Poissonian, using the Γ function, for various values of μ. The dashed curves are Gaussians with the same μ, and variance set to $\sigma^2 = \mu$

5.7.3 *Using the Gaussian as an Approximation*

It may take an infinite number of trials for the binomial to become indistinguishable from the Gaussian, but for practical purposes it is a good approximation already by $n = 20$. This is very useful, because the factorials involved in the binomial formula are very tedious to calculate for large values of n. If we believe a situation should produce a binomial distribution, then we know the mean and variance expected; for calculation purposes we can use the Gaussian with the same mean and variance. Of course, the Gaussian is a PDF, rather than a discrete PD; so we have to be careful to think about the range of r we are considering (see the exercises). Likewise, for a Poisson situation with known μ, we know the variance will be $\sigma^2 = \mu$ and we can approximate with the appropriate Gaussian. As well as detailed calculations, such Gaussian approximations are often used in quick rule-of-thumb judgements. If we see a "3 sigma event" then we know that is a rare thing.

Approximating distributions with a Gaussian becomes particularly important in Chaps. 8 and 10 in Part III, when we are looking at parameter estimation and model fitting. It enables us to greatly speed up numerical computations when we are trying to get errors on estimated parameters.

5.8 Multi-variate Gaussians

Quite often in Physics we meet situations described by two or more variables each of which has its own Gaussian distribution, resulting in a *multivariate Gaussian*. Two important examples we will discuss later are the distribution of particle velocities (which we will look at in the next section) and the distribution of posterior probabilities in model fitting (which we will look at in Chap. 10). When discussing the univariate Gaussian, we found it instructive to consider the integrated amount of probability within a distance x from the mean, in units of the dispersion, $z = (x - \mu)/\sigma$. The equivalent for a multivariate Gaussian is to ask how much integrated probability is within a *radial distance* r of the mean. We will now look at how to calculate this quantity, starting with the bivariate or 2D Gaussian.

5.8.1 Bivariate or 2D Gaussian

To simplify, let us assume that both component Gaussians are zero-centred, with the same value of σ. We also assume that the two variables are independent, so that $p(x, y) = f(x)g(y)$. In this case each of $f(x)$ and $g(y)$ are Gaussians, and so the joint probability density at position x, y is

$$p(x, y) = \frac{1}{\sigma\sqrt{2\pi}} \exp\left[\frac{-x^2}{2\sigma^2}\right] \frac{1}{\sigma\sqrt{2\pi}} \exp\left[\frac{-y^2}{2\sigma^2}\right],$$

which gives

$$p(x, y) = \frac{1}{2\pi\sigma^2} \exp\left[\frac{-(x^2 + y^2)}{2\sigma^2}\right].$$

The quantity $p(x, y)$ is the joint probability density, defined such that $p(x, y)\mathrm{d}x\mathrm{d}y$ is the amount of probability in a small region within x to $x + \mathrm{d}x$, and y to $y + \mathrm{d}y$. What we want next is to define the *radial probability density* $p(r)$ defined such that $p(r)\mathrm{d}r$ is the amount of probability within a range r to $r + \mathrm{d}r$. To find this, we ask how much probability there is in an annulus of width $\mathrm{d}r$ at radial distance r, where $r^2 = x^2 + y^2$. Multiplying the joint probability density by the area of the annulus $2\pi r\mathrm{d}r$, we can see that the *radial probability density* is given by

$$p(r) = \frac{r}{\sigma^2} e^{-r^2/2\sigma^2}. \tag{5.4}$$

Notice that this is no longer a Gaussian shape, because of the factor of r outside the exponential. Unlike the Gaussian, this function is analytically integrable, which can be quite useful (see the exercises). The result is that the cumulative probability out to $r = R$ is

$$P(< R) = 1 - e^{-R^2/2\sigma^2}.$$

We can invert this to find how many multiples of σ we need to go out to in order to reach a given integrated probability P:

$$k = \frac{R}{\sigma} = \sqrt{-2\ln(1 - P)}.$$

Recall that for the 1D Gaussian, a distance from the mean of $k = 1$ i.e. within $\pm 1\sigma$, contains 68.3% of the integrated probability; for the 2D Gaussian however, to get 68.3% of the integrated probability we need to reach $k = 1.52$. Likewise, to reach integrated areas the same as we would get when reaching out to 2 or 3σ in the 1D case, for the 2D case we need to integrate out to $k = 2.49$ or $k = 3.44$ respectively.

5.8.2 3D Gaussian

For the 3D problem, we likewise assume that the three variables x, y, z are independent, and that the conditional distributions in each axis are Gaussian, each with the same σ. Then to find the radial density distribution we need to sum over the shell of thickness $\mathrm{d}r$ at r, so that we get

$$p(r) = \frac{r^2}{\sigma^3}\sqrt{\frac{2}{\pi}}e^{-r^2/2\sigma^2}.$$

This function, like the 1D Gaussian, is not analytically integrable. However it has of course been numerically integrated. To get the same integrated areas we would get when reaching out to 1, 2, or 3σ in the 1D case, we need to reach out to $k = 1.88$, $k = 2.84$, and $k = 3.58$ respectively. We will see in Chap. 10, Sect. 10.5.6 that the general problem of integrating an N-dimensional Gaussian is the same as integrating a χ^2 distribution with N degrees of freedom.

5.8.3 2D Gaussian: More General Case

Above, we assumed that the x and y distributions were independent, equal variance, and zero centred. Note that the contours of equal probability density will be circles centred on zero. The equations of such contours are then given by $x^2 + y^2 = k^2\sigma^2$ for various values of k, with $k = r/\sigma$ as above. If the distributions are not zero centred, we can of course just use new variables $x' = x - \mu_x$ and $y' = y - \mu_y$. Contours will then be circles centred on μ_x, μ_y. Suppose next that the distributions have differing dispersions, σ_x and σ_y. Then the probability density is

$$p(x, y) = \frac{1}{2\pi\sigma_x\sigma_y}\exp\left(-\frac{1}{2}\left[\frac{x^2}{\sigma_x^2} + \frac{y^2}{\sigma_y^2}\right]\right).$$

The iso-density contours will then be ellipses, given by

$$\frac{x^2}{\sigma_x^2} + \frac{y^2}{\sigma_y^2} = k^2.$$

The major axis of the ellipse will be parallel to either the x or y axis, depending on whether σ_x or σ_y is larger. Once again, we can reduce this situation to the simple case by transforming one of the variables, e.g. such that $x' = x\frac{\sigma_y}{\sigma_x}$.

Finally however, suppose that the variables are not independent. Then it can be shown that the PDF is

$$p(x, y) = \frac{1}{2\pi\sigma_x\sigma_y\sqrt{1-\rho^2}} \exp\left(-\frac{1}{2(1-\rho^2)}\left[\frac{x^2}{\sigma_x^2} + \frac{y^2}{\sigma_y^2} - 2\rho\frac{x}{\sigma_x}\frac{y}{\sigma_y}\right]\right),$$

where x and y are the mean-centred variables, and

$$\rho = \sigma_{xy}/\sigma_x\sigma_y,$$

where σ_{xy} is the covariance of the PDF, as defined in equation (2.5). The quantity ρ is a normalised version of σ_{xy} known as the *correlation coefficient*, which we will discuss further in Chap. 9. The iso-density contours are given by

$$\frac{x^2}{\sigma_x^2} + \frac{y^2}{\sigma_y^2} - 2\frac{\rho xy}{\sigma_x\sigma_y} = k^2.$$

This is the equation of a *rotated ellipse*. With some longwinded algebra you can find that the rotation angle is given by

$$\theta = \frac{1}{2}\cot^{-1}\left(\frac{A-C}{2B}\right) \quad \text{where} \quad A = 1/\sigma_x^2, C = 1/\sigma_y^2, B = 2\rho/\sigma_x\sigma_y.$$

Once again, you can recover the simple case by a suitable transformation of co-ordinates:

$$\begin{pmatrix} x' \\ y' \end{pmatrix} = \begin{pmatrix} \cos\theta & \sin\theta \\ -\sin\theta & \cos\theta \end{pmatrix} \begin{pmatrix} x \\ y \end{pmatrix}.$$

This problem is closely related to how we estimate the errors on correlated parameters in model fitting. We will take another look in Chap. 10.

5.9 Gaussians in Disguise

Sometimes in Nature one sees frequency distributions that don't look at all Gaussian, but which turn out to have an underlying Gaussian cause; it is just that our measured data is in terms of a variable which is a transformation of the underlying variable which does have a Gaussian distribution. We will look briefly at two examples.

5.9.1 The Maxwell–Boltzmann Speed Distribution

In a gas in thermal equilibrium, the speeds of particles follow a distribution which rises initially as v^2 and then declines exponentially. However, in three dimensions, each component of velocity v_x, v_y, v_z has a Gaussian distribution. Collisions between molecules continually re-distribute momentum. Consider the x-component of momentum. Each collision adds or subtracts momentum with equal probability. The final momentum is then the sum of many random variables, and has a probability given by a Gaussian distribution. (In fact, the molecule will perform a *random walk* in momentum space, as described in Chap. 6, Sect. 6.2.4). This logic applies separately to each velocity co-ordinate, so that the result is a trivariate Gaussian. The distribution of *speeds* is then given by the 3D radial formula above, but with $r = v = (v_x^2 + v_y^2 + v_z^2)^{1/2}$.

Note that each component Gaussian is zero centred—each particle is just as likely to be moving in the positive x-direction as the negative x-direction. The variance must be related to the temperature of the gas—the hotter the gas the larger the velocity dispersion. However, more massive particles will be slower moving, so variance may depend inversely on particle mass m. We could guess that $\sigma^2 = kT/m$ where k is a constant of Nature that relates these quantities. If we make this substitution we find

$$p(v) = 4\pi \left(\frac{m}{2\pi kT}\right)^{3/2} v^2 e^{-mv^2/2kT},$$

which is precisely the *Maxwell–Boltzmann* velocity distribution law, and the constant k is the well known Boltzmann constant.

How does this result compare to the Boltzmann energy distribution we derived in equation (3.6)? It is essentially the same thing. Given that particle energy is $E = \frac{1}{2}mv^2$, we can re-express the Maxwell–Boltzmann formula in terms of E, also multiplying by $\mathrm{d}E/\mathrm{d}V$ so that the result is probability density per unit energy rather than per unit speed, and find that

$$f(E) \propto E^{3/2} e^{-\beta E}.$$

The exponential term is the usual *Boltzmann factor*. The $E^{3/2}$ term is an example of *degeneracy*. The energy states are not of equal probability before we consider the effect of the total-energy constraint—there are more ways to have higher energy. In this classical model, we do not have discrete energy states to count, but there is more *phase-space volume* per unit energy at higher energy.

5.9.2 Stellar Masses: An Example of a Log-Gaussian

The left-hand side of Fig. 5.8 shows the distribution of the masses of the stars found in two different young star clusters. They agree well—it seems there is a universal

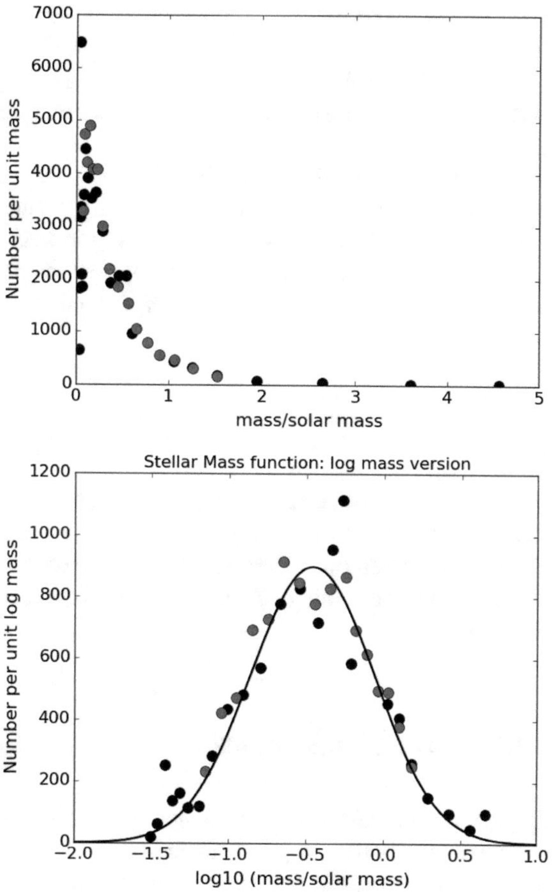

Fig. 5.8 **Upper**: The distribution of stellar masses in two young star clusters, α Per (black symbols) and Praesepe (grey symbols), from Lodieu et al. (2012) and other references therein. The final few low points at small masses are very likely to be due to incompleteness of the sample (measuring stellar masses is very tricky), so overall there is an inexorable rise in numbers towards small masses. **Lower**: The same data, but grouped in bins of log mass, and expressed per unit log mass. The curve shows a function which is Gaussian in log mass

stellar mass function. The distribution does not look Gaussian—it increases steeply towards small masses, with a long tail towards high masses. Note that this graph shows us the relative number of stars per unit mass. Suppose instead we calculate the log of the mass of each star, and we make a histogram of the numbers per unit log mass. We do this by grouping the stars in bins of log mass, and in each bin dividing by the width of the bin in log mass, to give the density per unit log mass.

The result is shown on the right-hand side of Fig. 5.8. It looks Gaussian. The figure shows an example curve plotted through the points (its not a formal fit).

Log-Gaussians occur quite frequently in Nature—why does this happen? Recall that we argued that the Gaussian results from the combination of many variables. Suppose rather than considering $x + y$, suppose we consider their multiplication xy. In the additive case, to get any net outcome we need this-or-this-or-this. In the multiplicative case we need this-*and*-this-*and*-this. But if we consider $\log xy = \log x + \log y$ we return to the additive case—so we should get a Gaussian distribution in the log. Why this happens in the forming of stellar masses is not fully understood. Star-formation is a very difficult and messy problem, usually studied by numerical simulation. The best recent simulations make a good job of reproducing the stellar mass function, but the existence of a log-Gaussian form hints that something much simpler is going on.

How do we recognise a log-Gaussian in linear form? Suppose $z = \log_{10} x$ and the density per unit z is $p(z)$. Then the density per unit x is $f(x) = p(z)\mathrm{d}z/\mathrm{d}x$, and $\mathrm{d}z/\mathrm{d}x = 1/x \ln 10$. Of course if were using natural logs throughout we wouldn't need that ln 10 factor, but it is usual in physical science to work with data in base 10 logs. So if $p(z)$ is Gaussian in z, but we plot a density histogram as a function of x, we will see

$$f(x) = \frac{1}{x}\frac{1}{\sigma\sqrt{2\pi}}\frac{1}{\ln 10}\exp\left[-\frac{1}{2}\left(\frac{\log_{10} x - \mu}{\sigma}\right)^2\right],$$

i.e. to first approximation we get a $1/x$ distribution. Note that μ and σ are the mean and dispersion of $z = \log x$, not of x. The $1/x$ is what gives the histogram its characteristic steep rise to low values. It also makes qualitative sense. If to get large values of x you need this-*and*-this-*and*-this things to happen, large values get rapidly more unlikely.

5.10 Error Distributions

A common problem for experimental physics is to understand the errors on our measurements, and to propagate them in analysis of the measurements. For example, the distributions in Fig. 5.1 are not perfectly smooth. How do we assess the errors on a binned sample histogram like this? Common practice is to assume that the individual binned values themselves can be seen as having a Gaussian distribution. Suppose in some specific bin x_i the true value is f_i. If we imagine making the same measurement repeatedly, we would get various values g_i, centred on f_i, but with some spread described by an *error variance* σ_{err}^2. But how do we know what the value of σ_{err} is? There are three important cases.

5.10.1 Poisson Errors

Quite often the measurement we make involves a *count* of some kind—e.g. how many two-photon events did the ATLAS experiment see in that energy range? The number of counts N that we observe can be seen as an estimate of the true value f, but N will follow a Poisson distribution with $\sigma_{\mathrm{err}} = f^{1/2}$. We can take N as an estimate of f and use $\sigma_{\mathrm{err}} = N^{1/2}$. For large N the distribution will be closely Gaussian and our estimate of the error will also be good. For small N we have to be much more careful, both because the Gaussian approximation may be poor for small N, and because N will be a poor estimate of f. Note also that the value we have in our histogram may not be the count N itself, but some quantity derived from N. In that case, we can use the error propagation formulae from Chap. 2 to calculate the appropriate σ.

Our phone-number example illustrates the point. Looking at Fig. 5.1 you can see that the histogram bar heights jump about somewhat—the distribution is not smooth. Consider the maximum bin, with 51 counts. The "rule of thumb" error on that value is $51^{1/2} = 7.1$, so we should see a bin to bin scatter of around that size, and this is indeed pretty much what we see.

5.10.2 Theoretically Understood Errors

Sometimes from the physics of the experiment we know exactly what the sources of noise are, and can calculate what the errors ought to be. Poisson errors are an example of this, but there are others. For example, when we measure a voltage, it will be subject to thermal fluctuations or *Johnson noise* for which there is a well known formula depending on the temperature of the system.

5.10.3 Empirically Derived Errors

Sometimes, we simply don't know enough about the system in question to reliably calculate what the errors ought to be, and so need to estimate them from the data itself. We could do this by repeating the experiment several times, and estimating $\sigma_{\rm err}$ from the scatter within the repeat values. Alternatively, we could take several neighbouring bins, and on the assumption that the true value is not changing too fast, estimate $\sigma_{\rm err}$ from the bin-to-bin scatter. Later we will look more closely at how we can best estimate mean and variance from experimental data.

5.10.4 Propagation of Gaussian Errors

In Chap. 2 we looked at how errors propagate, deriving the general formula

$$\sigma_z^2 = \sigma_x^2 \left(\frac{\partial z}{\partial x}\right)^2 + \sigma_y^2 \left(\frac{\partial z}{\partial y}\right)^2,$$

and then using this to find handy formulae for various cases. The general formula is a correct statement about how variances combine. If the PDFs for both x and y are

Gaussian, then if $z = x + y$ or something similar, such as an average of the two, then the convolved PDF will as we have seen also be Gaussian. This makes the combined σ_z easy to interpret—we know where 90% probability will be, or 95% and so on. But beware! If x and y are not normally distributed, or if we combine them in a more algebraically complicated way, then in general the PDF of z will *not* be Gaussian. This can lead to serious mistakes when interpreting discrepant data points.

5.11 Key Concepts

Some of the key concepts from this chapter are:

- The ubiquitous Nature of the Gaussian distribution
- The use of a standard form for the Gaussian, with $\mu = 0$ and $\sigma = 1$
- The physical origin of the Gaussian from situations where many factors are combined
- The idea of the convolution of two functions
- The result that if you add two random variables, the PDF of the new variable is the convolution of the two original PDFs
- The result that the convolution of two Gaussians makes another Gaussian, so that the Gaussian is a convergent form for the convolution process
- The fact that most other distributions tend towards the Gaussian form, for large N
- The result that the radial PDF resulting from a bivariate Gaussian is *not* Gaussian
- The fact that the Maxwell–Boltzmann speed distribution is really just a trivariate Gaussian in disguise
- The idea that error distributions are *usually* Gaussian, but not always!

The key formulae from this chapter are: the formula for the Gaussian distribution ((5.1) left) and its standard form ((5.1) right); the formula for calculating the convolution of two functions (5.2) and the result of convolving two Gaussians (5.3); and the radial PDF resulting from a bivariate Gaussian (5.4).

5.12 Further Reading

The normal distribution was first clearly defined by Gauss in 1809, but an equivalent mathematical form was derived by De Moivre in 1738, as an approximation to binomial coefficients. However, it was Laplace, in a series of works, who had the key insight that the distribution arose from aggregating observations, and who explicitly showed that the binomial tended towards the normal form. This can be seen as the first version of the *Central Limit Theorem*, and indeed is known as the De Moivre–Laplace Theorem. The more complete modern version of the Central Limit Theorem, that adding random variables with a wide variety of PDFs produces a convergence on the normal form, was proved by Lyapunov in 1901. The historical development, as well as much technical detail, is in the excellent wikipedia pages on the Normal Distribution, and on the Central Limit Theorem. These pages also show a nice simulation of how the distribution of coin tosses evolves towards the normal.

Most advanced statistics textbooks will have some kind of derivation of the Central Limit Theorem—for example Miller and Miller (2013). The typical approach involves re-expressing a PDF as a set of "characteristic functions" which is essentially a Fourier Transform of the PDF. A good introduction to Fourier Series analysis is Folland (1992)

As discussed in this chapter, the stellar mass function seems to be a good example of a normal distribution in disguise, but it took some time to recognise. The mass function was usually thought to be first a power-law, then a segmented power-law, before being recognised finally as a log-Gaussian (Miler and Scalo 1979; Chabrier 2003). However, well before these key observational papers, Larson (1973) had already recognised that a simple probabilistic approach to fragmentation of the molecular clouds in which stars form, would lead naturally to a log-Gaussian form. However, these matters are still controversial. A good review is in Bastian et al. (2010).

5.13 Exercises

5.1 Distribution A is a Gaussian PDF $p_A(x)$ with $\mu = 0$ and $\sigma = 1$. Distribution B is a Gaussian $p_B(x)$ with $\mu = 3$ and $\sigma = 1$. For what values of x does p_B exceed p_A?

5.2 Show that for a Gaussian, FWHM$=2.355\sigma$. What is the Full Width at 20% of maximum?

5.3 Show that one standard deviation from the mean is where the Gaussian has the steepest slope. If you draw a tangent to the curve at this point, where does it intercept the x-axis?

5.4 What is the probability of getting 6 heads and 10 tails in 16 tosses of a coin? First calculate the exact answer, then try using the Gaussian approximation and compare the answer. (The Gaussian is of course a distribution for a continuous random variable. Consider a range of x that approximates the relevant value.)

5.5 What is the probability that at least 70 out of 100 mosquitos will be killed by a new insect spray, if the probability that any one mosquito will be killed by the spray is 0.75? Find an approximate answer, rather than trying to find the exact answer.

5.6 Derive the result quoted in equation (5.4), that for a bivariate Gaussian, the integrated probability to radial distance r is

$$P(< R) = 1 - e^{R^2/2\sigma^2}$$

where σ is the variance of the conditional distribution for each of x and y. Compare the multiples of σ you have to go to in the 1D and 2D cases to reach integrated probabilities of 68, 90, and 95%.

5.7 A chemist sells on average 150 tubes a week of a particular brand of toothpaste. If she wants to make sure that there is at most a 5% chance of running out in a given week, how many should she keep in stock?

5.8 An astronomer is estimating the number density of a certain type of star by counting stars of that type within a limited volume. It is desired that the number counted should be within 10% of the "correct" number. How many stars should be in the sample if there is to be a 95% chance of achieving this accuracy?

5.9 Show that the mean absolute deviation (MAD) for a Gaussian distribution is $\sqrt{\frac{2}{\pi}}\sigma$.

5.10 For a gas in thermal equilibrium, in terms of the temperature T and Boltzmann constant k: (a) what is the most probable speed v, (b) what kinetic energy E does this correspond to? and (c) What is the probable value of E? Are the answers to (b) and (c) the same?

References

Bastian, N., Covey, K.R., Meyer, M.R.: A universal stellar initial mass function? A critical look at variations. Ann. Rev. A&A **48**, 339 (2010)
Chabrier, G.: Galactic stellar and substellar initial mass function. PASP **115**, 763 (2003)
Drukier, G.A., et al.: The global kinematics of the globular cluster M92. Astron. J. **133**, 1041 (2007)
Folland, G.B.: Fourier Analysis and Its Applications. Brooks/Cole Publishing Co., Pacific Grove (1992)
Galton, F.: Regression towards mediocrity in hereditary stature. J. Antropol. Inst. G. B. Irel. **15**, 246 (1886)
Larson, R.B.: A simple probabilistic theory of fragmentation. MNRAS **359**, 211 (1973)
Lodieu, N., Deacon, N.R., Hambly, N.C., Boudreault, S.: Astrometric and photometric initial mass functions from the UKIDSS Galactic Clusters Survey - II. The Alpha Persei open cluster. MNRAS **426**, 3403 (2012)
Miller, I., Miller, M.: John E. Freund's Mathematical Statistics with Applications, 8th edn. Pearson, Upper Saddle River (2013)
Miler, G.E., Scalo, J.M.: The initial mass function and stellar birthrate in the solar neighborhood. ApJ Supp. **41**, 513 (1979)

Websites (all accessed March 2019):

Wikipedia page on the normal distribution: https://en.wikipedia.org/wiki/Normal_distribution
Wikipedia page on the central limit theorem: https://en.wikipedia.org/wiki/Central_limit_theorem

Chapter 6
Distributions Arising from Random Processes in Time

6.1 Outline of Content

- Random walks in 1D and 3D
- The Poisson Process
- The waiting time or exponential distribution
- The Cauchy/Lorentz distribution
- Power law tails

A stochastic process is something that evolves in time, with a random element. Examples might be the diffusion of one gas through another, the change of stock market prices with time, or the flickering light curve of a quasar. The sequence of values $x(t)$ that arises is known as a *time series*. We will study the characteristics of the time series themselves in Chap. 12. In this chapter we look at the resulting *distribution of values* $f(x)$. If the process produces a distribution of data values that is the same whenever you look at it, we say that the process is *statistically stationary*. However, this is not always the case. In the problem of diffusion, the distribution of particles depends on the length of time since the initial release of particles. The diffusion of particles is a good example of a *random walk*. Let's look at that concept first.

6.2 Random Walks

Imagine some evil-smelling gas released at some point in space. How long before the smell reaches your nose? The molecules of the evil-smelling gas start to travel through the air but suffer repeated collisions, changing direction. They undertake a *random walk*. Each molecule will take a different path, ending up, after many steps,

A. Lawrence, *Probability in Physics*, Undergraduate Lecture Notes in Physics,
https://doi.org/10.1007/978-3-030-04544-9_6

at some x, y, z position. How many particles end up near a given spot in space? In other words, what is the joint probability distribution for x, y, z? Lets start by considering a simplified case, where the steps are in one dimension rather than three, and are of fixed size.

6.2.1 1D Random Walk

Suppose that the molecule goes through n steps, each one of length a, but that these steps can be either forward or backward. At each step the molecule moves either $+a$ or $-a$, with equal probability. A simple computer simulation of a hundred runs of this process is shown in Fig. 6.1, illustrating how the track produced gradually wanders away from the starting point. Every track is different, but it is clear that on average they slowly diverge. This is captured by the right-hand side of Fig. 6.1, which shows the distribution of data values $f(x)$ resulting from a large number of runs, at three different time steps. Moving forward in time by a factor of four seems to produce a distribution that is wider by roughly a factor of two. We can show that in fact this is exactly what we expect.

We can think of getting a positive step as a "success" and so after n steps the number of positive steps r will be given by the binomial distribution with $p = 1/2$. However r positive steps means $n - r$ negative steps, and so the *net displacement* is $d = ra - (n - r)a = (2r - n)a$.

The mean of r is $\mu_r = np = n/2$, so the mean displacement is $\mu_d = 0$. The variance in r is $\sigma_r^2 = np(1 - p) = n/4$, so by the usual variance propagation formula, $\sigma_d^2 = (2a)^2\sigma_r^2 = na^2$. In general, the distribution of the number of positive steps will be given by the binomial distribution for r. The distribution of displacements will be given[1] by transforming to $d = (2r - n)a$. For large n the distribution of displacements will approximate to the Gaussian, and the probability density for a distance travelled near x will be given by

$$p_{1D}(x) = \frac{1}{\sigma\sqrt{2\pi}} e^{-x^2/2\sigma^2} \quad \text{with} \quad \sigma^2 = na^2. \tag{6.1}$$

This distribution has mean $\mu = 0$ at all times, but the dispersion depends on the number of time steps—$\sigma = a\sqrt{n}$. It is a normal distribution, but it is spreading with time, as the number of steps increases.

[1]This simple fixed step-size simulation actually gives a distribution which oscillates on a fine scale—an even/odd number of steps can only result is an even/odd net displacement. However, for small step sizes this effect smooths out.

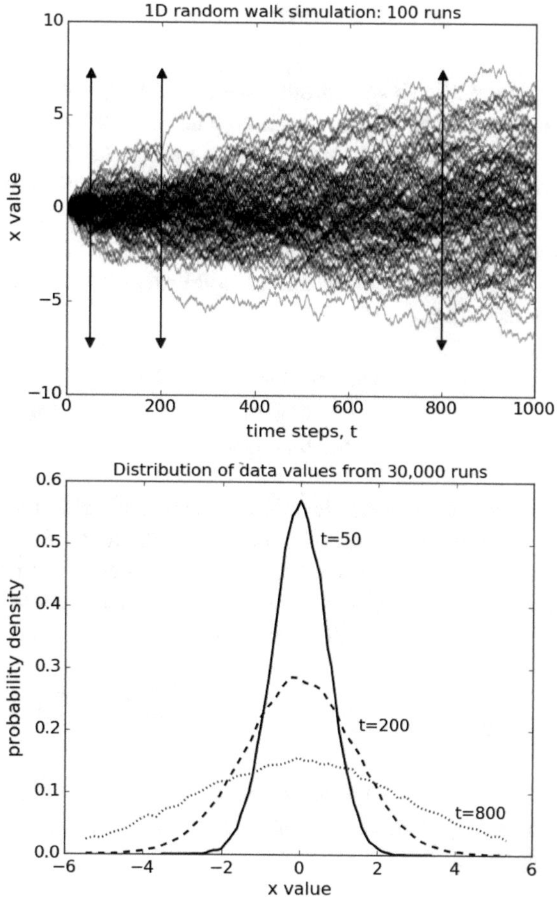

Fig. 6.1 **Upper**: Simulation of random walk in 1D. At each time step the step size is $a = 0.1$, and the simulation is continued for $n = 1000$ time steps. The figure shows the tracks of 100 such simulation runs. The vertical lines show the positions of the vertical slices used for the neighbouring figure. **Lower**: At each of the indicated times, the resulting x value is collected for each simulation run. The curves show the resulting (normalised) distributions $f(x)$

6.2.2 3D Random Walk

In three dimensions, each collision scatters the molecule into a random direction. The result will be a *trivariate Gaussian*, i.e. a Gaussian independently in each of x, y, z. In Chap. 5, Sect. 5.8.2, we saw how we can find the radial density distribution by considering the probability in a radial shell. We thus find that

$$p_{3D}(R) = \frac{4\pi R^2}{\left(2\pi\sigma^2\right)^{3/2}}\ e^{-R^2/2\sigma^2} \quad \text{with} \quad \sigma^2 = na^2. \tag{6.2}$$

Note that this equation gives the probability density per unit radius. (We have used capital R here to avoid confusion with the number of forward steps r from the previous section.)

6.2.3 Spreading Rate

As the particles spread out, how far do they get as a function of time? There is no unique answer to this question. The particle distribution is always centred on the origin, but develops into a gradually wider distribution. However, we could interpret the question as "what is the typical radial distance reached by a particle?". This is a better posed question, as the radial density distribution $p(R)$ at first rises as R^2, coming to a peak, and then declines exponentially. However, we could choose a location measure in several different ways, which give somewhat different answers, for example:

$$\begin{array}{lll} \text{peak of PDF} & R_{\text{pk}} & = a\sqrt{2n} \\ \text{mean value of R} & \bar{R} = \int Rp(R) & = a\sqrt{8n/\pi} \\ \text{root-mean-square R} & R_{\text{rms}} = (\bar{R^2})^{1/2} & = a\sqrt{3n} \end{array}$$

Let us take the peak-location version, and interpret this formula in a gas physics context. If we take the step length a as the mean-free-path between collisions λ, and the molecules are travelling at typical speed v, then the time between collisions is $\tau = \lambda/v$. Then in time t the number of steps is $n = tv/\lambda$. Finally then the typical distance the smell will travel in time t is

$$l = \sqrt{2v\lambda t}, \tag{6.3}$$

i.e. the distance goes as the square root of time. Of course in a real gas, the steps are not of a fixed length; they have their own distribution. However the basic results stay the same.

6.2.4 Random Walks in Momentum Space

Collisions between molecules also exchange momentum and energy, and so are undergoing a kind of random walk in momentum space. Does this mean they get hotter and hotter? Imagine a single slow molecule dropped into a hot gas. At first it does undergo a random walk in momentum, but gradually it gains less and less on each collision on average as it speeds up, until it is similar to the other molecules, in which case it is just as likely to donate *them* energy. This kind of random walk with decreasing step length can of course be mathematically modelled, but we won't pursue that here.

6.3 The Poisson Process and Its Derivatives

Quite often in Nature we see discrete events that seem to occur at a constant rate on average, but at random intervals. One example would be disintegrations of atoms in a radioactive source. Another example would be the noise we sometimes see in electronic components when single electrons strike the component. Yet another might be the arrival of customers at a queue. Such a process is known as a *Poisson Process*, and can be seen as a sequence of points on a timeline, as illustrated in Fig. 6.2. In a similar manner, we could consider a distribution of points in space, such as the position of trees in a forest, or stars in a galaxy. How do we model such a process? To simplify, we will consider only the 1D Poisson process, and think of the points as distributed in time. To clarify our thinking, it will be useful to refer to Fig. 6.2, which defines three different time intervals that we will use.

6.3.1 Distribution of Events Versus Time

We assume that the events occur at a rate λ per unit time. Over an interval of time T the expected number of events will on average be $\mu = \lambda T$. If T is not very large, then the actual observed number n will fluctuate noticeably from bin-to-bin. You can probably guess that the fluctuations follow the Poisson statistics that we discussed in Chap. 4, but we will come to that in a moment. However for large enough T, the number of counts can be considered constant, $n \sim \mu$—a uniform distribution in time.

Now consider a very small time bin, Δt. The expected number of events $\mu << 1$. Note that we can't have fractional events—each such bin Δt either contains an event or it doesn't. However, we can see that

$$p = \lambda \Delta t$$

is the probability that a given bin contains an event. The probability of two events in the same time interval is p^2. As long as Δt, and hence p, is small enough, this

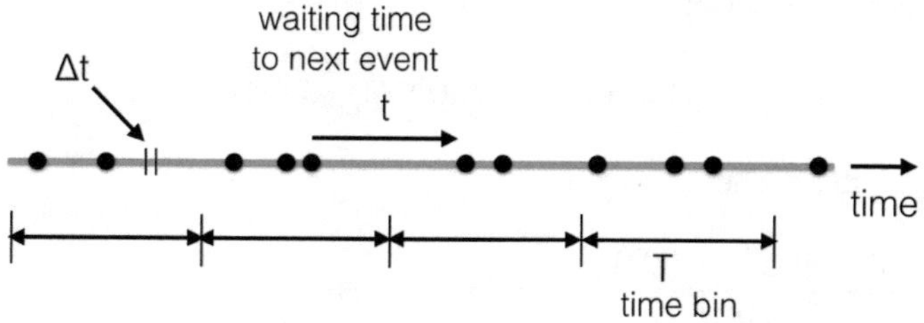

Fig. 6.2 Illustration of a one-dimensional Poisson process in time. Each event is given by a separate symbol, and occurs at random but at a constant average rate. The various time intervals shown are explained in the text

probability is negligible compared to the probability of a single event. Numerically then, we could construct the *sequence of arrival times* by taking one Δt bin at a time and using a random number generator to decide whether that bin has an event or not.

6.3.2 Bin-to-Bin Fluctuations

Returning to the larger time bins of size T, suppose we pick T such that the expected number per bin is say $\mu = 3$. How often do we get $n = 2$, or $n = 5$, and so on? The answer is of course that $f(n)$ is given by the Poisson distribution, but lets step through the logic. We can divide up T into smaller bins Δt. In each small time-step we have a success/failure trial with probability $p = \lambda\Delta t$. Overall we have $N = T/\Delta t$ trials. The expected number of successes n is then given by the binomial distribution. As we let $\Delta t \to 0$ and so $N \to \infty$, this tends towards the Poisson distribution. The count fluctuations are therefore given by

$$f(n) = \frac{\mu^n e^{-\mu}}{n!} \quad \text{with} \quad \mu = \lambda\Delta t.$$

6.3.3 Waiting Time Distribution

A striking feature of Fig. 6.2 is that although the events arrive at random, they appear to be clumped—sometimes there is a long gap, and sometimes several in a row. What is the distribution of waiting times from one event to the next? This calculation is relevant to many familiar circumstances—for example, if I just missed a bus, how long am I likely to wait for the next one?[2] To treat this problem, it is easier to think in terms of a lack of events in a given time interval.

The probability of there being no event in time Δt is $1 - \lambda\Delta t$. Suppose now that after an accumulated time t there has been no event. We denote the probability of this by $P_0(t)$. What is the probability that there is *still* no event by $t + \Delta t$? We need to have no event by time t (probability $P_0(t)$) and no event in Δt (probability $1 - \lambda\Delta t$). Multiplying these together we get

$$P_0(t + \Delta t) = P_0(t)\ (1 - \lambda\Delta t).$$

Re-arranging this we get

$$\frac{P_0(t + \Delta t) - P_0(t)}{\Delta t} = -\lambda P_0(t) \quad \text{that is} \quad \frac{\mathrm{d}P_0}{\mathrm{d}t} = -\lambda P_0,$$

which has the solution

[2] Of course, its not obvious that buses do actually arrive at random...

$$P_0(t) = e^{-\lambda t},$$

where $P_0(t)$ is the probability of no event by time t. We would also like the probability density that the next event will occur at time t. This must be

$$\text{Distbn of waiting times} \qquad f(t) = \lambda e^{-\lambda t}. \tag{6.4}$$

You can see this because the integral $\int_0^T f(t)$ is $1 - e^{-\lambda T}$, which is the probability that the next event occurs somewhere in the range 0 to T, as opposed to $P_0(T) = e^{-\lambda T}$ which is the probability that the next event occurs beyond T. The distribution $f(t)$ is known as the *waiting time distribution* or sometimes just as the *exponential distribution*.

6.4 The Cauchy Distribution

Figure 6.3 shows the profile of a resonance feature measured in a nuclear physics experiment. The core of the feature looks roughly Gaussian. However on the high energy side we a long tail inconsistent with a Gaussian distribution. The right hand side of the figure shows a log-log version of the same dataset. The extended tail is roughly a straight line in this log-log plot, with a slope of 2, showing that the profile is following a *power law form* with $y \sim x^{-2}$. Distributions showing $1/x^2$ tails occur in various circumstances in Nature. The simplest mathematical form that gives the right shape is the *Cauchy distribution*

$$f(x) = \frac{1}{\pi(1 + x^2)}. \tag{6.5}$$

Below we shall look at two examples of how such a shape can arise naturally through random processes. But first, lets look at the properties of this distribution, which are somewhat strange.

6.4.1 Moments of the Cauchy Distribution

The zeroth moment of the Cauchy distribution is 1.0, so that $f(x)$ can represent a properly normalised probability density function. To see this, note that the derivative of $\arctan x$ is $1/(1 + x^2)$. The half-integral of $f(x)$ from $x = 0$ to $x = X$ is therefore

$$I(X) = \frac{1}{\pi}\big[\arctan x\big]_0^X = \frac{1}{\pi} \arctan X.$$

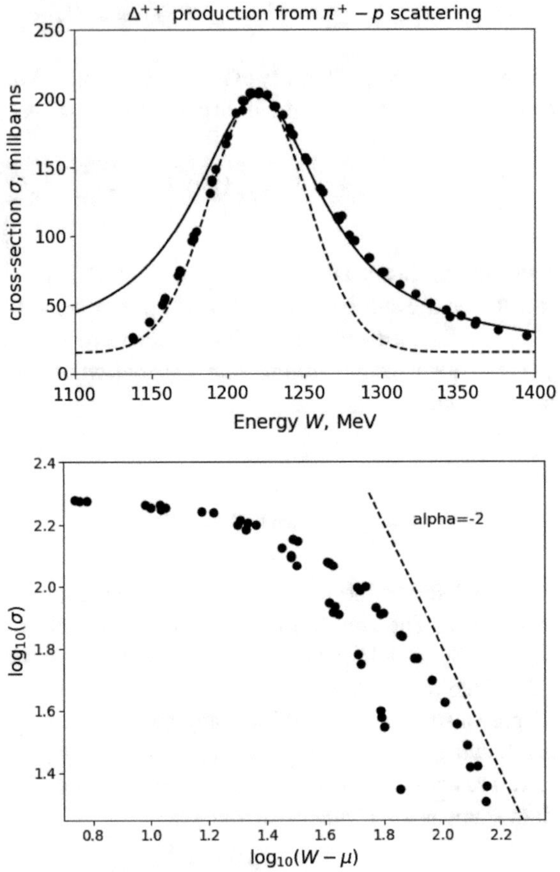

Fig. 6.3 **Upper**: Cross-section of Δ^{++} production from $\pi^+ - p$ scattering. Data kindly provided by Mikhail Bashkanov. For comparison, the dashed line shows a Gaussian profile and the solid line shows a Lorentzian profile. For a little more explanation, see the Further reading section. **Lower**: Log-Log version of the plot, recentred on the mean. The extended wing tends towards an x^{-2} shape at high energies

Then of course $\tan \pi/2 = \infty$ and so we get

$$I(\infty) = \frac{1}{\pi}\frac{\pi}{2} = \frac{1}{2}.$$

However, what we just calculated was the half integral from zero upwards, so the full integral is 1.0, as required.

The mean, defined as the expectation value $E[x]$ is formally an undefined quantity. We won't prove that here, but it is not too disturbing, as $f(x)$ has a well defined mode or median, which is clearly at $x = 0$, as the function is symmetrical in x. What about the variance? Given that the distribution is centred at $x = 0$ we simply want the expectation value of x^2, and find

$$E[x^2] = \frac{1}{\pi}\int \frac{x^2}{1+x^2}\mathrm{d}x = \frac{1}{\pi}\int\left(1 - \frac{1}{1+x^2}\right)\mathrm{d}x$$
$$= \frac{1}{\pi}\left(\int 1.\mathrm{d}x - \int \frac{1}{1+x^2}\right)\mathrm{d}x$$
$$= \frac{1}{\pi}(\infty - \pi)$$
$$= \infty.$$

Disturbingly, the Cauchy distribution has infinite variance. It can be shown that all higher moments are also infinite.

6.4.2 Lorentzian Form

Physicists normally use a slightly different formulation of the Cauchy distribution, known as the *Lorentzian distribution*, or sometimes the *Breit–Wigner profile*. The first step is to recentre at a value $x = \mu$ rather than $x = 0$. Next, although the dispersion σ is undefined, we can still objectively characterise the width of a given distribution. The usual way to do this is to quote the Full Width at Half Maximum (FWHM), usually denoted by the symbol Γ. Adjusting the normalisation to make sure the integrated probability is 1.0, we find that the required Lorentzian form is:

$$f(x) = \frac{1}{\pi}\frac{\Gamma/2}{(x-\mu)^2 + (\Gamma/2)^2}. \tag{6.6}$$

This is a symmetrical distribution with mode $= \mu$, and a maximum value of $2/\pi\Gamma$. At $(x - \mu) = \Gamma/2$ we get $f(x) = 1/\pi\Gamma$, i.e. half the maximum value, as required. You can see that the standard Cauchy form is a Lorentzian with $\mu = 0$, $\Gamma = 1$.

6.4.3 Angular-to-Linear Transformations: The Rotating Gun Problem

A Cauchy–Lorentz distribution can arise when a phenomenon involves a transformation from an angular co-ordinate to a linear co-ordinate. If the expected distribution is uniform in the angular co-ordinate, it will be Cauchy-like in the linear co-ordinate. As an example, consider a gun firing bullets at random, but at a uniform rate, i.e. as a Poisson process in time. The distribution of bullets as a function of time will be uniform on average. Now imagine the gun to be rotating, with uniform angular velocity. The distribution of bullets with angle will now be uniform. Finally, imagine a wall some distance from the rotating gun. What is the probability of a bullet hitting a particular position on the wall?

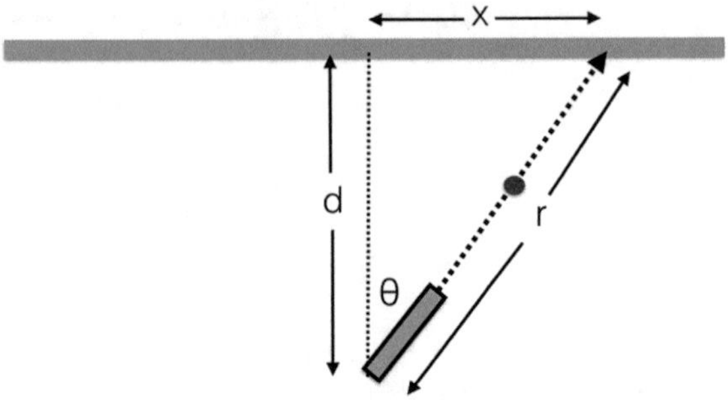

Fig. 6.4 Illustrating the geometry of the rotating gun problem

Figure 6.4 shows the geometry of this situation. If we care only about the forward travelling bullets, which are capable of hitting the wall, the probability density as a function of angle θ is $p(\theta) = 1/\pi$. To transform to the density per unit x we note that $p(x)\mathrm{d}x = p(\theta)\mathrm{d}\theta$ and so get

$$p(x) = \frac{1}{\pi}\frac{1}{\mathrm{d}x/\mathrm{d}\theta}.$$

From the figure we have

$$x = d\tan\theta \quad \text{so} \quad \frac{\mathrm{d}x}{\mathrm{d}\theta} = d\sec^2\theta \quad \text{and} \quad \sec\theta = \frac{(d^2+x^2)^{1/2}}{d}.$$

and so we get

$$p(x) = \frac{1}{\pi}\frac{d}{d^2+x^2},$$

which you can see is a Lorentzian with $\Gamma = 2d$.

6.4.4 *Resonance and the Lorentzian*

Another way to achieve a Cauchy–Lorentz distribution is through the phenomenon of resonance. This crops up in a variety of places—the shape of absorption lines in stars, the dependence of nuclear cross-sections on energy, or the Rayleigh scattering of light by molecules in the atmosphere. In each case the problem takes a probabilistic form—for example, what is the probability a given photon will be scattered? However, the essence of any such process is analogous to the problem of a damped driven oscillator.

If we have a spring with spring constant k this produces a returning force $F = -kx$, and if a mass m is attached this has a natural frequency of oscillation with angular frequency $\omega_0 = (k/m)^{1/2}$. Now suppose we try to force an oscillation at frequency ω_f with driving force $F = F_0\cos\omega_f t$, and in addition we have a damping term proportional to velocity, $F = -\lambda v$. Then, as you will find in standard texts on vibrations and waves, you can show that the energy of the resulting oscillation you get at a chosen driving frequency ω_f is

$$E(\omega_f) = \frac{F_0^2/2}{(\omega_f - \omega_0)^2 + m^2\lambda^2}.$$

When the driving frequency is near the natural frequency, you get a large output—i.e. ω_0 is the resonant frequency. However, far from the resonant frequency, the output goes as $E \propto 1/\omega_f^2$. You can see that, apart from a normalising factor, the expression above is identical to the standard Lorentzian form.

6.4.5 Comparison to Gaussian

Figure 6.5 compares the Lorentzian and Gaussian distributions. In order to compare them, because the Lorentzian has no defined value of σ, they have been calculated to have the same value of FWHM, Γ. For the Gaussian, $\Gamma = 2.355\sigma$ (see Exercise 5.2). The vertical normalisation has been done in two different ways. In the first method the two curves are both properly normalised probability density functions, i.e. they integrate to 1.0. This emphasises that the Lorentzian has a substantially lower peak probability density, and very extended tails. The second version re-normalises the Lorentzian so the peak probability density is the same as for the Gaussian. What this reveals is that within the core region, the Gaussian and the Lorentzian are closely similar. This is of very practical concern. The usual practice when analysing data is to assume that observed peaks are of Gaussian shape. A common problem (especially in statistical inference) is to estimate the probability of unusual events, e.g. a data point seen at $>2.5\sigma$. Examining the core region, a value of σ may have been estimated on the assumption that the peak is of Gaussian shape; if in truth it is Lorentzian, then the probability of events in the tail can be enormously larger (see exercises).

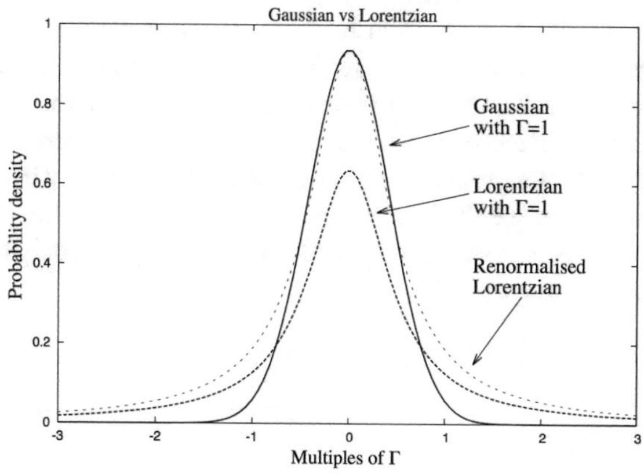

Fig. 6.5 Comparison of Gaussian and Lorentzian functions. See text for details

Another key point is that because of the pathological Nature of the Cauchy–Lorentz distribution, with no defined variance, it does not converge to a Gaussian on repeated convolution. Convolution of a Lorentzian with another Lorentzian gives a third Lorentzian; it is a separate convergent point of the convolution process. This is another reason why it is a common and important distribution.

6.5 Power Law Tails

The left-hand side of Fig. 6.6 shows a histogram of the number of US cities with a given population. It is strongly peaked towards low values—there are many small cities, and fewer and fewer large ones. There is no obvious "typical" city size. This is rather reminiscent of the situation for the distribution of stellar masses (see Chap. 5, Sect. 5.9.2). In the stellar mass case we found that if we looked at the distribution of values of log mass rather than mass, there is after all a typical value (of log mass), and the distribution looks Gaussian. That is not the case here—the distribution of the log of city size does not look Gaussian. Instead, what is revealing is to stick with the number per unit city size versus city size, but to plot the log of each of these values. This is done for the city size data in the right hand side of Fig. 6.6, and shows a straight line in this log-log plot. In others, the frequency of city size seems to follow a *power law*, i.e.

$$f(x) = Ax^{-\alpha} \quad \text{so that} \quad \log f = \log A - \alpha \log x. \tag{6.7}$$

In the city-size case, it seems that $\alpha \sim 1.5$. Note that at this point, our frequency is not a normalised probability density. In fact, this is problematic, as we shall see shortly.

Power law distributions, often extending over many decades of the relevant variable x, are quite common in Nature. We see them in the distribution of cosmic ray energies, the spectra of solar flares, the distribution of incomes, the relative popularity of web sites, and many other places. In some areas of science, especially in economics, the power-law distribution is known as a *Pareto distribution*. A variety of values for the power law index α is seen, ranging from $\alpha = 0.5$ to $\alpha = 3$ or even steeper. Before we look at how such power law distributions might be generated in Nature, we should look more carefully how they can be mathematically characterised. In fact, as we shall see below, a power law cannot truly be the correct description of a probability distribution for all values of x; it can only be an approximate description of a *power law tail*.

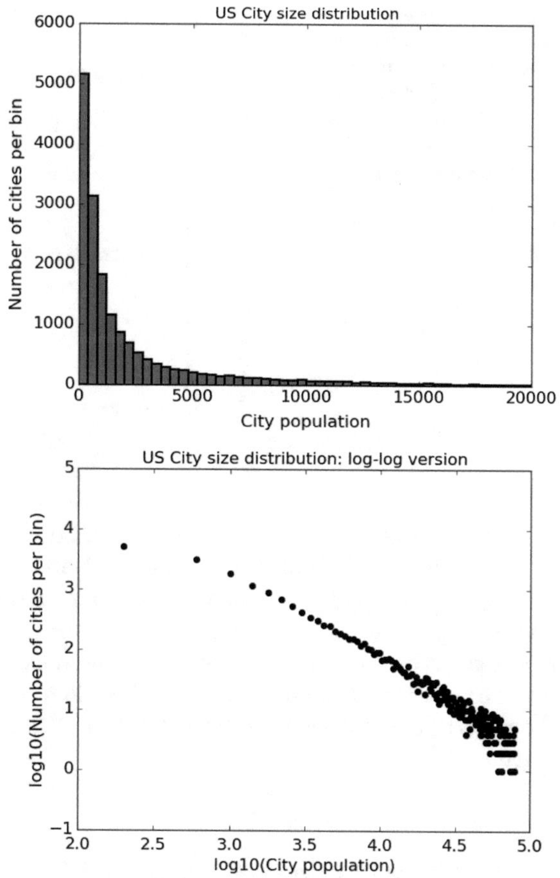

Fig. 6.6 **Upper**: Frequency distribution of US city sizes as of the year 2000, taken from Clauset et al. (2009). **Lower**: Same data points, but plotting the log of the frequency versus the log of the city size

6.5.1 *Normalising a Power Law Distribution: Power Law Tail Approximations*

Suppose we want to normalise the frequency distribution of equation (6.7), to produce a probability density function. Integrating equation (6.7) gives

$$\int_{-\infty}^{+\infty} f(x) = \frac{A}{1-\alpha}\left[x^{(1-\alpha)}\right]_{-\infty}^{+\infty}.$$

If the power law is steep, $\alpha > 1$, the integration diverges towards small values of x. A steep power law therefore cannot be a strictly true mathematical description at low values. Indeed, in real world examples, we always see a flattening of the slope towards low values. You can see this in the right hand of Fig. 6.6, but it is even more obvious in Fig. 6.7, the distribution of earthquake magnitudes. In fact, the lower end

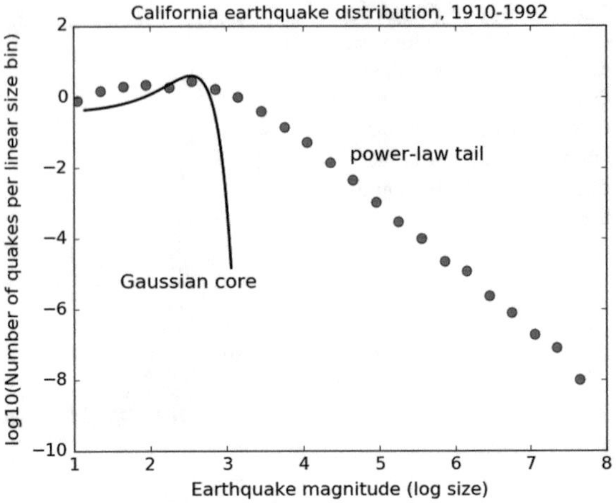

Fig. 6.7 Frequency distribution of earthquake magnitudes, taken from Clauset et al. (2009). Note that the "magnitude" of an earthquake is already effectively a logarithmic quantity, but that the frequency is terms of the implied linear size. The strange looking curve on the left is simply a normal Gaussian, but plotted in log-log space

is roughly consistent with a Gaussian distribution. The power-law phenomenon is usually about extended tails to more normal distributions.

Suppose we recognise this and consider only values at $x > x_{\min}$. Unfortunately, towards large values of x, the integral still blows up if $\alpha < 1$. Therefore, just as a steep power law cannot be a true description at low values of x, a flat power law cannot be a true description at high values. Real world distributions do indeed quite often show two different slopes at high and low values, or a slow curvature in logarithmic space.

6.5.2 Moments of a Power Law Distribution

If we restrict our power law distribution to $x > x_{\min}$, and also require $\alpha > 1$, then requiring the zeroth moment to be 1.0, gives us the normalisation factor A, in terms of $x_{\min}$, and a useful way to express a power law distribution:

$$A = (\alpha - 1)x_{\min}^{-(1-\alpha)} \quad \text{and so} \quad p(x) = \frac{\alpha - 1}{x_{\min}}\left(\frac{x}{x_{\min}}\right)^{-\alpha} \quad \text{for} \quad x > x_{\min}.$$

Is there a "typical" value? What is the first moment? We find

$$E[x] = \int_{x_{\min}}^{\infty} xp(x) = \frac{A}{2-\alpha}\left[x^{2-\alpha}\right]_{x_{\min}}^{\infty}.$$

If $\alpha > 2$ this produces a sensible result.

$$E[x] = x_{\min}\frac{\alpha - 1}{\alpha - 2}.$$

If $\alpha < 2$ the mean value is technically infinite. However, it is worth remembering that this is the first moment of a mathematical expression; the sample mean of any sample of data which is approximately described by a power law with $\alpha > 2$ will still produce a sensible finite value. However, for such a power law, you will find that the sample mean fluctuates rather wildly from sample to sample. In a similar way, as we look at higher moments, we find that the m'th moment is only defined for $\alpha > m + 1$, but if that holds we have

$$E[x^m] = x_{\text{min}}^m \frac{\alpha - 1}{\alpha - 1 - m}.$$

6.5.3 *Lorentzians and log Gaussians in Disguise*

If a dataset shows a power law tail with $\alpha = 2.0$, then it may well simply be a Lorentzian distribution; at large x this follows $p(x) \propto 1/x^2$. If a dataset looks power-law like, but you only have data values over a limited range of x, say one or two decades, then the true distribution could be log-Gaussian. The log-Gaussian has a very "stretched out" distribution, and over a modest range of x will produce a straight-ish line in a log-log plot; however, the slope in log-log space slowly changes, getting steeper at larger values of x. As described in the "Further reading" section of Chap. 5, early measures of the stellar mass function were fitted with power law functions; a wider range of measurements led to the popularity of two power law forms; but now it is conventional to model the stellar mass function with a log-Gaussian form. Note that the larger the value of σ (i.e. the dispersion in $\log x$) the larger the range over which x will look like a single power law.

6.5.4 *Power Laws from Inverse Quantitities*

In this and the next two sections I follow the excellent discussion of Newman (2005) in examining ways that approximately power law distributions can arise in Nature. The first and simplest is by looking at the distribution of inverses of a quantity. Suppose some quantity y has a known PDF $f(y)$, but that the quantity we have actually measured and found the distribution of, is its inverse, $x = 1/y$. Then the PDF of x will be

$$p(x) = f(y)\frac{\mathrm{d}y}{\mathrm{d}x} = -\frac{f(y)}{x^2}.$$

Suppose $f(y)$ is zero centred, and we are looking at large values of x that correspond to small values of y, where $f(y) \sim$ const for any reasonable distribution, be it Gaussian or whatever. Then to a good approximation, $p(x) \propto x^{-\alpha}$ with $\alpha = 2$. Positive values of y correspond to negative values of x and vice versa. More generally if $x = 1/y^\beta$ we can get a power law with $\alpha = 1 + 1/\beta$.

This method may seem artificial, but in fact it crops up quite often, especially when we look at *fractional changes* in a quantity. Suppose we measure a quantity y repeatedly and look at the relative change $x = \Delta y/y$. The absolute values Δy may have a Gaussian distribution, but the distribution of x will show a power law tail with $\alpha = 2$.

6.5.5 *Power Laws from Growth and Survival Processes*

Another simple mathematical trick to generate a power law is to take the ratio of two exponentials. Suppose the quantity has a PDF $f(y) = e^{ay}$, but the quantity x that we measure is related to y as $x = e^{by}$. Then the PDF of x will be given by

$$p(x) = f(y)\frac{\mathrm{d}y}{\mathrm{d}x} = \frac{p(y)}{\mathrm{d}x/\mathrm{d}y} = \frac{e^{ay}}{be^{by}} = \frac{\left(e^{by}\right)a/b}{be^{by}} = \frac{x^{a/b}}{xb}.$$

but the last expression is of the form

$$p(x) \propto x^{-\alpha} \quad \text{with} \quad \alpha = 1 - a/b.$$

Combining exponentials can occur in a variety of circumstances, but probably the most important class of phenomena involves the competition between exponential growth and an exponential decline in survival rate. As a specific example let us examine the acceleration of extremely energetic particles in the cosmos. This can happen in a variety of places—solar flares, supernova remnants, quasar jets—but the process is always roughly similar. The particles enter a region where they are colliding with much more massive entities—for example the upstream and downstream shock fronts in an expanding blast wave. (The word colliding here is a bit of an oversimplification of course.) In a normal gas, when particles collide with each other, they share energy, both gaining and losing energy, rapidly establishing a Gaussian. When however the dynamics of a particle is dominated by collisions with much more massive objects, it gains energy in every collision, and the result is an exponential growth of particle energy, on some growth timescale T_g. At the same time, there is a small chance in each unit of time that the particle will escape the acceleration region. This leads to a exponentially decreasing probability of survival with time, on some escape timescale T_e. Put the exponential growth of energy and the probability of escape together we have

$$E = E_0 e^{t/T_g} \quad \text{and} \quad p(t) \propto e^{-t/T_e}.$$

We have deliberately left the normalisation of the PDF undefined, because of the ambiguities discussed previously. This pair of expressions is just like our discussion above, with $y = t$, $b = 1/T_g$, and $a = -1/T_e$. We therefore predict the particle energy distribution

$$p(E) \propto E^{-\alpha} \quad \text{with} \quad \alpha = 1 + T_g/T_e. \tag{6.8}$$

If the growth timescale is much shorter than the escape timescale, $T_g << T_e$, we expect an energy distribution following E^{-1}; if the timescales are comparable, we have $\sim E^{-2}$; if escape dominates, i.e. $T_e << T_g$, the power law tail will become very steep and so unimportant. Other possible examples of competing growth and death processes include population sizes, and the distribution of personal wealth. In a capitalist economy, rich people get richer, but they are also subject each year to a finite risk of a catastrophe occurring and becoming bankrupt.

6.5.6 Other Power Law Generation Methods

We will briefly mention a few other processes thought to lead to power law distributions.

(i) The *Yule process* was invented to explain the distribution of the number of biological species per genus, but has been applied elsewhere. Over time, species split and make new species, growing the population in that genus, but every so often a new species is different enough that it starts a new genus.

(ii) Earlier in this chapter we discussed the random walk. Every so often by chance the process will return to the starting point—sometimes quickly, sometimes after a long time. It can be shown that the distribution of return times is a power law. (This is sometimes known as the *gambler's ruin* problem.)

(iii) When a substance undergoes a phase transition from one state to another, it often does so in localised regions. As these increase in number density, they start to connect up. At a certain critical density, the size distribution of connected regions is a power law. Although we have talked in terms of phases in a substance, analogous situations hold in a variety of circumstances. This arrangement requires fine tuning, but it has been argued that many systems naturally end up at a critical point. This idea is known as *self organised criticality*.

6.6 Key Concepts

Some of the key concepts from this chapter are:

- How a random walk produces a binomial distribution of displacements
- How a 3D random walk in practice produces a trivariate Gaussian

- How in diffusion, distance travelled goes as the square root of time
- The idea of a Poisson process, and how it leads to an exponential distribution of waiting times
- The Cauchy(Lorentz) distribution, and how it formally has an infinite variance
- How a Cauchy distribution can be produced by resonance process
- How power-law tails frequently occur in Nature
- How power-law tails can be produced by inverse processes, and by growth-and-survival processes.

The key formulae from this chapter are: the distribution of displacements in a random walk, in the 1D (6.1) and 3D (6.2) cases, and the formula for distance travelled versus time (6.3); the formula for the distribution of waiting times in a Poisson process (6.4); the formula for a Cauchy distribution (6.5), and its Lorentzian form (6.6); the general form of a power-law tail (6.7); and the formula for a power-law tail produced by competing growth and death processes (6.8).

6.7 Further Reading

Diffusion and the simplest kinds of random walk are discussed in many basic textbooks on thermal physics. Likewise resonance processes and Lorentzian distributions are covered in textbooks on nuclear physics and elsewhere. To get a deeper appreciation of random walks, Poisson processes and shot noise, there are many books on stochastic processes. Two books that are focused specifically on stochastic processes in Physics are Lemons (2002) and Jacobs (2010). We will look at these issues more closely in Chap. 12.

Its harder to find good treatments of power law distributions in basic textbooks, despite their ubiquity in Nature. A large number of intriguing examples are collected in Newman (2005) and Clauset et al. (2009), and both those papers have a clear approach to explaining the underlying processes. There is a very interesting and useful web page associated with the Clauset paper (see web references). Power-law distributions are better known in social sciences, where they are known as Pareto distributions; the wikipedia page is a good overview. An interesting blog which looks at Economics from a Physicist's point of view is "Economics as Classical Mechanics" by Ivan Klitov.

Power-law distributions are closely related to fractals. The classic (and very readable) book which started the subject is Mandelbrot (1982).

6.8 Exercises

6.1 Perfume is released from a small point and diffuses through the air. If the typical mean free path of the molecules is 1.3×10^{-7} m, and they have typical velocity $490\,\mathrm{m\,s^{-1}}$, how long does it take before the peak of the radial density distribution of perfume molecules reaches 1 m away?

6.2 Equation (6.2) gives an expression for the radial density profile resulting from a 3D random walk. Verify that this is the same as the equation for the radial profile of the 3D Gaussian equation in (5.5), if $\sigma^2 = na^2$ where n is the number of steps, and a is the size of each step. If we express the radial distance in units of σ, so that $R = z\sigma$, show that the radial density profile is

$$\rho(z) \propto z^2 e^{-z^2/2}$$

Show that the peak of the density profile is at $z = \sqrt{2}$.

The expression above is for the density per unit radius, i.e. summed over a radial shell. How does the density per unit volume depend on z? How does the density at $z = 5$ compare to the density at $z = 0$? Given that the human nose can detect very small numbers of molecules, why does this tell us that the calculation in the previous problem significantly overestimates how long it takes for the smell to detected?

6.3 The previous question proves one of the results of Sect. 6.2.3, that the peak of $p(R)$ for a 3D random walk is at $R = a\sqrt{2n}$, where a is the step length, and n is the number of steps. Can you prove the second result, that the mean of $p(R)$ is at $R = a\sqrt{8n/\pi}$? (Hint: you need the third "error integral" $I_3 = \int_0^\infty x^3 e^{-hx^2} = 1/2h^2$.)

6.4 In our analysis of Poisson processes, we assumed that we can ignore the chance of two events in the same time period Δt. Supposed that events are detected by a particle counter at a rate of 10 per second. How small a time interval do we need to consider so that the probability of two events is no more than 1% of the probability of a single event?

6.5 The table below shows the frequency of time intervals between successive disintegrations of Thorium atoms in a radioactive sample, measured over the space of 2496 s. (This is real data from Marsden and Barratt 1911.) Does this match what you would expect? Either plot the data with a theoretical curve over the top, or calculate the corresponding expected numbers for each bin.

Time interval (s)	Observed frequency
0–1/2	101
1/2–1	98
1–2	159
2–3	114
3–4	74
4–5	48
5–7	75
7–10	59
10–15	32
15–20	4
20–30	2
Over 30	0

6.6 If W_{50} means Full Width at 50% of peak, i.e. FWHM, and W_{20} means Full Width at 20%, compare the ratio W_{20}/W_{50} for the Lorentzian and the Gaussian.

6.7 A particle physics experiment measures the energy at which a specific type of collision event occurs. After measuring a few tens of collisions, it looks like the distribution of collision energies is roughly Gaussian, with FWHM = 0.86 GeV. An event is then seen at an energy 1.16 GeV away from the mean energy. One of the scientists points out this is at several sigma from the mean, with a probability of less than one in a thousand, and therefore suggests a physically different kind of event has been seen, which should be published quickly. If the distribution is actually Lorentzian rather than Gaussian, how big a mistake would this scientist be making? (Hint: you can approximately integrate the Lorentzian by noting that it roughly follows x^{-2}.)

6.8 It has been suggested that the number of individuals with wealth w follows a power law density distribution $n(w) = Aw^{-\alpha}$, i.e. $n(w)\mathrm{d}w$ is the number of people with wealth between w and $w + \mathrm{d}w$. This is assumed to apply above some minimum value $w_{\min}$, and $\alpha > 1$ in order for the total number of people to be finite. Show that the richest half of the population is located at $w > w_{1/2} = w_{\min}2^{1/(\alpha-1)}$. In the USA, current evidence is that $\alpha \sim 2.1$. Show that the richest half of the population own 94% of the wealth.

References

Clauset, A., Shalizi, C.R., Newman, M.E.J.: Power-law distributions in empirical data. SIAM Rev. **51**(4), 661–703 (2009)

Jacobs, K.: Stochastic Processes for Physicists: Understanding Noisy Systems. Cambridge University Press, Cambridge (2010)

Lemons, D.S.: An Introduction to Stochastic Processes in Physics. Johns Hopkins University Press, Baltimore (2002)

Mandelbrot, B.: The Fractal Geometry of Nature. W.H. Freeman Company Ltd., New York (1982)

Marsden, E., Barratt, T.: The probability distribution of the time intervals of a particles with application to the number of a particles emitted by uranium. Proc. Phys. Soc. **23**, 367 (1911)

Newman, M.E.J.: Power-laws, Pareto distributions, and Zipf's law. Contemp. Phys. **46**, 323, (2005)

Websites (all accessed March 2019):

SAID database at George Washington University: http://gwdac.phys.gwu.edu

Web page associated with Clauset et al 2009: http://tuvalu.santafe.edu/~aaronc/powerlaws/data.htm

Discussion of Pareto distributions at Wikpedia: https://en.wikipedia.org/wiki/Pareto_distribution

"High income distribution from the IRS", a posting in "Economics as classical mechanics" by Ivan Klitov: http://mechonomic.blogspot.co.uk/2012/02/high-income-distribution-from-teh-irs.html

Part III
Probabilistic Inference: Reasoning in the Presence of Uncertainty

In Part II, we looked at how frequency distributions arise in natural circumstances. This can be seen as the "forward machinery" of the handling of probability—if we know the circumstances that pertain, we can turn the handle on our machinery and predict the distribution of events that we should see in an experiment. Now in Part III, we look to see if we can run the machinery backwards—that is, given some events that we observe, can we deduce what the physical circumstances must be? This is the problem of *Statistical Inference*. For example, given a peak in the counts at some channel number in our particle detector, can we say whether the particle that caused this peak has a mass of 128 GeV or 134 GeV? If we want certainty, the answer is a clear "no"—we cannot be sure which mass is correct. However, this doesn't mean we can say nothing.

The key concept behind all the methods in Part III is the idea of *likelihood*—if I assume such-and-such, what is the probability I would have seen what I saw? In using the idea of likelihood, there are two paths we can follow, that correspond to the two different uses of the term "probability" that we met in Part I—the frequentist path, and the Bayesian path. Historically, these have been often seen as philosophically warring. In fact, they are both valid approaches to statistical inference, either of which can be used as long as we are careful about the statements we make.

In Chap. 7, we look at the basics of *hypothesis testing*. We first see how we can use likelihood together with Bayes's formula to adjust our initial (prior) degree of belief in a hypothesis to get an updated (posterior) belief, after taking the observed data into account. Alternatively, we can make an absolute test of our *confidence* in a hypothesis, using the observed data. When we extend our ideas to multi-value datasets, this turns out to be simple for Bayesian methods, but for confidence methods requires us to combine our data points into a single *test statistic*. We introduce the most important example, the χ^2 test statistic, which quantifies the degree of scatter in a dataset.

In Chap. 8, we look at how to use the ideas of hypothesis testing to estimate the "best" value of a parameter, and the uncertainty on our estimate, by seeing the possible values of the parameter as a kind of tuneable hypothesis. We explore several different (closely related) ways to do this—by finding maximum likelihood, maximum posterior, and minimum χ^2.

In Chap. 9, we look at how we deal with two variables, and how to test whether or not there is evidence for a *correlation* between them. This involves clarifying the ideas of independence and covariance. We then look at how we test the mathematical relationship between two variables, using the *least squares method* to find the parameters of the best fitting line that goes through the data points.

Finally, in Chap. 10, we look at *model fitting* in general. We have most of the apparatus we need from the previous chapters, but some extra subtleties arise when we construct credibility or confidence intervals for multiple parameters, depending on which parameters are considered "interesting", and whether the parameters are correlated. We also look briefly at practical numerical issues of how to find the minima/maxima of multidimensional surfaces, and characterise their curvature, in order to find best parameter values and their uncertainties.

Chapter 7
Hypothesis Testing

7.1 Outline of Content

- The idea of data likelihood
- Using Bayes's formula to update hypothesis probabilities
- Confidence testing and P-values
- Pitfalls of P-values
- Bayesian inference with compound datasets
- Confidence methods with compound datasets: using test statistics
- The χ^2 test statistic

A common activity in scientific life is the making of *hypotheses* followed by their testing. For example, a theorist may state that "my theory implies that there should be a particle of mass such and such; if you make so and so measurements you should see a spike in the data". Some experimenters then make the necessary measurements. There is a small spike in the right place, but a few other spikes that look almost as big. Have we seen the theorised particle? Such theoretical predictions can seem rather complicated. To get at the essence of this problem, lets look at something more artificial but much simpler.

Consider the urn so beloved of statisticians, containing coloured balls. If we *know* that our urn contains four red balls and six black balls, we can calculate the probability of picking three red balls in a row. Now suppose however we do not know the number of red balls and black balls, but have just picked out three balls and found them all to be red; can we now tell what the urn contains? The answer of course is that we cannot make a unique and unambiguous deduction. But we can make probabilistic statements. Suppose two colleagues make differing hypotheses. Colleague A suggests that the urn contains exactly three red balls and a hundred black balls; Colleague B suggests that there are five of each. If we have just picked out three red balls, then instinctively the first hypothesis (three red balls and a hundred black balls) seems much less likely to be right than the second (five red balls and five black balls). How do we firm up and quantify this instinctively reasonable feeling?

A. Lawrence, *Probability in Physics*, Undergraduate Lecture Notes in Physics,
https://doi.org/10.1007/978-3-030-04544-9_7

7.2 Data Likelihood

The key concept is *likelihood*—the probability of getting the data we have seen, given the hypothesis we are assuming. To use this idea, we need a *well framed hypothesis*. As well as stating the hypothesis unambiguously, we need to state it in a manner that allows us to perform a calculation of the probability. For example, suppose we have tossed a coin $n = 10$ times and found heads to come up $x = 4$ times. If our hypothesis is "this coin is a fair one, which comes up heads or tails equally often", then we can calculate the likelihood $L(x)$ from the binomial distribution for $p = 0.5, n = 10, x = 4$. An alternative hypothesis could be "this coin is not fair; it is weighted". This is a perfectly valid logical hypothesis, but it doesn't allow us to calculate a likelihood. We would need to be more specific—for example, "this coin is weighted and typically comes up heads eighty percent of the time".

The *experiment* that produces the data also needs to be carefully defined. Our coloured-balls problem is a good example. If we just say "we got three reds in a row", this is insufficient information—did we replace each ball before we picked the next one, or did we put it to one side? Suppose we take the hypothesis that the urn contains three red and three black balls. If we replace the balls, then the probability of getting exactly three reds in a row is $\frac{3}{6} \times \frac{3}{6} \times \frac{3}{6} = \frac{1}{8} = 0.125$. If we do not replace the balls, the probability of getting exactly three reds in a row is $\frac{3}{6} \times \frac{2}{5} \times \frac{1}{4} = 0.05$.

In general an experiment might produce a set of numbers $D = x_1, x_2, x_3, \ldots$. If we have an unambiguous hypothesis H, and we carry out a well understood experiment, then we can calculate the joint probability of getting the complete set of numbers D in those circumstances. We then define likelihood as:

$$\text{likelihood} \quad L(D) = P(D|H) \tag{7.1}$$

where D is understood as shorthand for $x_1, x_2, \ldots$. For the first part of this chapter, we will simplify matters by assuming that the experiment gives a single data value, x. For example, x could be the number of red balls found after n picks, or the voltage measured in some piece of equipment. Then we have $L(x) = P(x|H)$. If X is a discrete random variable (such as the number of red balls observed), then $L(X)$ is a simple probability. If x is a continuous random variable (such as a voltage value), then $L(x)$ is a likelihood density, i.e. probability per unit x.

Clearly, getting a smaller likelihood reduces the believability of a hypothesis. How do we sharpen that idea? There are two ways, depending on the question we are asking.

7.2.1 Comparing Hypotheses Using Relative Likelihoods

We might have several hypotheses and ask "which of these hypotheses is more believable?" Given a number of hypotheses H_A, H_B, H_C, ... we can calculate each of

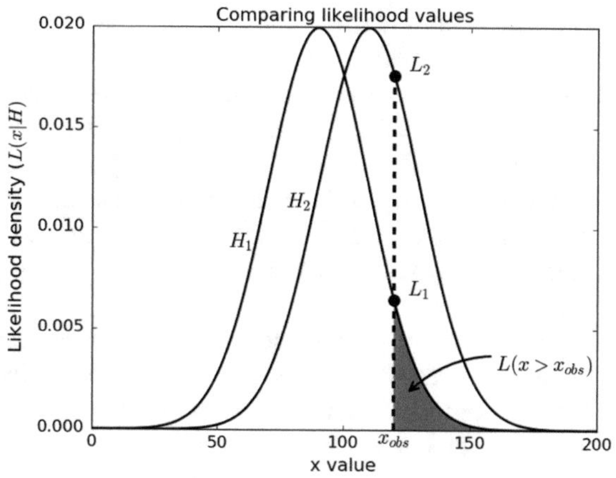

Fig. 7.1 The two ways of using likelihood values. Given a hypothesis H, we can calculate the likelihood distribution $L(x|H)$ for the possible values of the measured data value x. Given a specific observed data value x_{obs}, we could compare the predictions of two different hypotheses. In the illustration above, hypothesis H_2 is clearly a better explanation of the data than hypothesis H_1, because $L_2 > L_1$. Alternatively, we could consider the hypothesis H_1 in isolation. The data value x_{obs} seems rather on the high side. The shaded area shows the integrated probability of getting a value as large as x_{obs} or larger. The smaller that area, the less confidence we would have in H_1

$L_A(x), L_B(x), L_C(x), \ldots$ in turn. It seems reasonable to choose the *best* hypothesis as the one that gives the *maximum likelihood*. The idea is illustrated in Fig. 7.1.

Note that the absolute values of L do not matter. We need only the relative values, to tell us the relative preferences of our hypotheses. We can do better if we give each hypothesis a quantitative probability value $P_A, P_B, \ldots$ (i.e. credibility, or degree-of-belief, as discussed in Chap. 1). We might start by assigning "initial guess" probabilities. Then after performing the experiment, we can use the relative likelihood values $L_A, L_B, \ldots$ to adjust our hypothesis probabilities in the light of the data. In Sect. 7.3 we will carefully step through the method for implementing this idea.

7.2.2 Assessing a Single Hypothesis Using Absolute Likelihood

Alternatively we might ask "how much confidence do I have in this hypothesis"? In our earlier example, when we see three red balls in a row, we instinctively feel dubious about the hypothesis that there are three red balls and a hundred black ones, regardless of what hypothesis somebody else might suggest. But how small a value of $L(X)$ should make us suspicious?

The difficulty, as we discussed in Chap. 1 (Sect. 1.4.4) is that absolute values of $L(X)$ for specific values of X are not very helpful. For example, consider our example from Chap. 4 (Sect. 4.5), of a department store selling diamond rings. The hypothesis might be that we are selling μ rings per week. If in any given week we actually sell X rings, the likelihood of that happening will be given by the Poisson distribution with μ and X. So for example if $\mu = 20.7$, the likelihood of selling exactly $X = 25$ rings is $L(X) = 0.0523$—it is fairly unlikely. But this shouldn't make us doubt the hypothesis. We would have been equally as impressed by selling 26, or 27. The likelihood of selling *at least* 25 rings comes from summing all the individual likelihoods, and comes to $L(X \geq 25) = 0.198$, a not very unlikely result at all. If we are using absolute likelihood to test our confidence in a hypothesis, that result would not impress us.

The situation is even more obvious for continuous variables such as a voltage, because the probability of exactly x is formally zero. When testing a hypothesis in isolation, we always need to consider likelihood integrated over a range. The likelihood integrated over some range tests our *confidence* in our hypothesis, given the data we have seen. A low value makes us trust the hypothesis less. The general idea is illustrated in Fig. 7.1. But what integration range do we use? The answer to this depends on the precise question one is asking. In Sect. 7.6 we will carefully step through the method for implementing the idea of confidence testing.

7.3 Updating Credibilities Using Bayes's Formula

Now we will take forward the idea of Sect. 7.2.1, comparing hypotheses using relative likelihoods. In the Bayesian approach, we start with an initial guess at relative degrees of belief in our hypotheses, then use data to update those beliefs. This corresponds to everyday reasoning. If I ask you "will it rain today?", in the absence of any information, you are likely to say "well I have no idea, so lets say 50-50". You may use background information to make a better starting guess. For example, if you live in Arizona, your first guess would be rather different than if you live in Glasgow. Then if you go outside and notice that the sky is dark grey and the atmosphere is oppressive, that experimental data collection will change your degree of belief that it will rain today. Of course this is all rather qualitative; how can we make this method objective and quantitative? We have already seen in Chap. 1 how we can attach a probability (credibility) to a hypothesis. But how do we use the data to do our updating?

7.3.1 Bayes's Formula for Hypothesis Probabilities (Credibilities)

You will recall from Chap. 1 (Sect. 1.5.3) that if we have two events A and B, then $P(A, B) = P(B|A) \cdot P(A) = P(A|B) \cdot P(B)$, which leads to Bayes's theorem

$$P(B|A) = \frac{P(A|B)P(B)}{P(A)}.$$

Once we realise that we can treat credibilities mathematically like probabilities, the trick is to consider Bayes's theorem with $B = H$ where H stands for a hypothesis, and $A = D$ where D is a dataset resulting from an experiment. We could now write Bayes's theorem as

$$P(H|D) = \frac{P(D|H)P(H)}{P(D)},$$

but it is instructive to re-write the equation as follows

$$P_2(H|D) = P_1(H)\,\frac{L(D|H)}{L(D)} \quad \text{or in short} \quad P = \pi\,\frac{L}{E}. \tag{7.2}$$

We can then interpret the formula as follows:

- $\pi = P_1(H)$ is the initial or *prior probability* of our hypothesis
- $L = L(D|H)$ is the *likelihood* of our data, given our hypothesis
- $E = L(D)$ is a normalising factor, and is the *overall likelihood* of the data
- $P = P_2(H|D)$ is the *posterior probability* of our hypothesis, i.e. the updated value.

In Sect. 7.2 we already looked at how to calculate likelihoods, given a carefully framed hypothesis, and a well defined experiment. Now lets look more carefully at how we deal with priors and posteriors.

7.3.2 Assigning Priors

We start by assigning an initial hypothesis probability or "prior" on a subjective but reasonable basis. When we do this, we need to consider a complete set of hypotheses under consideration, and make sure that the priors add up to 1.0. In our rain forecast example, there are just two hypotheses—either it will rain or it won't. We could just simply give each of these $\pi = 0.5$. Alternatively, if we happen to know that in our town it rains on average one day in four, we could start with $\pi(\text{rain}) = 0.25$ and $\pi(\text{not rain}) = 0.75$. For the various "balls in urns" problems, there could in principle be many different plausible hypotheses, but we need to start by clearly defining a set of hypotheses that we are choosing to compare. The Bayesian method is all about hypothesis comparison. Some scientists worry that choosing a set of hypotheses, and assigning them priors, is a subjective process. Bayesians on the other hand see this as a strength of the method, as it corresponds to normal scientific reasoning—we begin by using our judgement, and then proceed objectively.

7.3.3 Posteriors and How to Normalise Them

The posterior probability represents our degree of belief in the hypothesis, after we take the data into account. The formula tells us that we re-scale the prior by the likelihood, but normalised by the quantity $E = L(D)$. How do we find this normalising factor? To get this, we have to remember that although the formula shows us how to update a single hypothesis, we should always be comparing a set of hypotheses. Suppose we have two hypotheses A and B. Then we have

$$P_A = \pi_A \frac{L_A}{E} \qquad P_B = \pi_B \frac{L_B}{E}.$$

However, if we have a complete set of hypotheses, then we require $P_A + P_B = 1$. Substituting from above and re-arranging we find

$$E = L(D) = \pi_A L_A + \pi_B L_B. \tag{7.3}$$

More generally we find $E = \sum \pi_i L_i$. The normalising factor $E = L(D)$ is therefore the *marginal likelihood*, i.e. the sum of the likelihoods for each hypothesis, weighted by the prior probability for each. This can be interpreted as the overall likelihood of the data, regardless of the hypothesis, but it is really just a normalising factor which gives us properly normalised posteriors. The symbol E is used because marginal likelihood is sometimes referred to as the "Bayesian Evidence" (see Chap. 10).

7.4 Bayesian Inference Worked Examples

7.4.1 Testing Spin State Theories

Suppose we have an experiment where we apply a magnetic field and then measure the spin states of two particles; each could be found to spin-up or spin-down with respect to the magnetic field. The possibilities are uu, dd, ud, du. In theory A, the particles are indistinguishable, so ud and du are the same. The states occur at random. Theory A therefore predicts that uu occurs a third of the time. However in Theory B, ud and du are not the same, but again the states occur at random. Theory B therefore predicts that uu occurs a quarter of the time. Before doing any experiments, we have no obvious reason to prefer one theory over the other, so should assign a prior credibility of 0.5 to each. Now suppose we run the experiment 25 times, and find we get uu 11 times. How does that change our degree of belief in the two theories? Summarising what we have:

Hypotheses	$A: p(uu) = 1/3$	$B: p(uu) = 1/4$
Priors	$\pi_A = 0.5$	$\pi_B = 0.5$
Data	$x = 11, n = 25$	

The likelihood of getting $x = 11$ is given by the binomial distribution for $n = 25$, with p given by the appropriate value for each hypothesis.

$$L_A = f_{\text{bin}}(n, x, p = 1/3) = 0.0862 \qquad L_B = f_{\text{bin}}(n, x, p = 1/4) = 0.0189.$$

The marginal likelihood is given by

$$E = \pi_A L_A + \pi_B L_B = 0.5 \times 0.0862 \; + \; 0.5 \times 0.0189 \; = 0.0526.$$

Finally, we get the posterior probabilities

$$P_A = \pi_A \frac{L_A}{E} = 0.5 \times \frac{0.0862}{0.0526} = 0.82 \text{ and } P_B = 0.5 \times \frac{0.0189}{0.0526} = 0.18.$$

Note that P_A and P_B add up to 1.0. Taking the data into account makes Theory A considerably more likely to be correct than Theory B, but it is far from certain. What if we performed another experiment? We could then use the first experiment as background information, and set the priors as $\pi_A = 0.82$, $\pi_B = 0.18$, and repeat the calculation using the new priors and new data.

7.4.2 Particle Mass Test

Now we will look at an example using probability densities rather than discrete probabilities. Suppose two rival theories predict the mass of a new particle as $m_A = 223\,\text{GeV}$ and $m_B = 260\,\text{GeV}$ respectively. We can reasonably start by giving these hypotheses equal priors, $\pi_A = \pi_B = 1/2$. Now suppose an experimental measurement gets a value $x = 256 \pm 12\,\text{GeV}$, with the error taken to represent a Gaussian distribution—in other words, we take x to be drawn from a Gaussian distribution with $\mu = m_A$ or $\mu = m_B$, and $\sigma = 12$. Which theory is best? For theory A, the measurement 256 ± 12 is at a Gaussian deviation $z = (223 - 256)/12 = -2.75$. The likelihood density is given by the Gaussian probability density at 2.75σ, which is $L_A = 0.0091$. For B we get $+0.33\sigma$ which gives $L_B = 0.378$. Note that we don't have to worry about the units of the probability densities, as long as they are both the same—its only the relative values that matter. The marginal likelihood is

$$E = L_A \pi_A + L_B \pi_B = \frac{1}{2}(L_A + L_B).$$

and so we get the updated probabilities

$$P_A = \frac{1}{2}\frac{L_A}{E} = \frac{L_A}{L_A + L_B} \qquad P_B = \frac{1}{2}\frac{L_B}{E} = \frac{L_B}{L_A + L_B}.$$

Putting the numbers in we find $P_A = 0.029$ and $P_B = 0.971$.

7.5 The Posterior Odds Ratio

When we are considering just two hypotheses, we could examine the ratio of the two posteriors, in which case the marginal likelihood E cancels out:

$$\frac{P_A}{P_B} = \frac{\pi_A}{\pi_B}\frac{L_A}{L_B}. \qquad (7.4)$$

If all we want is the relative probability of our two hypotheses, then we don't need to explicitly calculate E. If we re-express the posterior ratio as the ratio of two integers, then we get a traditional betting style *odds ratio*, where we compare the chance of something happening to the chance of it not happening. For example, if $P_A = 0.8$ and $P_B = 0.2$ so that $P_A/P_B = 4.0$ then we could refer to this as "4:1 on". On the other hand, if $P_A/P_B = 0.3$ this is the same as "7:3 against".

7.6 Confidence/Significance Testing: *P*-Values

Now we return to the other method of using likelihoods—assessing our confidence in a single hypothesis using absolute likelihood. As discussed in Sect. 7.2.2, for this to make sense, we always need to consider likelihood summed or integrated over a range. Suppose for example our hypothesis H implies some probability density function $p(x)$ for measured values of x, and that we have obtained an unexpectedly large value of x. Then what we want is the "P-value"

$$P(>x) = \int_x^\infty p(x). \qquad (7.5)$$

We can then accept or reject H based on the value of P. The intellectually healthy thing to do is to *decide in advance* what value of $P = \alpha$ would make us decide to reject H. The value α is known as the **significance level**, and $1 - \alpha$ is the **confidence level**. For example, if we pick $\alpha = 0.05$, then any P value less than α is said to be a result at "5% significance", or conversely would allow us to reject at 95% confidence.

Another way of looking at this is as follows. Having chosen our significance level α, we can find a critical value of $x = x_{\rm crit}$ where $P(> x_{\rm crit}) = \alpha$ exactly. Then when we make the measurement, any value with $x > x_{\rm crit}$ would cause a rejection. We can divide all possible points x into those in the *rejection class* with $x > x_{\text{crit}}$ and those in the *acceptance class* with $x < x_{\text{crit}}$.

7.6.1 The Idea of a Null Hypothesis

Why this emphasis on rejection rather than acceptance? Note that in this approach, we are avoiding giving our hypothesis a sliding-scale numerical credibility, but just looking at how likely our data is. If in fact, a hypothesis is correct, we don't necessarily expect a large P value—we will get x values spread either side of the mean and can easily get $P = 0.5$ or 0.2 or 0.7 etc. So it makes sense to provisionally accept a hypothesis as true, unless we see some data that causes us doubt. However it would be a mistake to automatically provisionally accept a controversial or interesting hypothesis, such as "I propose there is a previously unsuspected particle", or "Lima beans cause cancer". That would give equal credence to too many potential ideas. So instead, the normal procedure is to identify a boring or *null hypothesis* that we must reject before we consider other hypotheses. For example, to test the Lima bean hypothesis, the null hypothesis would be that eating Lima beans has no effect on cancer rates. Suppose that the proportion of people with cancer in the general population is well known, and also its variance. Then we take a sample of Lima bean eaters, and test them and get some number. The null hypothesis would predict the probability distribution for our observed value, and we can calculate the P value for our result. If P is small we reject the null hypothesis. This does not prove that the Lima bean hypothesis is correct; it is just that it can at least be considered.

7.6.2 One-Tailed Versus Two Tailed Tests

Above, we considered only large values of x, and rejected any value with $x > x_{\text{crit}}$. This is known as a "one-tailed test". However, in some cases small values of x might also be suspicious. In this case we will want to construct a rejection class split into two parts, at high and low values, as indicated in Fig. 7.2. We therefore pick two boundary values x_1 and x_2, chosen so that $P(< x_1) + P(> x_2) = \alpha$. This is known as a "two-tailed test". Unfortunately there are infinitely many ways to do this. If $p(x)$ is symmetrical, then we can at least place x_1 and x_2 an equal distance either side of the mean, and the problem is solved. For an asymmetric $p(x)$, the choice of x_1 and x_2 is arbitrary, but we can at least ask that they be decided before the test is conducted. One standard method is to set the boundaries equidistant from the mean, such that the integrated probability in the central acceptance region is equal to $1 - \alpha$. Note however that the upper and lower rejection regions will then not be the same size, as illustrated in Fig. 7.2.

How do we know when to use a one-tailed test, and when to use a two-tailed test? For this we do need to know not just the null hypothesis, but also the *alternative hypothesis*. Suppose we are tossing a coin, but suspect it is biased. The null hypothesis

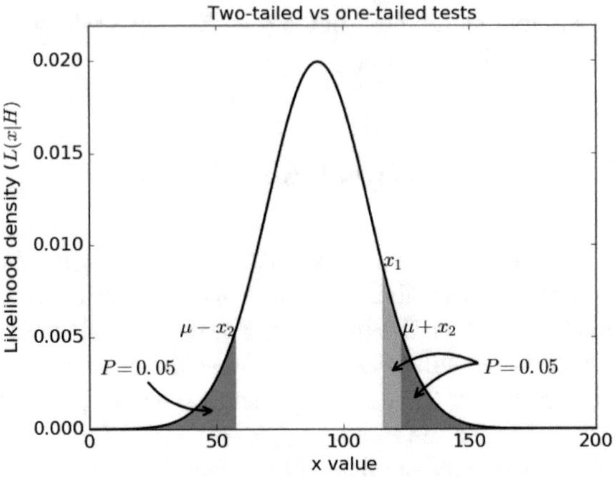

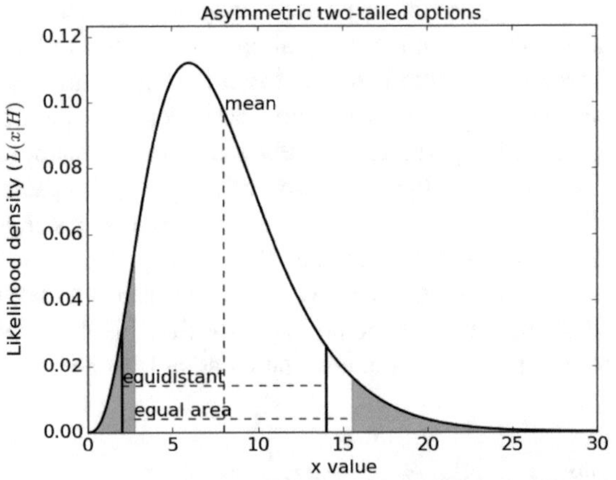

Fig. 7.2 Comparing one-tailed and two-tailed 10% significance tests. **Upper**: In this illustration the likelihood distribution is symmetrical. The shaded region at $x > x_1$ contains 10% of the probability, and so is appropriate for a one tailed test. The choice of x_1 is unique, given the requirement for $P = 0.1$. The two darker shaded regions contain 5% probability each, and so are appropriate for a two-tailed test. The choice of upper and lower values is arbitrary, but fixed if we require the two shaded regions to be the same size. **Lower**: In this case $L(x)$ is asymmetric, and it is much less clear how to choose upper and lower boundaries. We could require the upper and lower regions to have 5% probability each, which gives the shaded regions, or we could for example ask for the boundaries to be symmetrically placed either side of the mean, which gives the boundaries indicated by the vertical lines

is that the coin is fair. We toss the coin a number of times, and the null hypothesis predicts a probability distribution for the number of heads. If we simply have some reason to suspect the coin is not a fair one, we would consider both low and high points, and do a two-tailed test. If however we have a reason to suspect the coin is biased towards heads, we are not interested in low points, and should do a one-tailed test.

7.6.3 *Gaussian z-Test*

In very many cases, the null hypothesis will predict some true value x_t, with a Gaussian distribution of measured values x, perhaps because the measurements have normally-distributed errors. Assuming that we know both the mean μ and standard deviation σ for the predicted distribution of measured values x, likelihood values are completely determined by the normalised deviation from the mean, $z = (x - \mu)/\sigma$. How often will we get various values of z?

In Chap. 5 we saw that the area under a Gaussian within $z = \pm 1, 2, 3$ is 68.3%, 95.5%, and 99.7% respectively. (Often expressed as being "within 2σ" and so on.) So on a two-tailed test, if our data value gives say $z = 2.1$, we could reject the null hypothesis at 95% confidence, but not at 99% confidence.

Suppose we turn this round, and ask what are the *critical values* of z we need for 90, 95, or 99% confidence? This depends of course on whether we perform a one-tailed or two-tailed test. Standard tables show us that the answer is

	one-tailed	two-tailed
90%	$z = 1.28$	$z = 1.64$
95%	$z = 1.65$	$z = 1.96$
99%	$z = 2.33$	$z = 2.58$

7.7 Significance Testing Worked Examples

7.7.1 *The Dodgy Coin*

Suppose we have a coin that we suspect is biased towards heads. Then $p = 1/2$ is the null hypothesis, but we are interested in a suspiciously large number of heads, so we decide to do a one-tailed test, and look for 5% significance level, or 95% confidence. We toss the coin $n = 10$ times, and get $x = 8$ heads. The binomial formula shows that for $p = 1/2$ the value $x = 8$ has probability 4.4%. However, 9 or 10 heads have probability 1.0 and 0.1%, so the probability of $x = 8$ or larger is $P = 0.055$, and we cannot reject at 95% confidence. Only $x = 9$ or $x = 10$ would be in the "rejection class". Note that if we had tossed the coin 20 times and got 16 heads—exactly the same fraction as 8 out of 10 heads—the probability of $x = 16$ or worse would be 0.59%, and we would be able to reject the null hypothesis that the coin is fair at high confidence.

7.7.2 *The Flaring Star*

Some stars undergo rare outbursts lasting several days. One night, an observation of a suspected flaring star finds it to be somewhat bright. Should we devote telescope time to observing it repeatedly over the coming days? The star was observed with a photon counting detector over a one hour exposure, and 238 counts were recorded. Given the normal brightness of the star, 202 counts would have been predicted. Is the difference likely to be real? Only high points are interesting, so we do a one-tailed test, and because telescope time is expensive, we look for 99% confidence. The null hypothesis predicts a Poisson distribution with $\mu = 202$ and $\sigma = \sqrt{202} = 14.21$. What is the probability of $x = 238$ or worse? Rather than explicitly calculating the Poisson distribution values for 238, 239, ... we note that the mean of 202 is large enough that the Poisson distribution will be well approximated by a Gaussian. For a one-tailed test, 99% confidence needs $z > 2.33$. In this case, $z = (238 - 202)/14.21 = 2.53$. The observed brightening is therefore very significant.

7.7.3 *The Probability of Nothing*

Let us return to our sales-of-diamond rings problem. Suppose in any given week a department store sells on average 3.4 rings. One week they sell none at all. Does that indicate something has gone wrong? Or is it just a fluke? In this case, because numbers are small, we should certainly use the proper Poisson distribution. For a given μ, we simply have $p(x = 0) = e^{-\mu}$. So for $\mu = 3.4$ we get $P = 0.033$. That result is then significant at 95% confidence but not at 99% confidence. However, this conclusion could fall foul of the "multiple test trap" (see below). A value of $P = 0.033$ means one chance in thirty. So if we watch the figures every week, selling nothing is bound to happen once or twice a year...

The problem of seeing nothing is quite common, so its worth working out some standardised critical values. For a chosen significance level α, we can ask, what value of μ would give a chance that small of seeing nothing? This is just

$$\mu_{\text{crit}} = -\ln \alpha. \tag{7.6}$$

So for seeing nothing to be significant at 5%, i.e. to reject at 95% confidence, we need $\mu > 3.00$; for 99% confidence we need $\mu > 4.61$. Put another way, if we see nothing, then there is still a 1% chance that actually μ is as large as 4.61.

7.8 Pitfalls of Using *P*-Values

There are several common problems when using P values.

(i) Getting fooled by flukes. The decision from significance testing is always provisional. Getting a result at 5% significance does not mean you have disproved a null hypothesis. It means that if you reject it, you'd be right 19 times out of 20 and wrong 1 time out of 20. Rejecting the null hypothesis when in fact it turns out to be true is known as a "Type I error". Failing to reject when in fact the hypothesis is false is known as a "Type II error".

(ii) We learn nothing about the interesting hypothesis. Typically one starts with an "interesting" hypothesis, carefully constructs a corresponding null hypothesis, and then tests this. However, many different "interesting" hypotheses could correspond to the same null hypothesis. If we (provisionally) reject the null hypothesis, it does not tell us whether our original interesting hypothesis is correct; it just means it lives to fight another day.

(iii) The "wrong distribution" problem. If we believe that our data points are normally distributed, but in fact they turn out to be distributed in a Lorentzian fashion, we could badly underestimate the probability of a high point. This is important because we quite often use a Gaussian approximation to other distributions; this may be good at small multiples of σ but much poorer at higher multiples. Of course this problem also applies to Bayesian inference.

(iv) The multiple test trap. Suppose we have 20 data points, and find that one them is high, appearing at 2.5σ. This has only a 1% chance of happening for any one data point; but we have 20 shots at that 1% chance, so in fact seeing one high point is not particularly impressive. Sometimes you have implicitly done multiple tests without realising it. The diamond-ring problem above is a typical example. You only think to do the test when you see a strange result.

(v) The false positive problem. This is really a combination of problems (ii) and (iv). Suppose an airport security team are using face-recognition software to spot terrorists. The null hypothesis is that any specific passenger is not a terrorist, in which case there is, say, only a 1% chance that the software will think it has spotted a terrorist. So if the software raises the alarm, does this mean there is a 99% chance that person is terrorist? No. Suppose one passenger in 10,000 is a terrorist; but one (non-terrorist) passenger in every 100 is setting off the alarm. So as passengers stream through, the false alarms will be outnumbering the terrorists by a factor of 100.

7.9 Comparing Bayesian and Confidence Testing Results

We have looked at two different ways of using likelihood to perform statistical inference. How do they compare? They are consistent, but the Bayesian method gives us more information, especially with modest amounts of data. Let us look at a very simple example.

Suppose we are watching someone toss a coin, which we suspect might be doubled headed. Let us label the hypothesis of there being two heads as HH, of two tails as TT, and of a normal coin as HT. Our data result can be labelled h for heads and t for tails. If we toss the coin once and get h, what can we say about the hypothesis HH?

First, consider this problem with the significance testing method. Given that $L(h|HH) = 1$ this suggests that we have no information at all. We certainly cannot reject HH. But of course we should really propose that HT is the null hypothesis and test that. In this case $L(h|HT) = 1/2$, so we can't even reject the null hypothesis. So in this approach, we could conclude that we have no useful information at all. Now lets try the Bayesian method. If we assume we start with ignorance so that HH, TT, and HT are all equally probable, then $\pi(HH) = \pi(TT) = \pi(HT) = 1/3$. In this case getting a result of heads certainly *does* give us information. The likelihoods are $L(h|HH) = 1$, $L(h|TT) = 0$, $L(h|HT) = 1/2$, and we get $E = 1/2$; updating using Bayes's formula we get $P(HH) = 2/3$, twice as likely as before the experiment. So in this simplest case the significance and Bayesian methods give very different answers.

Now suppose we toss the coin twice and get heads both times. Using the posteriors from the one-toss experiment as our new priors (noting $\pi(TT) = 0$), we can update again, and we find $P(HH) = 4/5$, $P(HT) = 1/5$. For the significance analysis, we would consider the probability of getting two heads out of two tosses, which is given by the binomial distribution: $L(2h|HH) = 1$ and $L(2h|HT) = 1/4$. So now the significance with which we could potentially reject HT, $P = 1/4$, is close to, but not identical to, the Bayesian probability of HT, $P(HT) = 1/5$. This difference becomes smaller with more tosses. If we get 4 heads out of 4 tosses, $L(4h|HT) = 1/16$ whereas the Bayesian analysis gives $P(HT) = 1/17$.

In general, the significance and credibility methods give qualitatively similar but not quantitatively identical answers. This is because the methods are asking related but different questions, both of which are reasonable. The credibility method is asking "what is my relative degree of belief in this hypothesis, compared to the other possible hypotheses?" whereas the significance method is asking "if I were to bet on this hypothesis, how often would I lose?"

7.10 Hypothesis Testing with Multi-value Datasets

So far in this chapter, we have implicitly or explicitly assumed that the data concerned is a single event, or a single data value x, or that a single data value for testing can very simply be extracted from the data—for example, testing a single discrepant high point. But what if the data concerned is a collection of two or more data values? How do we use these to do our testing? This is an area where the Bayesian and significance techniques are rather different.

7.10.1 Bayesian Inference with Compound Datasets

Because we simply use the point likelihood, or the likelihood density, with no need to integrate over a range, the Bayesian method deals with compound datasets quite simply, using the joint likelihood or likelihood density. Continuing the particle-mass example from Sect. 7.4.2, suppose we have the same two hypotheses $m_A = 223\,\text{GeV}$ and $m_B = 260\,\text{GeV}$. However, now measurements have been made by three different labs, with results $x_1 = 256 \pm 12$, $x_2 = 216 \pm 16$, $x_3 = 238 \pm 13$. Given hypothesis A we can calculate the *joint likelihood density* $L_A(x_1, x_2, x_3) = L(x_1|m_A) \cdot L(x_2|m_A) \cdot L(x_3|m_A)$, and similarly for L_B where the individual likelihoods are just given by the Gaussian probability density at the relevant z-deviation. Each of the joint probabilities L_A and L_B may be quite small, and their absolute values, because they are densities, are dependent on the units being used; but none of this is a problem, because all that matters is the relative values of likelihood density for the various hypotheses under consideration. The marginal likelihood is given by

$$E = L(D) = L_A(D)\pi(A) + L_B(D)\pi(B),$$

where D is shorthand for the complete compound dataset x_1, x_2, x_3. We then get the posteriors just as before, with

$$P(A) = \pi(A)\frac{L_A(D)}{E} \qquad P(B) = \pi(B)\frac{L_B(D)}{E}. \tag{7.7}$$

The ability to deal with compound datasets in a simple way, and indeed the same way each time, is a big advantage of the Bayesian method.

7.10.2 *Significance Testing with Compound Datasets: Test Statistics*

In discussing significance testing, we saw the necessity of integrating likelihood over a range of data values, both to get sensible values we can compare from one situation to another, and to get the sense we wanted of "this unlikely or worse". However, with a compound dataset $x_1, x_2, x_3, \ldots$ its not at all obvious how to pick a sensible integration region. Even for two variables, there are many different reasonable ways you could do this, as illustrated in Fig. 7.3.

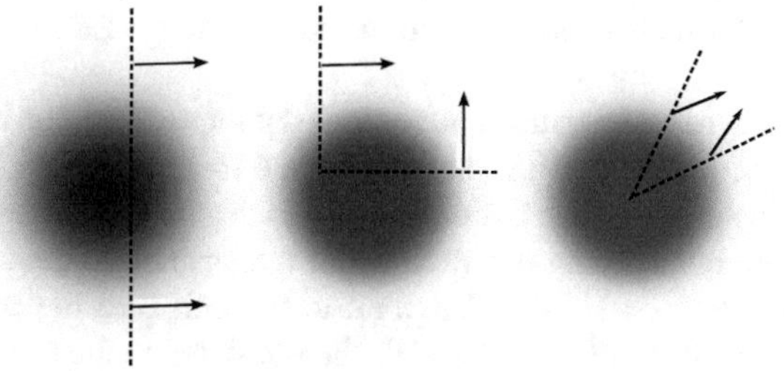

Fig. 7.3 Illustrating how one could integrate over a two dimensional probability density function in several different ways. In each part of the figure, x and y represent the random variables, with the grey scale representing the joint probability density. The dashed lines indicate regions we could choose for summing the probability

The solution in traditional statistical inference is to combine the data values into a single number **test statistic** K. For example, we could find the mean of all our data values; or we could look at the deviation of each number from some prediction, and find the average deviation. Then we could ask the question "assuming the null hypothesis, if I ran this experiment many times, what distribution of K values would I get?". This is known as the *sampling distribution* for K. Finally, we can compare our observed value of our test statistic with the expected distribution in order to decide whether to accept or reject the null hypothesis. A large part of traditional statistics consists of deriving and studying the distributions of such test statistics.

7.10.3 Sample Mean z-Test

Perhaps the simplest example of a test statistic is the mean value of a set of data values, which we could then compare to a hypothesised true value. If there are N original data points, and they are each normally distributed with variance σ^2, then the average of the N points should also be normally distributed with variance $\sigma_K^2 = \sigma^2/N$. (You can see this from the standard error propagation formulae; consider adding N variables. See also Chap. 8, Sect. 8.3.) We can then revert to our standard z-test using this new σ_K.

Suppose we don't know the value of σ—can we still use the sample mean test? This requires the use of the t-statistic, which we shall discuss in Chap. 8, Sect. 8.3.4.

7.11 The χ^2 Test Statistic

By far the most important test statistic is χ^2. It is the single most common test statistic you will encounter in the statistical and scientific literature. The general idea is to test the *scatter* of our data points compared to some prediction. Suppose we have a set of N measurements x_i, and our hypothesis is that the true value is x_t. We further suppose that the measurements all have the same Gaussian error distribution, with standard deviation σ_t. In other words, we assume that the x_i are independent random variables, each of which is drawn from the same parent distribution with mean x_t and variance σ_t^2.

Our first guess for a statistic representing data spread might be to sum all the values of $x_i - x_t$, but that will typically average to zero, which isn't helpful. We might consider using $|x_i - x_t|$, but this is hard to deal with mathematically. So, just as when we were considering the variance of a distribution, the simplest practical answer is to use $(x_i - x_t)^2$. We also want to normalise to the expected standard deviation, just like with the z statistic. We therefore define the test statistic

$$\chi^2 = \frac{1}{\sigma_t^2} \sum_{i=1}^{N} (x_i - x_t)^2. \tag{7.8}$$

7.11.1 χ^2 as Excess Variance

Notice that the summed expression is very similar to the standard formula for variance, $\sigma^2 = (1/N)\sum(x_i - \mu)^2$. If we think of $(1/N)\sum(x_i - x_t)^2$ as the "observed variance" $\sigma^2_{\rm obs}$ then we can see that

$$\chi^2 = N\frac{\sigma^2_{\rm obs}}{\sigma_t^2}.$$

The χ^2 statistic is therefore telling us how much the data points scatter about the prediction, compared to how much we would expect them to scatter. If the hypothesis is good, then we should normally have $\chi^2 \sim N$. If we divide by N, this "normalised" χ^2 should typically be equal to 1. The normalising factor is known as the "degrees of freedom" and is given the symbol ν. In our first simple case here, we just have $\nu = N$. Later on, when we talk about parameter estimation and model fitting, we will see that the effective number of degrees of freedom is $\nu = N - m$ where m is the number of parameters fitted. In general, it is customary to define the "reduced chi-squared" as $\chi^2_\nu = \chi^2/\nu$ and this will in general be close to 1.0 for a good hypothesis.

7.11.2 Distribution of χ^2

Typically we should have $\chi^2 \sim N$, but how do we tell what a "bad" value of χ^2 is? Let us re-iterate the standard procedure we set out earlier. The null hypothesis is that all the N values x_i are drawn from the same underlying Gaussian with mean x_t and variance σ_t^2. If we were to repeat the experiment many times, what distribution of χ^2 values would we get? In other words, what is the *sampling distribution* for χ^2? Then if we see an unexpectedly large value of χ^2, we know how often this should happen by chance.

Let us start by asking, if a random variable x has a Gaussian distribution, what will be the PDF of the variable $y = x^2$? To conserve probability, we need $p(x)\mathrm{d}x = f(y)\mathrm{d}y$ and so the probability density for y is given by $f(y) = p(x)/(\mathrm{d}y/\mathrm{d}x)$ which gives

$$f(y) = \frac{1}{\sqrt{8\pi}}y^{-1/2}e^{-y/2}.$$

Generalising further, if we have ν independent random variables $x_1, x_2, \ldots, x_\nu$ and each of these is normally distributed with means μ and variance σ^2, then the quantity we want is

$$y = \chi^2 = \sum_{i=1}^{\nu}\left[\frac{x_i - \mu}{\sigma}\right]^2.$$

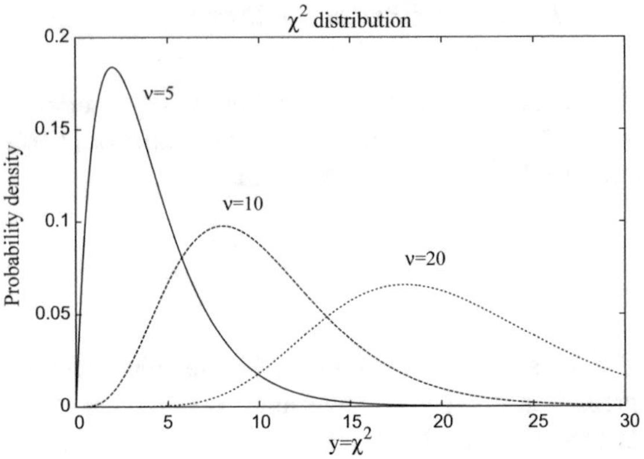

Fig. 7.4 Illustrating the χ^2 distribution for three different values of the degrees of freedom ν

Calculating the PDF of this sum of random variables is a little involved so we won't derive it here. The result is that the PDF for our new variable y is given by

$$f_\nu(y) = f_\nu(\chi^2) = \frac{1}{2^{\nu/2}\Gamma(\nu/2)} y^{\frac{\nu}{2}-1} e^{-y/2}. \tag{7.9}$$

Note that this defines a PDF $f(y)$ for the variable $y = \chi^2$, so the density is per-unit-χ^2 not per-unit-χ. We have written $f_\nu(y)$ to show that the PDF is a function of $y = \chi^2$, but there is a family of curves for different values of the parameter ν. Here Γ is the usual gamma-function, and ν is the *degrees of freedom* as discussed above. This distribution has mean ν and variance 2ν, but the shape of $f(y) = f_\nu(\chi^2)$ depends quite strongly on ν, as shown in Fig. 7.4. At small values of ν the distribution is very asymmetric; at large ν it tends towards a Gaussian shape.

7.12 Using χ^2

Just like with other test statistics, what is really interesting is not the probability density $f(y)$, but its integral $P = \int_y^\infty f(y)\mathrm{d}y$. A good hypothesis will give $\chi^2 \sim \nu$. A bad hypothesis will normally give a *larger* value of χ^2. In other words, the question we want to answer is "on the null hypothesis, what fraction of times would we get a value of χ^2 that large or larger by chance?". Unfortunately the integral of χ^2 does not have an analytic solution, so like with the Gaussian z statistic we use tables or computer routines. Note that when looking up P values for χ^2 you have to look up both the right value of χ^2 and the right value of ν.

7.12.1 A Variable Star

Suppose we have measured the brightness of a star four times, and we got 256, 239, 237, and 278 counts respectively. Is there any evidence that this star is variable, or are the differences just random? A theorist has predicted that we ought to see a constant value of 251 counts. We can use Poisson statistics to find that we ought to have $\sigma = \sqrt{251} = 15.84$. Count values of this size are large enough that a normal distribution should be a good approximation. Using these predicted values of $\mu = 251$ and $\sigma = 15.84$ we can compare to each of the four observed values. This gives us a value of $\chi^2 = 4.36$, compared to the expected mean of $\chi^2 = 4.0$. Looking up in our table, we find that $P(\chi^2 > 4.36) = 0.36$ for $\nu = 4$—in other words a χ^2 value that big or bigger occurs 36% of the time. The spread of points we got is therefore nothing remarkable. The star could still be variable, but we have no evidence for it.

7.12.2 Using Individual Errors

Above, we assumed that σ is the same for all our data points x_i. However, a quite common situation is that the hypothesis that we are testing specifies x_t, but each data point has a different error given by σ_i. We then calculate

$$\chi^2 = \sum_{i=1}^{N} \frac{(x_i - x_t)^2}{\sigma_i^2}. \tag{7.10}$$

This makes no difference to the theoretical PDF for χ^2, but sometimes could make quite a big difference to the observed value of χ^2, so we get a more sensitive test if we do know individual errors.

7.13 Key Concepts

Some of the key concepts from this chapter are:

- The idea of likelihood as the probability of the data given the hypothesis
- The difference between comparing hypotheses using point likelihood or likelihood density values, and testing confidence in a single hypothesis using integrated likelihood
- The use of Bayes's formula to update hypothesis probabilities (credibilities) given the data likelihood
- The importance of background information in setting prior credibilities
- The importance of choosing a significance level in advance before carrying out a test

- The idea of trying to reject a null hypothesis, rather than trying to prove an interesting one
- The idea of a test statistic
- The use of the Gaussian z-statistic
- The use of the Poisson "nothing seen" test
- The idea of the χ^2 statistic to test the scatter of data points compared to that expected under a null hypothesis
- The behaviour of the χ^2 distribution—asymmetric at low ν, converging towards symmetric Gaussian form at high ν.

The key formulae from this chapter are: the definition of likelihood (7.1); the formula for updating hypothesis probabilities using data likelihoods ((7.2) and (7.7)); the formula for calculating the marginal likelihood or Bayesian evidence (7.3); the formula for comparing hypotheses using the posterior ratio (7.4); the formula for a P-value by integrating over a PDF (7.5); the formula for the "see nothing" critical value of the mean in a Poisson process as a function of significance level (7.6); the definition of the χ^2 test statistic ((7.8) and (7.10)); and the formula for the distribution of χ^2 (7.9).

7.14 Further Reading

Traditional statistics textbooks (see Chap. 1 references) are very strong on significance/confidence testing, and discuss a bewildering variety of test statistics. Most of the time all you need is either the Gaussian z-test or the χ^2 test. Recently a lot of scepticism has emerged about the use of P-values (see for example Baker 2016 and the references and links therein). As far as I can see, it all centres round one or more of the pitfalls described in Sect. 7.8. There is nothing wrong with P-values as long as you know what you are doing, but they are certainly mis-used. By far the worst problem I think is the unconscious multiple test trap—only the interesting results get published.

Bayesian techniques have a long history, starting with Bayes (1763), but for most of the nineteenth and twentieth century statistics was dominated by significance testing. Slowly over the last few decades, Bayesian methods have become very popular, but most of the development was in rather advanced research papers or specialist books (see Chap. 1 references). It has taken a while to sink in how *simple* the Bayesian method is. Sivia and Skilling (2006) was the first really good simple book. Another really good short book that is well suited to both beginners and specialists is Stone (2013).

7.15 Exercises

7.1 A lab measures the mass of the Yotta particle as a test of their equipment, as it should give $m = 1375\,\text{GeV}$, and if their equipment is working as expected, each measurement should have a Gaussian error $\sigma = 65\,\text{GeV}$. They make four measurements, which give 1326, 1376, 1422, and 1512 GeV respectively. Consider these two different scenarios, using a 5% significance test:

(a) Before looking at the results, one of the scientists involved has claimed that there was a problem with the equipment on the fourth measurement, likely to make the value too high. When the results come out, do they support this suggestion?

(b) After the results come out, the high value on the fourth measurement is spotted, which makes the experimental team concerned that there is a problem with the equipment. Do the data values support this suggestion?

7.2 For an exercise class in statistics, a technician has fetched from the lab an urn containing coloured balls. You know that there are two of these urns—urn A has 10 red balls and 20 black balls; urn B has 15 red balls and 15 black balls—but you don't know which one the technician has fetched. One of the tutors, Fred, claims with confidence that the technician fetched urn A; but Jane claims with equal confidence that urn B was fetched. You have no particular reason to trust Jane more than Fred or vice versa. You do a test run, picking out three balls in succession, replacing the ball each time. All three picks give red balls. On a Bayesian basis, how does this change your view of whether Fred or Jane is most likely to be right? Fred then insists you try again, and you pick out two black balls in succession. What now is the probability that Jane is right?

7.3 A student is fitting a curve to some data points. (We will learn how to do this in Chap. 8.) They are not sure how to calculate their error bars, but notice that five out of six points lie above the line, which makes them suspicious. Is this in itself enough to make them reject the fit? Is there a Bayesian way to approach this problem?

7.4 A chemical spray is known to kill 75% of mosquitos. A medical worker is testing a batch which they suspect to be faulty. They spray 100 mosquitos and find that only 65 of them die. Are their suspicions justified? What if they had found 60 to die? What if they tested on 1000 and 650 died?

7.5 An experiment is measuring Thorium decays. The standard theory predicts a decay rate of $0.35\,\mathrm{s}^{-1}$. A new theory predicts $0.37\,\mathrm{s}^{-1}$. How long do we need to run the experiment to distinguish the two theories at 95% confidence?

7.6 We normally expect that a poor hypothesis, or incorrect parameter value, will give a value of χ^2 larger than the mean value expected for a good hypothesis, $\chi^2 \sim \nu$; values smaller than ν will just occur by chance sometimes for any acceptable hypothesis. However, experimenters sometimes find that consistently small values of χ^2 keep appearing in their data analysis, more often than ought to happen by chance. What do you think might be causing this problem?

7.7 For an underlying Gaussian with mean μ and standard deviation σ, what is the likelihood density of a given data point x? If instead you have a collection of N data points x_i, each with its own error σ_i, what is the joint likelihood of this set of values? You can consider each point to be drawn from a different Gaussian, all with the same μ, but with different σ_i. Also, we can consider the σ_is to be fixed, and so think of the likelihood as a function of μ. Use this to show that $\ln L = \text{const} - \chi^2/2$

7.8 An experiment is undertaken measuring the time T it takes for a metal ball to fall 2.00 m. The measurement is done 50 times, as precisely as possible, but the results are then grouped in to bins of size 0.01 s. The results are shown in the table below.

Estimate the observed sample mean T_{obs} and standard deviation σ_{obs}. From your knowledge of physics, calculate what the correct answer for flight time T_f should be. Then test the observed distribution of measurements against the expectation of a Gaussian distribution with $\mu = T_f$ and variance σ^2_{obs}. Do this by making a prediction for each bin, calculating an appropriate error for that individual bin as a separate data point, and finally using the χ^2 statistic.

Time (s)	Observed frequency
0.59–0.60	2
0.60–0.61	2
0.61–0.62	11
0.62–0.63	6
0.63–0.64	12
0.64–0.65	8
0.65–0.66	4
0.66–0.67	3
0.67–0.68	1
0.68–0.69	1

References

Baker, M.: Statisticians issue warning over misuse of P values. Nature **531**, 151 (2016)

Bayes, T.: An essay towards solving a problem in the doctrine of chances. Philos. Trans. R. Soc. **53**, 370 (1763)

Sivia, D.S., Skilling, J.: Data Analysis: A Bayesian Tutorial, 2nd edn. Oxford University Press, Oxford (2006)

Stone, J.V.: Bayes Rule: A Tutorial Introduction to Bayesian Analysis. Sebtel Press, Sheffield (2013)

Websites (all accessed March 2019):

Online version of Baker (2016), with useful links: http://www.nature.com/news/statisticians-issue-warning-over-misuse-of-p-values-1.19503?WT.mc_id=SFB_NNEWS_1508_RHBox

Chapter 8
Parameter Estimation

8.1 Outline of Content

- Sample-based estimators for mean and variance
- Unbiased variance estimator
- Confidence intervals from z and t statistics
- Maximum Likelihood estimates
- Maximum Posterior estimates
- Constructing posterior probability intervals
- Minimum χ^2 estimates
- Constructing χ^2 confidence intervals

Rather than getting a yes-no decision on a hypothesis ("is this dataset consistent with the existence of a boson?"), we often want to estimate the value of one or more parameters ("based on this data, what is the mass of the boson?"). The problem of parameter estimation is closely related to that of hypothesis testing. You can think of there being a family of hypotheses, with some tuneable parameter θ. If we try out different values of θ, how do we decide which value is "best"? And what *range* of θ values do we find acceptable?

There are two routes forward. Sometimes, the Nature of the parameter we wish to estimate suggests a simple sample-based method, giving us some kind of formula or algorithm for calculating our estimate. An example would be taking the mean of a series of N measurements as our estimate of the true value that we have repeatedly tried to measure. But then, what is the uncertainty on that process? To answer that, we imagine repeating our experiment of N measurements many times and looking at the distribution of answers.

The second route forward is to try out a range of θ values, and for each value to calculate a number which quantifies the agreement between the hypothesis and the data. The usual choices for such an agreement-testing number are the likelihood L, the posterior probability P, or the scatter χ^2. We can then examine the resulting function—either $L(\theta)$ or $P(\theta)$ or $\chi^2(\theta)$—and choose the θ value which maximises/minimises the function as our "best" value. This gives us a *point estimate*. We can also use our function to answer a question such as "what range $\theta_{\rm min}$ to $\theta_{\rm max}$ gives me 90% of the posterior probability?". This provides us with a *parameter interval*.

A. Lawrence, *Probability in Physics*, Undergraduate Lecture Notes in Physics,
https://doi.org/10.1007/978-3-030-04544-9_8

In this chapter we will normally be considering a single parameter θ, or at most two parameters. In Chap. 10, we will see how these methods generalise to multi-parameter model fitting.

8.2 Estimating Mean and Variance of a Normal Population

The methods we will discuss in the sections below are quite general, but in order to develop them we will repeatedly consider variations of the same simple problem—how to estimate mean and/or variance from a set of data values, believed to be drawn from a normal distribution. There are three main variants to consider.

(i) *Unknown* μ, *known* σ. For example, we might be attempting to measure the mass m of a particle, and so make multiple measurements x_i, each of which has a known experimental error σ. Then each x_i can be seen as a random drawing from a normally distributed population with mean m and variance σ^2, and the problem is to find an estimate of m.

(ii) μ *and* σ *both unknown*. For example, we might wish to find both the typical height of a particular species of animal, and the spread in this quantity. Alternatively, in the context of the mass-measurement example above, it might be that we don't know what the experimental error is, and wish to estimate it from our dataset at the same time as estimating the mass.

(iii) *Unknown* μ, *individual errors*. A common situation is that each measurement x_i is supposed to be a measurement of the same thing, for example particle mass m, but has a different experimental error σ_i. Then each x_i is assumed to be drawn from a normal distribution with the same mean, but with a different variance for each data point.

In the worked examples below, we repeatedly use the same example dataset—$N = 6$ measurements which give values $x_i = 256, 239, 237, 278, 266, 241$.

8.3 Methods Using Sample Mean and Variance

If we had a single value x, that would clearly be our best estimate of μ. When we have a series of equivalent measurements, it would seem obvious that the thing to do is to average the estimates. Likewise, we might consider the spread of data values around the average to be a reasonable estimate of the variance of the population from which the data values have been drawn. So in other words, we estimate μ and σ using the sample mean and variance:

$$\hat{\mu} = \bar{x} = \sum x_i/N \qquad \hat{\sigma}^2 = s^2 = \sum (x_i - \bar{x})^2/N. \tag{8.1}$$

Note that here we use the notation $\hat{\mu}$ to mean "estimate of μ". Sometimes we need to distinguish carefully between the *estimator*, which refers to the algorithm or formula used, and the *estimate*, which refers to the estimated value. It is (hopefully) clear from context which we mean in various cases.

Applying this method to the example dataset at the end of Sect. 8.2 we get $\hat{\mu} = \bar{x} = 252.83$ and $\hat{\sigma} = s = 15.26$. However, there is a problem in using sample variance to estimate the population variance, which we discuss next.

8.3.1 Bias in Sample Variance

Using sample variance as an estimator for the true population variance, together with the sample mean as an estimator for the true mean, produces a **biased estimate for variance**. To see why, imagine trying various different values μ' for the mean, and seeing what sample variance we would get if that was actually the true value: $s^2 = \sum (x_i - \mu')^2/N$. What value of μ' would give us the smallest value of s^2? This comes from

$$\frac{\mathrm{d}s^2}{\mathrm{d}\mu'} = 0 = \frac{1}{N}\sum (x_i - \mu')^2 = \frac{2}{N}\sum (x_i - \mu') = \frac{2}{N}\left(\sum x_i - N\mu'\right)$$

and therefore

$$\frac{2\sum x_i}{N} = 2\mu' \qquad \text{and so} \qquad \mu' = \frac{\sum x_i}{N} = \bar{x}.$$

In other words, the usual sample mean is always exactly the value which gives us the smallest sample variance. But remember that the sample mean will not in general be quite the same as the *true* population value. If we actually used the true value of μ, the analysis above shows that this would in fact *always* give us a larger value of s^2 than the standard sample variance; the sample variance is therefore a biased estimate. Note by the way, that if we used a fixed hypothesised value for μ, and estimated variances as $\hat{\sigma}^2 = \sum (x_i - \mu)^2/N$, that would not give a biased estimate; the problem is only when we use the same data for both $\hat{\mu} = \bar{x}$ and $\hat{\sigma} = s$.

How different is the sample variance s^2 from the true variance σ^2? Let us work out the expectation value of $\sigma^2 - s^2$.

$$\begin{aligned}
E[\sigma^2 - s^2] &= E\left[\tfrac{1}{N}\sum (x_i - \mu)^2 - \tfrac{1}{N}\sum (x_i - \bar{x})^2\right] \\
&= \tfrac{1}{N}E\left[\sum (x_i^2 - 2x_i\mu + \mu^2) - \sum (x_i^2 - 2x_i\bar{x} + \bar{x}^2)\right] \\
&= \tfrac{1}{N}E\left[N\mu^2 - N\bar{x}^2 - 2N\mu\bar{x} + 2N\bar{x}^2)\right] \\
&= E\left[\mu^2 - 2\bar{x}\mu + \bar{x}^2)\right] = E\left[(\bar{x} - \mu)^2\right].
\end{aligned}$$

The expression we have ended up with is the expected value of the square of the difference between the true population mean μ and our estimate $\bar{x}$. But this is just the sampling distribution variance for the mean, which we saw above is σ^2/N. So now we can see that

$$E[s^2] = \sigma^2 - \frac{\sigma^2}{N} = \frac{N-1}{N}\sigma^2.$$

In other words, the sample variance is on average smaller than the true variance by a factor $(N-1)/N$. If we want an unbiased estimate of the variance, we should therefore use

$$\hat{\sigma}^2 = s_{un}^2 = \frac{1}{N-1}\sum (x_i - \bar{x})^2. \tag{8.2}$$

For small samples, the difference is quite substantial. For our $N = 6$ example in Sect. 8.2, the standard sample variance gives $s = 15.26$, but the unbiased estimate is $s_{un} = 18.31$. For large N of course there is little difference. Both the standard sample variance and the unbiased version tend towards the true value σ as N increases.

8.3.2 Error on Mean from Sampling Distribution

What is the uncertainty on our estimate $\hat{\mu} = \bar{x}$? One possible answer comes from doing the following thought experiment. Suppose we hypothesise that the underlying population has true mean μ and variance σ. We imagine doing an experiment, getting N data values x_i, and using these to calculate $\hat{\mu}$. We then imagine repeating this experiment many times and getting many different values of $\hat{\mu}$. These values will have some probability distribution $p(\hat{\mu}|\mu)$, which we call the *sampling distribution* for $\hat{\mu}$.

The standard deviation of the estimator sampling distribution, σ_μ, can be used as the uncertainty on our estimate.[1] You might think that to calculate σ_μ we have to imagine many repeats of our experiment with N samples. In fact, it is simpler to think of splitting our sample into N groups of 1. Each data point gives us a separate estimate of the mean, with error σ. Let us first suppose that we know the value of σ, as would be the case if this is a known experimental error. We can then use our standard error propagation formulae to see the effect of adding many variables. (This assumes that the points are independent.) If $f = ax + by$ then $\sigma_f^2 = a^2\sigma_x^2 + b^2\sigma_y^2$ so with $\hat{\mu} = \sum x_i/N$ and all the x_i having the same $\sigma_i = \sigma$, we get $\sigma_\mu^2 = \sum \sigma_i^2/N^2 = N\sigma^2/N^2$ and so

[1] Keep a mental distinction between σ, s, and σ_μ. Generally σ refers to the point-to-point dispersion in the parent population; s refers to the point-to-point dispersion in the observed sample of N points; and σ_μ refers to the sample-to-sample dispersion you would get from repeating the experiment.

$$\sigma_\mu^2 = \frac{\sigma^2}{N}. \tag{8.3}$$

So reassuringly, the larger our sample, the more accurate is our estimate of μ. Using once again our example from Sect. 8.2 with $N = 6$ points, if we know the experimental error to be, say, $\sigma = 16$, then $\sigma_\mu = 16/\sqrt{6} = 6.53$, and we can say that our estimate of the mean is 252.83 ± 6.53. Suppose on the other hand we don't know the value of σ; should we use our sample-variance estimate? This needs a little more careful thought, as described in the next two sub-sections.

8.3.3 Confidence Intervals for the Mean: Known σ Case

The idea of a sampling distribution allows us to construct any confidence-range we wish for μ, using the significance-testing techniques of Chap. 7. If we propose a particular value of μ, then we can in principle derive the probability distribution $p(\hat{\mu}|\mu)$. Now we compare our actual value $\bar{x}$ with this distribution. If $\bar{x}$ is a rare value, we would reject the hypothesised μ value. Finally, we imagine varying μ until we find critical values $\mu_{\rm min}$ and $\mu_{\rm max}$, such that values outside that range would be rejected based on our data point $\bar{x}$, at our chosen confidence level.

How this works depends on whether we know σ or not. If all the data points are drawn from the same distribution with known σ, then $p(\hat{\mu}|\mu)$ will be a Gaussian with mean μ and variance σ_μ^2. We can then use the *sample mean* z-statistic from Sect. 7.10.3:

$$z = \frac{\mu - \hat{\mu}}{\sigma/\sqrt{N}}, \tag{8.4}$$

which has a standard Gaussian distribution with $\mu_z = 0$ and $\sigma_z = 1$. By finding appropriate critical values of z, we can then set any desired *confidence interval*. If we want say 90% confidence, and are doing a two-tailed test, then we include values with $|z| < 1.64$. Using our example from Sect. 8.2, with $N = 6$, $\hat{\mu} = 252.83$, $\sigma_\mu = 6.53$, then for the 90% range we want $252.83 \pm 1.64 \times 6.53 = 252.83 \pm 10.71$.

8.3.4 Confidence Intervals for the Mean: Unknown σ Case

Now suppose we don't know the error on each of our N estimates. We can proceed by estimating $\hat{\mu} = \bar{x}$ as before, but using the sample standard deviation $\hat{\sigma} = s$. Instead of z we then use the *t-statistic*:

$$t = \frac{\mu - \hat{\mu}}{s/\sqrt{N}}. \tag{8.5}$$

However, this statistic does *not* follow a normal distribution, essentially because, as we saw in Sect. 8.3.1, s is a biased estimate of the variance. The t statistic can be shown to follow the PDF

$$f_\nu(t) = \frac{\Gamma[(\nu+1)/2]}{(\pi\nu)^{1/2}\Gamma(\nu/2)}\left[1 + \frac{t^2}{\nu}\right]^{-(\nu+1)/2}.$$

where the degrees of freedom is $\nu = N - 1$. Like χ^2 this is a family of curves for different values of ν, with mean 0 and variance $\nu/(\nu - 2)$. The formula is rather ugly, but just as with z and χ^2, standard tables and computer routines make it easy to look up any given level of confidence. For large ν the t-distribution becomes close to the standard normal, i.e. with $\mu = 0$ and $\sigma = 1$, but for small ν it is significantly different. It gives a more accurate test for whether a sample is consistent with a proposed mean, and therefore also a better confidence interval for μ.

Returning once again to our example data set of $N = 6$ numbers from Sect. 8.2, we have $\hat{\mu} = 252.83$ and $s = 15.26$. The number of degrees of freedom is $\nu = N - 1 = 5$. Looking up a standard t-table for $\nu = 5$ we find that to get 90% confidence we want $|t| < 2.02$. Our desired confidence interval is therefore $252.83 \pm 2.02 \times 15.26/\sqrt{6} = 252.83 \pm 12.58$, a little larger than the known-error case. This makes intuitive sense—if we are using the data to estimate two things rather than just one, each of them will be more uncertain.

8.4 Maximum Likelihood Estimates

In the simple case we have discussed above, estimating the mean and variance of a normally distributed population, the natural estimator is fairly obvious, but what do we do in more subtle or complicated cases? If we have a range of different possible values for a parameter θ, we can calculate the *likelihood function* $L(\theta)$. Here we are are using $L(\theta)$ as shorthand for $L(D|\theta)$, i.e. the probability of the dataset D given the hypothesis with a specific θ value, but holding D fixed and varying the parameter value θ. A reasonable suggestion is that the "best" value of θ is the one that makes the dataset we have observed most probable—i.e. that *maximises the likelihood function*.

We can use this idea in two ways. (i) In some cases we may be able to solve for the maximum analytically. Below we will look at one of the most important examples—finding the maximum likelihood estimates of the mean and variance of a dataset with Gaussian errors. (ii) More often, there is no analytic solution and so we find the maximum numerically or graphically. The techniques for doing that are just the same as when we calculate "maximum posterior", so we will put off looking at the numerical technique until Sect. 8.5.

So first, lets see how the maximum likelihood method works by repeating the same example as in the previous section—estimating mean and variance.

8.4.1 *Maximum Likelihood Solution for Mean and Variance*

Suppose we have N data points x_i all drawn from a Gaussian with the same μ and σ. We can think of μ and σ as variables, and then ask what values will jointly give us the maximum likelihood for our dataset. The likelihood density for a specific value x_i is

$$L_i(\mu,\sigma) = \frac{1}{\sigma\sqrt{2\pi}} \exp\left[-\frac{1}{2}\left(\frac{x_i-\mu}{\sigma}\right)^2\right].$$

The probability of obtaining all N values is then

$$\begin{aligned} L(\mu,\sigma) = \textstyle\prod_{i=1}^{N} L_i(\mu) &= \left(\tfrac{1}{\sigma\sqrt{2\pi}}\right)^N \exp\left[-\tfrac{1}{2}\textstyle\sum_i \left(\tfrac{x_i-\mu}{\sigma}\right)^2\right] \\ \text{and so} \quad \log L &= -\tfrac{1}{2}N\log 2\pi\sigma^2 - \tfrac{1}{2\sigma^2}\textstyle\sum_i (x_i-\mu)^2. \end{aligned}$$

We wish to find the overall minimum as we vary μ and σ^2, and so we need to find both $\partial\log L/\mathrm{d}\mu = 0$ and $\partial\log L/d\sigma^2 = 0$. The first of these simultaneous equations gives

$$\frac{\partial\log L}{\mathrm{d}\mu} = 0 = \frac{\sum(x_i-\mu)}{\sigma^2},$$

but here we are treating σ as constant, and assuming $\sigma^2 \neq 0$ and so we get

$$\sum x_i - N\mu = 0 \quad \text{and so} \quad \hat{\mu} = \mu_{\mathrm{ML}} = \frac{\sum x_i}{N},$$

where we have written μ_{ML} to emphasise that this value is our ML estimate of μ. So the ML estimate of the population mean is just the same as the sample mean $\bar{x}$. Substituting this value into the second of our simultaneous equations we get

$$\frac{\partial\log L}{\mathrm{d}\sigma^2} = 0 = \frac{-N}{2\sigma^2} + \frac{\sum(x_i-\hat{\mu})^2}{2\sigma^4} \quad \text{and so} \quad \hat{\sigma}^2 = s^2_{\mathrm{ML}} = \frac{\sum(x_i-\hat{\mu})^2}{N}.$$

So the ML estimate of variance is just the sample variance. However, as we saw earlier, this is a biased estimate. It can be shown that in general ML estimators give the minimum variance, i.e. the smallest error, but they are not always unbiased. Here we have a useful example of the recurring lesson that in statistics there is rarely a unique answer to the question "which method is best?".

8.4.2 Weighted Mean Estimate

With the concept of maximum likelihood, we can address the third variation described in Sect. 8.2—the case where each measurement x_i has a different error σ_i. The σ_i values are assumed known and so can be considered fixed; we just want to vary μ and see what value gives us the maximum likelihood. The joint likelihood is

$$\left(\frac{1}{\sqrt{2\pi}}\right)^N \prod_i \left(\frac{1}{\sigma_i}\right) e^{-\frac{1}{2}\sum\left(\frac{x_i-\mu}{\sigma_i}\right)^2}.$$

The first two terms are constant with respect to μ. Maximising the third term is equivalent to minimising whats inside the exponential

$$K = -\frac{1}{2}\sum\left(\frac{x_i-\mu}{\sigma_i}\right)^2.$$

Differentiating, we get

$$\frac{\mathrm{d}K}{\mathrm{d}\mu} = 0 = \sum\left(\frac{x_i-\mu}{\sigma_i^2}\right) = \sum\frac{x_i}{\sigma_i^2} - \mu\sum\frac{1}{\sigma_i^2},$$

and so finally we find our *weighted estimate* for μ

$$\hat{\mu}_{\text{weighted}} = \frac{\sum x_i/\sigma_i^2}{\sum 1/\sigma_i^2}. \tag{8.6}$$

What is the uncertainty on our weighted mean estimate? We can repeat the trick we used in Sect. 8.3, considering our sample of N to be N samples of 1. Sample-1 has error σ_1, Sample-2 has error σ_2 etc. Using our standard error propagation formulae, you can see fairly easily that the net error on our estimate $\hat{\mu}$ is

$$\sigma_{\hat{\mu}}^2 = \frac{1}{\sum(1/\sigma_i^2)}. \tag{8.7}$$

Using the weighted mean as opposed to the normal mean can make a big difference. Suppose we measure the mass of a particle twice, with the first measurement being more accurate than the second. The first time we get 252 ± 3 and the second time we get 274 ± 14. The standard mean would give us 263.0, which is discrepant with the first measurement by more than 3σ. The weighted mean on the other hand gives 252.97, which is nicely consistent with the better measurement, but not quite the same. The poor measurement still adds some information, but not as much as the good measurement.

8.4.3 Discarding Dodgy Data Points: Sigma Clipping

In the example use of weighted mean above, we assumed that although some data points have bigger errors than others, they are all fair estimates of the true mean. The correct approach is therefore to include all the information, but in a weighted manner.

However, in real life some measurements are simply wrong, or biased for some reason, and could bias our estimate. These incorrect measurements might stand out as being individually discrepant from the mean, given their error. Can we spot these and remove them? As ever in statistics, there is no iron-clad solution, but a reasonable procedure known as "sigma-clipping" is often followed. First, you decide a reasonable deviation at which to exclude points—typically those above/below $\pm 3\sigma$, which would exclude only 0.4% of non-bogus points. Next, you estimate $\mu \simeq \bar{x}$ and $\sigma \simeq s_{ML}$ from all the points in the sample, and use this to find the deviant points. You discard the deviant points and re-calculate $\bar{x}$ and s_{ML}; then repeat until there are no deviant points.

8.5 Maximum Posterior Estimates

The maximum likelihood technique takes no account of our existing knowledge or judgement. If we combine the likelihood of the data with the prior probability of a hypothesis, we can, just as in the previous chapter, put these together to find the posterior probability for our hypothesis. However, we now have a continuous range of hypotheses corresponding to trial values of our parameter θ. Correspondingly, we have to assign a continuous prior probability density distribution $\pi(\theta)$. The marginal likelihood comes as usual from summing over all the possible hypotheses, and the updated posterior probability density[2] is given by Bayes's formula at each θ value

$$P(\theta) = \pi(\theta)\frac{L(\theta)}{E} \quad \text{where} \quad E = \int_{-\infty}^{+\infty} L(\theta)\pi(\theta)d\theta. \tag{8.8}$$

Note that if we were to choose a constant value for $\pi(\theta)$, and then look for the maximum of $P(\theta)$, this would be exactly equivalent to the maximum likelihood solution, and like the ML method, there are analytic solutions for some simple cases.

[2]Note that for a continuous parameter both $\pi(\theta)$ and $P(\theta)$ will be densities with respect to θ; on the other hand $L(\theta)$ is not a density with respect to θ—it is a just function of θ, but may be a density with respect to the data.

More often however, the posterior method is implemented numerically. We will work through a simple example in Sect. 8.5.4.

A common problem is that calculating the normalisation factor E may be computationally expensive—in principle we have to perform the calculation over an infinite range of θ, and the integral may converge only slowly. This problem is much worse for the multi-parameter model fitting we will discuss in Chap. 10. However, if all we want is to numerically locate the maximum of P, we can just work with the relative values

$$P'(\theta) = EP(\theta) = L(\theta)\pi(\theta).$$

In Sect. 8.5.2 we discuss how to recover the normalisation using a Gaussian approximation. However, to proceed we first need to choose a prior.

8.5.1 Choosing a Prior

From the physical situation in question, a sensible prior will often suggest itself. For example, suppose we are measuring particle velocities in a gas and wish to estimate the temperature T of the gas. We will probably know roughly what the temperature is already, but want a more accurate estimate from the measured velocities. The simplest thing would be to limit T to some range $T_{\min}$ to $T_{\max}$, and set $\pi(T)$ to be uniform over that range. For the prior to be properly normalised, we need $\pi(T) = 1/(T_{\max} - T_{\min})$. Note that we cannot make π uniform over *all* values of T because this would make the marginal likelihood infinite, or equivalently would make the prior probability density zero everywhere. Limiting allowed values of T to some range is a simple and practical solution.

We might feel uncomfortable however with a hard rule that T cannot be outside this range. A better approach might be allow larger or smaller values to become gradually more improbable. A popular and reasonable approach is to allocate a Gaussian shape to $\pi(T)$, i.e. with a mean value set to a starting guess T_1, and a standard deviation σ_T chosen to give a range that we feel might include the right value a majority of the time. Another popular choice is to set $\pi(T)$ to be uniform in $\log T$ rather than uniform in T. You will hear this referred to sometimes as "the Jeffreys prior".

8.5.2 Gaussian Approximation to $P(\theta)$

In principle we could calculate $P(\theta)$, or the relative values $P'(\theta) = \pi(\theta)L(\theta)$, over a fine grid of θ values and numerically locate the maximum. However, we could locate the maximum much more cheaply if we interpolate between our grid values with some smooth function, in which case we need only a handful of points straddling the peak. A common practice is to assume that $P(\theta)$ has a **Gaussian shape**. This is quite well justified whenever the likelihood, and hence the posterior probability, involves a large number of independent data points. To locate the peak, we need only to calculate $P(\theta)$ at three values of θ and then solve for the Gaussian parameters which fit those data points. This is easy to see because whenever P is Gaussian, then $\log P$ is a parabola. Suppose we have

$$P(\theta) = \frac{1}{\sigma_\theta\sqrt{2\pi}} \exp -\frac{1}{2}\left(\frac{\theta-\theta_0}{\sigma_\theta}\right)^2.$$

where θ_0 and σ_θ^2 are the mean and variance of our assumed Gaussian probability distribution $P(\theta)$. In fact we can just work with the relative values $P'(\theta) = EP(\theta) = L(\theta)\pi(\theta)$. Taking logs, we find

$$y(\theta) = \log P' = A - \frac{1}{2}\left(\frac{\theta-\theta_0}{\sigma_\theta}\right)^2 \quad \text{where} \quad A = \log\left(\frac{E}{\sigma_\theta\sqrt{2\pi}}\right).$$

The quantity $y = \log P'$ is therefore a quadratic in θ with three parameters $A, \theta_0, \sigma_\theta$. Let us suppose that we have three evenly spaced θ values at $\theta_1, \theta_2 = \theta_1 + \Delta\theta, \theta_3 = \theta_1 + 2\Delta\theta$, for which we find values y_1, y_2, y_3. Then after some simple but very tedious algebra we can find that

$$\theta_0 = \theta_3 - \Delta\theta\left[\frac{y_3 - y_2}{y_1 - 2y_2 + y_3} + \frac{1}{2}\right]. \tag{8.9}$$

We therefore have a location of the best value of θ to arbitrary accuracy by calculating the posterior at just three test values. A little bit more algebra also gives us the error on our best estimate, and the normalisation factor E

$$\sigma_\theta = \Delta\theta\sqrt{(2y_2 - y_1 - y_3)^{-1}}, \tag{8.10}$$

$$\log(E) = y(\theta_0) - \log(\sigma_\theta\sqrt{2\pi}). \tag{8.11}$$

The assumption of a Gaussian form gives us another popular quick way to calculate σ_θ—the **curvature method**. Using the parabolic form in (8.9), differentiating twice will show that $\mathrm{d}^2y/\mathrm{d}\theta^2 = -1/\sigma_\theta$ and so we have

$$\sigma_\theta = -\left(\frac{\mathrm{d}^2 \log P}{\mathrm{d}\theta^2}\right)^{-1}. \tag{8.12}$$

A rough estimate of the error can therefore be obtained by double-differencing three neighbouring points.

8.5.3 Constructing Posterior Probability Intervals

If we have a normalised $P(\theta)$, then we can construct an interval θ_1 to θ_2 such that the integrated posterior

$$P_{\text{int}} = \int_{\theta_1}^{\theta_2} P(\theta)\text{d}\theta.$$

contains the desired total amount of probability—0.68, 0.9, 0.95, or whatever. If $P(\theta)$ is symmetrical, the usual method is to take θ_1 and θ_2 an equal distance from the mode. For an asymmetric probability density function, there is no unique choice of θ_1 and θ_2, so one has to be carefully explicit about how the choice was made.

There are three ways we might achieve normalised P values. First, if we are lucky we might have an analytic solution for $P(\theta)$—but this really is quite rare. The second way is what one might call the brute-force numerical method. We start by calculating $P'(\theta) = \pi(\theta)L(\theta)$, and then get an approximate normalising factor by numerical integration over some range θ_{min} to θ_{max}:

$$E_{\text{approx}} = \int_{\theta_{\text{min}}}^{\theta_{\text{max}}} P'(\theta)\text{d}\theta.$$

A typical procedure might be to gradually extend the upper/lower limits until our estimated value E_{approx} changes from the previous value by less than some pre-decided amount, such as say 1%.

The third way to achieve normalised values is to assume the Gaussian approximation of Sect. 8.5.2 and extrapolate, using standard Gaussian integral probability regions. For example, if we want to contain 95% of the probability, the range we want is $\theta_1 = \theta_0 - 1.96\sigma_\theta$ to $\theta_2 = \theta_0 + 1.96\sigma_\theta$. Furthermore, we don't need to calculate the marginal likelihood E by laborious integration; on the assumption of a Gaussian form, we can solve for E from our three points as described in the previous section.

How reliable is the Gaussian approximation method? There is no black and white answer to this. The more data points there are the better it will be. A rule of thumb would be that 10–20 or more independent data points makes it fairly reliable. However, it should be borne in mind that any discrepancy between the real $P(\theta)$ and the Gaussian approximation will be in the far wings of the function. If you are happy with a 68% interval, i.e. the equivalent of $\pm 1\sigma$, then it will usually be very good. However, if you are looking for a 99% interval, your limits could be badly underestimated. In Chap. 10, when we look at multi-dimensional model fitting, we find that the Gaussian approximation in the tails of $P(\theta)$ becomes far less reliable.

8.5.4 Maximum Posterior Worked Example

Let's use the same simple example dataset of Sect. 8.2. Suppose then we have six measurements of the mass m of a particle, with the data values being $x_i = 256, 239, 237, 278, 266, 241$, and the error being $\sigma_x = 16$. We imagine many

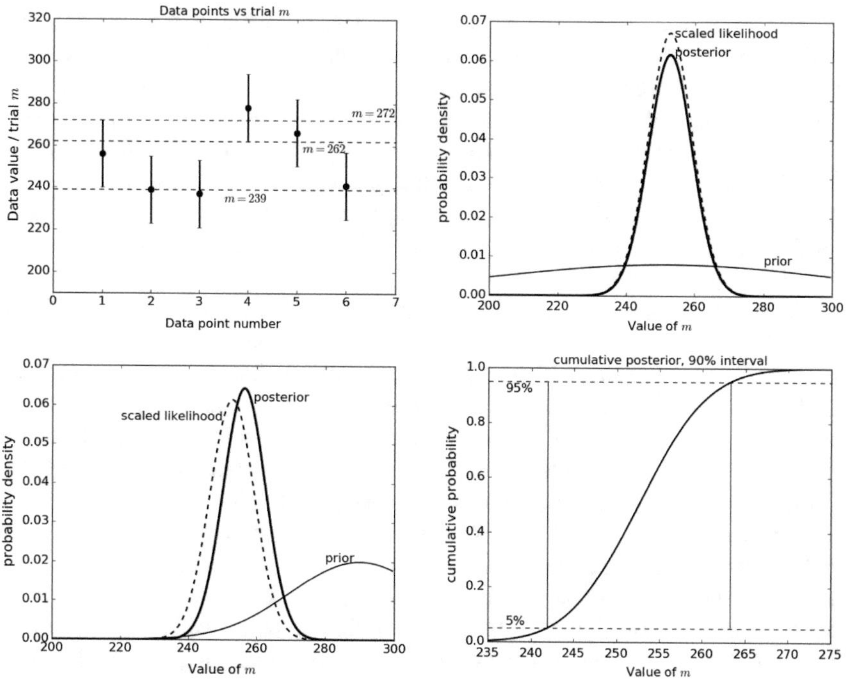

Fig. 8.1 Illustration of the use of Bayesian credibility methods to estimate a parameter, as described in the text. **Top left**: Six data points x_i compared to three different trial values of the parameter m. Note that the horizontal axis doesn't have a physical meaning; its just the index number for each of the six values. **Top right and bottom left**: Comparing the prior distribution $\pi(m)$, the joint data likelihood $L(m)$, and the resulting posterior $P(m)$, for two different priors. Note that both the prior and the posterior are normalised credibility density distributions, so that the *y-axis* units would depend on the units of m, but both curves integrate to 1.0. The joint likelihood is actually a very small number; an arbitrary scale factor has been applied in order to make a visual comparison with the other distributions. **Bottom right**: Ilustrating how to derive an uncertainty interval of 90%, based on the cumulative posterior distribution, derived using the first of the two priors

different possible values of m; for each hypothesised value, we can consider each data value to be drawn from a Gaussian distribution with mean m and standard deviation $\sigma = 16$. Figure 8.1, top-left, illustrates our six data values, compared to three different trial values of m. By eye its clear that $m = 262$ looks reasonable and $m = 239$ and $m = 272$ less so.

At a given value of m we can calculate the likelihood of each data value, which is given by the Gaussian probability density. For example, for $m = 239$, the value $x_1 = 256 \pm 16$ gives $L_1 = 0.0142$. The joint likelihood of the whole dataset is $L = L_1 \times L_2 ... \times L_6 = 1.6607 \times 10^{-12}$. Note that such joint likelihoods are often very small numbers, which needs some numerical care. Repeating the exercise for different values of m gives us $L(m)$, as illustrated in Fig. 8.1, top right.

Next we need to assign a prior probability density distribution $\pi(m)$. In a real-world example, this would be based on some physical judgement, or theoretical guess. As an illustration, we try two different priors, each of which has a Gaussian shape. In Fig. 8.1, top right, the prior has mean $m = 252$, and is quite broad, with $\sigma = 50$. In Fig. 8.1, bottom left, the prior is centered at $m = 290$, and is somewhat narrower, with $\sigma = 20$. We then combine our priors with $L(m)$ to get the un-normalised posterior $P'(m) = L(m)\pi(m)$. With this simple example, numerical integration is quite easy. For the first prior, we get $E = 2.0256 \times 10^{-12}$. We can then make the normalised posterior $P(\theta) = EP'(\theta)$.

The results are shown in Fig. 8.1, top right and bottom left. For the first prior, $P(m)$ is not much different from $L(m)$, apart from a normalisation factor. The best value is at the maximum of $P(m)$, which we find at $m = 253.0$. For the second prior, $P(m)$ is a kind of compromise between the prior and the likelihood. This is instinctively reasonable; our final conclusion involves both what our initial judgement told us, and what the data says. Does the choice of prior make a big difference to the final results? It always makes *some* difference, but how much depends on both the prior concerned and the data. If the information we have is such that the parameter is already fairly well constrained, then the precise prior may make little difference; over the range of interest, the behaviour of the prior may be close to flat for any reasonable choice of shape. Likewise, if the dataset is highly constraining, the choice of prior is not too important. Overall, the issue is which of $L(\theta)$ and $\pi(\theta)$ is changing faster with respect to θ. If $L(\theta)$ is dropping fast over a small range of θ then different choices of $\pi(\theta)$ will make little practical difference. However, note the converse—in cases where the data input is relatively modest, the choice of prior can make a *big* difference. People who don't like Bayesian methods see this as a weakness; fans of Bayesian methods see this same issue as precisely their strength—scientific *judgement* is required, and it makes a quantitative difference.

Next, let us examine how to construct an uncertainty interval. One way to do this is by using the cumulative posterior

$$P_{\mathrm{cum}}(<\theta) = \int_{-\infty}^{\theta} P(\theta')\mathrm{d}\theta'.$$

where of course in practice we start the integration at some suitably small θ_{min}. Depending on the physical meaning of the parameter in question, sometimes of course we may integrate from $\theta = 0$ and so on. The result in our case is shown in Fig. 8.1, bottom right. Suppose we want an interval to contain 90% of the probability for m. Then we simply read off where $P_{\mathrm{cum}} = 0.05$ and 0.95, giving a range of 241.91–263.23.

Finally, lets take a look at the three-point Gaussian solution. We can calculate our prior and our likelihood at say $m = 249.0, 253.0, 257.0$, i.e. with $\Delta m = 4.0$, giving $P(m) = 0.05176, 0.06157, 0.05001$ respectively. Taking logs and using equation (8.9), we find $m_0 = 252.82$ and $\sigma_m = 6.48$. For this simple example, the Gaussian approximation method gives the same answer as the full $P(m)$ method—but it ought to, because we have assumed that all our data points are drawn from a single Gaussian distribution. Note however, its computational efficiency.

8.6 χ^2 Parameter Estimation

In Chap. 7 we looked at how we can test either the relative believability of a set of hypotheses using priors and likelihood, or test our absolute confidence in a single hypothesis using a test statistic, the most popular example of which is the χ^2 statistic of Sect. 7.11, which measures the scatter of data values compared to what the hypothesis would suggest. If we have a tuneable parameter θ, we can do a χ^2 significance test for each value of θ in turn, and construct the function $\chi^2(\theta)$. Then, just as we can search for the θ value that maximises likelihood or maximises posterior probability, we can search for the value that minimises the value of χ^2. This is an extremely popular and widespread technique, and dominates much of the statistical testing used in the scientific literature, although Bayesian methods are slowly increasing in popularity.

8.6.1 Relation Between χ^2 and Likelihood

For normally distributed data points, the minimum χ^2 and maximum likelihood methods are essentially the same. Suppose we have N data values x_i and we believe that these are drawn from a parent population with mean μ and dispersion σ. Then we have

$$\chi^2 = \frac{1}{\sigma^2} \sum (x_i - \mu)^2,$$

Recall that in Sect. 8.4 we looked at the joint likelihood of N normally distributed data points, and then took logs, finding that

$$\log L = -\frac{1}{2} N \log 2\pi\sigma^2 - \frac{1}{2\sigma^2} \sum_i (x_i - \mu)^2,$$

We can see that the last term is just half of χ^2, and so we have

$$\log L = -\frac{1}{2} N \log 2\pi\sigma^2 - \chi^2/2. \quad (8.13)$$

This relation is generally true with respect to any dataset with normally distributed errors—minimising χ^2 is the same as maximising $\log L$.[3] The idea of maximum likelihood is more general, and can be applied in other cases; however a huge amount of literature on practical techniques is couched in terms of minimising χ^2.

[3] Sometimes you see people using "negative log likelihood" as a test statistic.

8.6.2 Finding Minimum χ^2

Just as with the maximum posterior method, there may occasionally be an analytic solution, but more often we will need to calculate $\chi^2(\theta)$, and find its minimum, numerically.

To locate the minimum of χ^2, we can use the same Gaussian/parabolic method as we used for locating the maximum posterior (Sect. 8.5.2). Note that because χ^2 is essentially the log of likelihood, if the likelihood function is Gaussian, then $\chi^2(\theta)$ will be a parabola. By analogy with the posterior probability case, we can see that the minimum χ^2 is located at

$$\theta_0 = \theta_3 - \Delta\theta \left[\frac{\chi_3^2 - \chi_2^2}{\chi_1^2 - 2\chi_2^2 + \chi_3^2} + \frac{1}{2} \right]. \tag{8.14}$$

8.6.3 Goodness of Fit

The main advantage of the χ^2 method is that it gives a way to quantify the absolute *goodness of fit* of our hypothesis. Once we have located the value θ_0 that gives us the minimum value of $\chi^2 = \chi^2_{min}$ over a tested range of θ values, we can ask "now we have found that best hypothesis, if it is correct, what is the probability of getting a value of χ^2_{min} or worse?". However, as noted in Sect. 7.11.2, because we have used the data to find the best-fit value, the degrees of freedom we use in interpreting χ^2 is $\nu = N - 1$ rather than N.

This ability to reject even the best value can be important because every hypothesis has a broader context. Suppose some theorist says "I propose there is a new particle. You will be able to see the effect of this particle if you make such-and-such measurements. I don't know what the mass m is, but my theory predicts what you should see depending on the mass." We could then make the measurements, and compute the likelihood of our observed data points, for each of a range of m values. We could get our "best" value of m from maximum likelihood, from maximum posterior, or from minimum χ^2. The good Bayesian at this point might say "I can tell you how believable different values of m are, but you should know that if the theory about the new particle isn't true anyway, then none of this means anything." However from the χ^2 test, we might end up saying "Well this value of m is the best of a bad bunch, but actually even for that best value, there is a low probability of getting the data we saw, so the whole idea might be nonsense."

8.6.4 Confidence Intervals Based on χ^2 and $\Delta\chi^2$

Once we have calculated the function $\chi^2(\theta)$, and have located its minimum at some value $\theta = \theta_0$, how do we find an uncertainty interval? Because χ^2 tests the significance of a result, what we can hope to get is a *confidence interval*. There are three ways of going about this.

The first way is to calculate an **acceptance interval**. Any given value of θ can be seen as an independent hypothesis, and based on the value of $\chi^2(\theta)$ we can accept or reject that hypothesis, using our chosen confidence level. Note that because we are considering a fixed hypothesis, the number of degrees of freedom is $\nu = N$ where N is the number of data points. We can then divide all the possible points into an acceptance class and a rejection class. The critical values of θ where we switch from acceptance to rejection define an *acceptance interval*.

The second way is to estimate a **Gaussian likelihood interval**. For a reasonable number of data points the likelihood function will be Gaussian, and because of (8.13), the function $\chi^2(\theta)$ will be a parabola. As well as allowing us to estimate the best value of θ as in Sect. 8.6.2, we can also use the "curvature method" of Sect. 8.5.2 to get the width of the function, giving us

$$\sigma_\theta = -\frac{1}{2}\left(\frac{\mathrm{d}^2\chi^2}{\mathrm{d}\theta^2}\right)^{-1}. \tag{8.15}$$

The third way to get a confidence interval is to use **the $\Delta\chi^2$ method**. This is related to the simple acceptance test method, but is more sensitive, using a new test statistic, $\Delta\chi^2$. To explain this, we first need a useful mathematical result. If we have a χ^2 variable χ_n^2 with $\nu = n$ degrees of freedom and a second variable χ_m^2 with $\nu = m$ degrees of freedom, then it can be shown that the difference between the two variables, $\Delta\chi^2 = \chi_n^2 - \chi_m^2$, is also distributed as χ^2, but with $n - m$ degrees of freedom. This is not too hard to prove, but we won't do so here. Lets look at how we use this theorem. The trick is to look at the change in χ^2 either side of the minimum. The probability distribution for the value of $\chi_{\min}^2$ will be given by χ^2 with $N - 1$ degrees of freedom, because we have determined our parameter θ from the data. Suppose now we try out a value of θ to one side of the minimum, and find a value χ_θ^2. This will be distributed as χ^2 with N degrees of freedom—each trial value of θ is fixed before the calculation. So the difference between this and the minimum value, $\Delta\chi^2(\theta) = \chi_\theta^2 - \chi_{\min}^2$ will be distributed as χ^2 with one degree of freedom. Then as above, we can choose a significance level α, and accept or reject each value of θ by testing whether the value of $\Delta\chi^2$ has a probability less than α. This will construct the $\Delta\chi^2$ confidence interval for θ.

The beauty of this method is that the required critical values of $\Delta\chi^2$ are always the same. Looking up values of χ_1^2 in a standard table, we find that 68.3%, 90%, 95% and 99% confidence requires $\Delta\chi^2 < 1.0, 2.71, 3.84, 6.63$ respectively.

8.6.5 Minimum χ^2 Worked Example

Let us once again use the simple example of Sect. 8.2—six data values $x_i = 256, 239, 237, 278, 266, 241$, all with error $\sigma = 16$, which are estimates of the mass

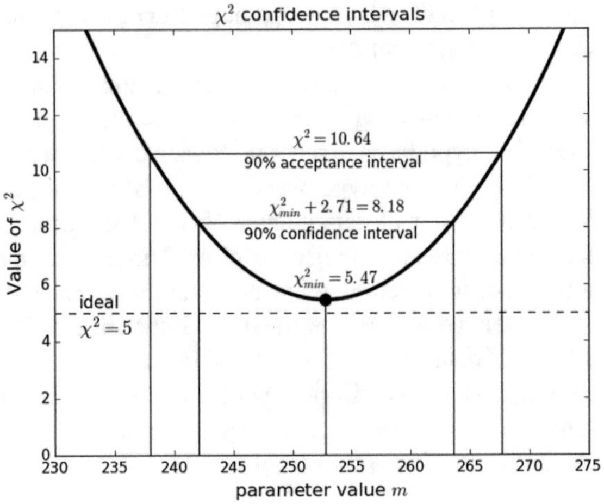

Fig. 8.2 Track of χ^2 versus parameter value m for the example data described in the text, together with two different 90% uncertainty intervals. The upper range is the acceptance interval, based on the critical value where $P(\chi^2 > \chi^2_{\rm crit}) = 0.1$ for $\nu = 5$. The lower value is the $\Delta\chi^2$ confidence interval, based on the critical value of χ^2 with $\nu = 1$

m of a particle. Figure 8.2 shows the result of calculating χ^2 for a range of values of m. The minimum can be located graphically to be at $m = 252.8$, where we find $\chi^2_{\rm min} = 5.47$. The value of $\chi^2_{\rm min}$ should be distributed as χ^2 with $\nu = N - 1 = 5$ degrees of freedom. From standard tables, we find $P(>^2_{\rm min}) = 0.36$, which shows that the hypothesis that all the data points are drawn from a single mean is acceptable.

From standard tables, we can also find $P(>\chi^2) = 0.1$ at $\chi^2 = 10.64$. The points where our $\chi^2(\theta)$ curve intersect that value therefore gives us the 90% acceptance range, which gives us $\mu = 251.9 \pm 14.9$, as illustrated in Fig. 8.2. On the other hand, if we use the $\Delta\chi^2$ statistic, to get 90% confidence, we need to go to $\chi^2_{\rm min} + 2.71 = 8.18$. Seeing where the curve hits that level, we find our 90% confidence range to be $\mu = 251.9 \pm 10.7$, which is closely similar to the the posterior probability interval for the case where the prior is broad.

The interval we get from $\Delta\chi^2$ also agrees closely with the result we get if we use the curvature method to get σ_θ, and extrapolate to 90% confidence assuming that this represents the standard deviation of a Gaussian distribution. For many practical purposes, the curvature and $\Delta\chi^2$ methods agree well, but the agreement will typically worsen as we go out into the wings of $\chi^2(\theta)$, i.e. if we want 95 or 99% confidence.

8.7 Key Concepts

Some of the key concepts from this chapter are:

- The idea of a parameter as representing a family of hypotheses
- How to get the error on an estimate from its sampling distribution
- The fact that the sample variance is a biased estimator of the population variance, and how to calculate an alternative unbiased estimate
- The use of the z-statistic for cases where σ is known in advance

- The use of the t-statistic for cases where σ is estimated from the data
- The idea of using maximum likelihood as a general technique for finding parameter estimates
- Maximum likelihood estimates for mean and variance, including the weighted-mean estimate
- The idea of using maximum posterior as a general technique for finding parameter estimates
- Techniques for getting maximum posterior estimates for parameters, including choice of prior, numerical techniques and the Gaussian approximation
- Techniques for constructing posterior uncertainty intervals
- The use of minimum χ^2 as a general technique for finding parameter estimates
- How to assess goodness of fit from minimum χ^2
- The use of $\Delta\chi^2$ to construct confidence intervals.

The key formulae from this chapter are: standard sample mean and variance estimators (8.1), and the error on the sample mean (8.3); the unbiased estimator for sample variance (8.2); definitions of the z and t statistics for constructing uncertainty intervals on the mean ((8.4), (8.5)); the maximum likelihood formulae for the weighted mean and its error ((8.6), (8.7)); the formula for constructing a posterior function, given a prior (8.8); the formulae for estimating maximum posterior, its error, and the posterior normalisation ((8.9), (8.10), (8.11)); the curvature formula for getting the error on a maximum posterior estimate (8.12); the relationship between χ^2 and likelihood (8.13); the parabolic approximation formula for getting a minimum χ^2 estimate (8.14); and the curvature formula for getting the error on a χ^2 estimate (8.15).

8.8 Further Reading

Traditional statistics textbooks (see the Chap. 1 references) usually have good treatments of parameter estimation using maximum likelihood and χ^2 techniques. A particularly practical guide which is popular with physical scientists is Bevington and Robinson (2002); as well as sound explanations, it also has many useful tables, and examples of computer code. As with hypothesis testing, Bayesian methods are becoming increasingly popular. You will find good simple treatments in Sivia and Skilling (2006) and in Stone (2013).

The t statistic we discuss in Sect. 8.3.4 is often known as "Student's t". This name is nothing to do with its use by students. The t-distribution was put forward in a paper in the journal *Biometrika* signed only with the name "Student". (Student 1908). This was a pseudonym used by William Sealy Gosset. He worked for the Guiness brewery, who did not allow their scientists to publish and risk breaching trade secrets; publishing under a pseudonym was the compromise Gosset came to with his employers. The story is nicely told in the Wikpedia page on Gosset.

8.9 Exercises

8.1 I recently measured the brightness of a star three times and got 225 ± 6, 218 ± 9, and 286 ± 36. (The units don't matter!) If I didn't have the errors, but only the brightnesses, what would I get for the estimated mean? With the errors known, what is the weighted mean and error? Calculate a 95% confidence range for the mean. Does it include the simple unweighted mean? Finally, how does our 95% confidence region compare to the 95% acceptance region?

8.2 How many data points do you need for the sample standard deviation to be the same as the unbiased standard deviation estimate, to within 5% or better?

8.3 A particle physics experiment has made five measurements of the mass of the Yotta particle, which gave m =83.6, 92.9, 77.3, 88.4, and 89.5 GeV respectively. From these measurements, estimate the sample mean, sample variance, and error on the mean. The standard theory predicts $m_s = 91.93$ GeV. Is this value consistent with the data at 95% confidence? Assume that the data points are normally distributed with mean m_s and variance σ^2. First, compare the prediction and the data using the standard z-statistic, using the sample variance as an estimate of σ. Then compare them using the t-statistic. How much difference does this make? Construct a 95% confidence interval for m using the t-statistic.

8.4 Show that for $\nu = 1$ the t-distribution is the same as the Cauchy distribution. (This will need some external mathematical investigation)

8.5 A radiation counter records a series of radioactive decays. The data saved is a set of N numbers t_i where t_i is the time between each event and the next one. Assuming the data follow the usual exponential (waiting time) distribution, derive a formula for the maximum likelihood estimate of the rate λ.

8.6 Show that the formula for the unweighted mean reduces to the usual formula for the mean if all the errors are the same.

8.7 (a) A coin is being tested for bias. This is quantified by p, the probability of landing heads up on any single coin toss. The test is to toss the coin n times and record the number of heads r. An initial test has 4 tosses and scores 3 heads. On this basis, what is the "natural" estimate of p? If the number of heads does indeed follow a binomial distribution with this value of p, what is the expected dispersion in r, σ_r? Use this to get an estimate of the error on our estimate of p, σ_p.

(b) The observation is 3 heads out of 4. Write down an expression for the likelihood of the data as a function of p. Assuming a uniform prior for p, calculate the maximum posterior estimate for p. Calculate the Full Width Half Maximum (FWHM) for p. By assuming the posterior function is roughly Gaussian, estimate an error σ_p. Do these estimates for p and σ_p agree with those from part (a) above?

(c) Suppose we use a prior function for p that is more weighted towards the middle, as seems instinctively reasonable. A good choice might be $\pi(p) = 6(p - p^2)$. You can confirm that this is centred on $p = 0.5$ and is well behaved in that it goes to zero at $p = 0$ and $p = 1$, and sums to 1.0. Using this prior, re-estimate p. Is it very different from the simple estimate?

(d) A fuller test makes 100 tosses and find 65 heads. For 100 tosses the Gaussian will be a very good approximation to the binomial. Use this approximation, and assume a uniform prior, to find a 95% posterior probability region for p.

8.8 A theoretical model involves a parameter Q. The theory predicts the outcome of some experiment which measures a quantity k. A number of measurements of k are made. Theorists have argued that the correct answer should be $Q \sim 17$. The experimenters therefore calculate the likelihood of the complete set of measured k values for trial values of $Q = 16, 0, 17.0, 18.0$. Applying a simple flat prior, they find (unscaled) relative posterior values of $P = 0.0372, 0.0489, 0.0416$ respectively. Using the Gaussian approximation to the posterior, estimate the value of Q and its error σ_Q. If you use the curvature method, what value of σ_Q does this give?

8.9 Consider N data points, all with the same error σ, assumed to represent a Gaussian standard deviation. The data points are thought to have a common mean μ, and you are using the data points to estimate μ. Show that the standard χ^2 statistic, the prior probablility π and the posterior probabiliy P are related by

$$\log P(\mu) = \text{const} + \log \pi(\mu) - \frac{1}{2}\chi^2(\mu)$$

In approximate terms, what would need to be the case for the minimum χ^2 estimate of μ to be close to the same as the maximum posterior estimate?

8.10 It is common to use the Gaussian approximation to the χ^2 distribution in order to construct confidence ranges. If we have N data points and use them to estimate m parameters, what value of χ^2 will correspond to 95% confidence, using this approximation? How accurate is this approximation for $m = 1$ and $N = 3$, 10 and 50?

References

Bevington, P., Robinson, D.K.: Data Reduction and Error Analysis for the Physical Sciences, 3rd edn. McGraw-Hill Education, New York (2002)

Sivia, D.S., Skilling, J.: Data Analysis: A Bayesian Tutorial, 2nd edn. Oxford University Press, Oxford (2006)

Stone, J.V.: Bayes Rule: A Tutorial Introduction to Bayesian Analysis. Sebtel Press (2013)

Student, : The probable error of a mean. Biometrika **6**, 1 (1908)

Websites (all accessed March 2019):

Wikipedia article on William Sealy Gosset ("Student"): https://en.wikipedia.org/wiki/William_Sealy_Gosset

Chapter 9
Inference with Two Variables: Correlation Testing and Line Fitting

9.1 Outline of Content

- The concept of correlation
- Covariance and the correlation coefficient
- Testing for correlation
- Rank correlation tests
- Correlation test pitfalls
- Least squares fitting
- Fitting to a straight line
- Testing the fit
- Curvilinear line fitting
- Arbitrary function fitting

A large part of science concerns looking for, and then trying to understand, causal relations between observed quantities. This is much harder where random variables are concerned. For example, take a look at Fig. 9.1. Each data point represents a galaxy where both the mass of the "bulge" component of a galaxy, and the mass of the central supermassive black hole it contains, have been estimated. It looks like these things are connected, which could be important. But the points are rather scattered, and have large error bars. Are we just being fooled by a chance distribution of points drawn from some random distribution? Note also that the authors have drawn a line going through the data, hopefully representing the true relationship between black hole mass and bulge mass. But is the slope of the line right? How do we decide what the "best" slope is? And what if a straight line isn't the right mathematical form? Can I test the prediction for my favourite theory?

When exploring possible correlations, we nearly always ask the same three questions. (i) Is it real, or just a fluke? (ii) What is the mathematical relationship between the variables? (iii) Does theory X get it right?

In the first half of this chapter we concentrate on the first question—is the apparent correlation real? We then take a simplified look at the second question, by looking at how to fit straight lines and polynomials to a set of data, using traditional least squares techniques. Full answers to the second and third questions will need the techniques of general model fitting, which we will tackle in Chap. 10.

A. Lawrence, *Probability in Physics*, Undergraduate Lecture Notes in Physics,
https://doi.org/10.1007/978-3-030-04544-9_9

9.2 Bivariate Data Sets and Their Origins

Before we set off, let us look carefully at bivariate datasets and how they come about. Earlier in the book we have mostly considered samples which are essentially lists of data points x_i which we take to be random drawings from some probability density function $p(x)$. When visually inspecting the data, we could draw each data point as a dot on the x-axis line, but more normally we would group them in bins and plot them as a histogram, for comparison to our expected $p(x)$ curve. In this chapter we are still dealing with a list of data points, but each point is an x_i, y_i pair. In the physical sciences, the usual way to visualise such data is to plot individual points on the x, y plane, as a "scatterplot", such as that seen in Fig. 9.1. We could also construct a two-dimensional histogram, but a scatterplot is more usual, because what we are exploring is the relation between x and y, rather than how often various x, y values happen. How do such x_i, y_i datasets relate to the underlying physical situation? It is useful to distinguish two different situations.

A: both variables random. Our x_i, y_i points might be random drawings from some underlying bivariate PDF, $p(x, y)$. Simulated examples are shown in Fig. 9.4. Something like this presumably underlies the data in Fig. 9.1, where both black hole mass and bulge mass have a distribution of values in the population of galaxies. Similarly, we might for example select people at random from a population, and for each of them measure both their height and their weight. In such cases both x and y are random variables. When we test for an underlying correlation between the two variables (see Sect. 9.5), we normally assume that $p(x, y)$ is a bivariate Gaussian (see Chap. 5, Sect. 5.8.1).

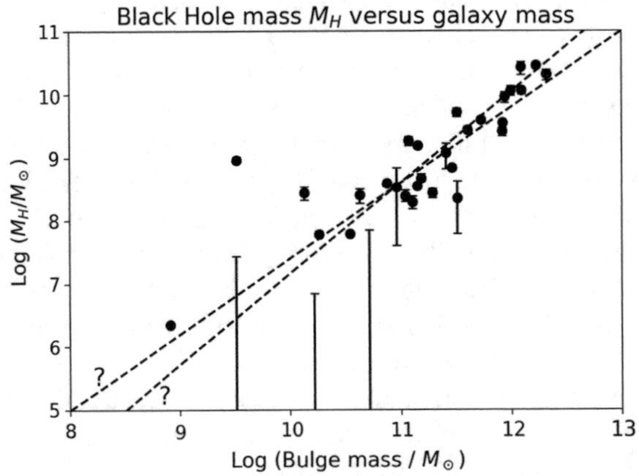

Fig. 9.1 Plot of black hole mass versus galaxy bulge mass, from the study by Magorrian et al. (1998). The "bulge" of a galaxy is the quasi-spherical component, as opposed to the flat disc component. The two dashed lines represent illustrate possible relationships—one is a one-to-one relationship, and the other is a line that one might think agrees with the data better. The vertical lines with no corresponding filled circles represent upper limits. (The original plot came from Magorrian et al. (1998), Astron. J, 115, 2285, and is copyright AAS, used with permission. We have taken data from Table 2 of that paper and re-drawn.)

B: only one variable random. In this situation, one of the variables, let us say x, is not random, but is varied in some controlled manner; then at each chosen x value we measure y, which is a random variable with some PDF $p(y)$. However, the PDF for y may be different at each x value, i.e. $p(y|x)$. For example, we might have an amplifier where we turn a dial to set a voltage, and then at each voltage setting we measure the output wattage. In such a case, x is known as the independent variable, and y is the dependent variable. Often, some theory predicts how y should depend on x, giving $y_{\text{pred}}(x_i)$, but the measurement has a random error σ. Then each y_i measurement is a random variable drawn from a PDF that is a Gaussian with mean y_{pred} and variance σ^2. When we fit lines to data (see Sect. 9.8), this is the situation we are assuming.

9.3 The Concept of Correlation

Looking at the points in Fig. 9.1, our instinct is that they look like they are probably "correlated". What do we mean by this? Lets first try to clarify the concept of correlation. Then we can discuss how to quantify the concept, and test hypotheses connected with correlation. Correlation is closely connected with the idea of statistical dependence but is not quite the same. Dependence, which we already met in Chaps. 1 and 2, is the more general concept, so lets recap that first.

9.3.1 Dependence: Shape Change

Consider the joint probability density function $p(x, y)$. At a given value of x, we can construct the conditional PDF for y, $g(y|x)$, and similarly we can construct $f(x|y)$. Graphically, if you think of $p(x, y)$ as a two dimensional surface, then you can visualise $g(y|x)$ as a vertical slice at position x. Note that in "Situation B", with only y being a random variable, we can still consider the distribution $g(y|x)$ at each x. If the two variables are independent, then the probability of y depends only on the value of y, and not on the value of x. In terms of our visualisation, the shape of a vertical slice, $g(y)$, is the same at all x. For "Situation A", both variables random, this also means that we can write the joint PDF $p(x, y)$ as the product of two distinct functions, $p(x, y) = f(x)g(y)$.

Examples of simulated PDFs are shown in Fig. 9.2, with the result of slices through y at three different x positions shown in the right hand panel in each case.[1] In the top row, the shape of the slice is the same at all three x positions. The normalisation differs, but not the shapes, as could be confirmed by re-normalising them to match. The x and y variables are *independent*. For both the middle row and the bottom row the y slices have shapes that clearly differ between the three x positions. The x and y variables are *dependent*.

[1]The simulations were all generated from bivariate Gaussan PDFs, but in the bottom row, we have made the ring shape by subtracting one Gaussian from another.

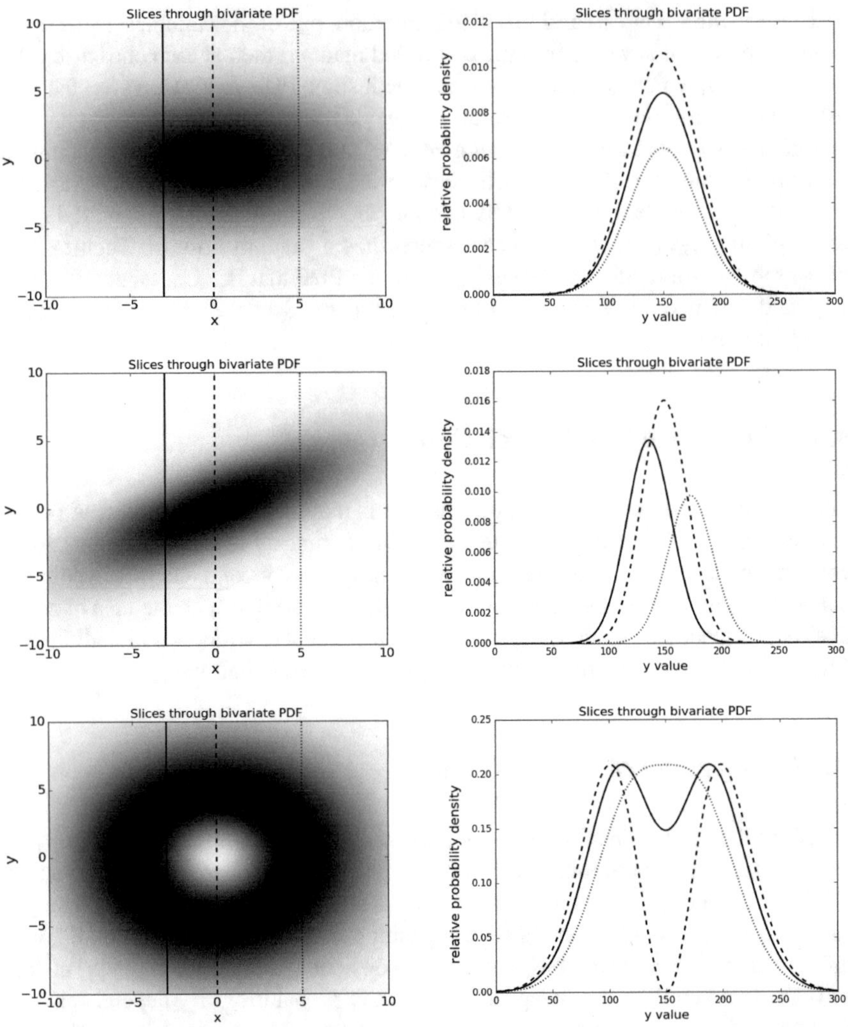

Fig. 9.2 Illustrating the difference between dependence and correlation using simulated PDFs, with the grey scale representing probability density. The top row shows a PDF for independent variables. Vertical slices at the indicated positions show shapes that are the same apart from a normalisation factor. The middle row illustrates variables which are both dependent and correlated. The shapes produced by vertical slices move systematically in y as the x position changes. The bottom illustrates variables which are dependent but not correlated. The shapes produced by vertical slices are quite different, but do not shift systematically in y as x changes

9.3.2 *Correlation: Systematic Drift*

When we say that two variables are *correlated* we mean that their *typical values* are connected—i.e. as we consider steadily larger values of x, the typical value of

y systematically drifts towards steadily larger values—or if the variables are *anti-correlated*, a larger value of x tends to mean a smaller value of y. Looking again at the examples in Fig. 9.2 we can see that the middle row fits this idea, but the bottom row does not—there is no systematic tendency for bigger x to mean bigger (or smaller) y.

Note that if two variables are correlated, they are also dependent—but it doesn't work the other way round. It is easily possible to be dependent but not correlated, as the bottom row of Fig. 9.2 shows.

9.4 Quantifying Correlation: Covariance

Above, we have talked about the drift of "typical" values. Can we make this idea more rigorous? Ideally, we might like to derive the mathematical form of any possible relationship between y and x. In Sect. 9.8 we shall examine how to do just that. But before doing anything that clever, it would be good to extract a single number from our dataset—a test statistic—that we can use to test whether there is any correlation at all, and if so, which measures the degree of correlation. The traditional way to quantify correlation is to consider the *covariance* of the joint distribution. Recapping the discussion of Chap. 2, Sect. 2.4.4, the sample covariance is

$$s_{xy} = \left[\frac{1}{N}\sum(x_i - \bar{x})(y_i - \bar{y})\right], \tag{9.1}$$

and the population covariance is

$$E[(x - \mu_x)(y - \mu_y)] = \mathrm{Cov}(x, y) = \sigma_{xy} = \lim_{N\to\infty} s_{xy}.$$

If the variables are independent then

$$E[(x - \mu_x)(y - \mu_y)] = E(x - \mu_x)E(y - \mu_y),$$

but each of those terms is zero by definition; so independent variables always have $\mathrm{Cov} = 0$. This is intuitively reasonable because for any given data point, whether $x_i - \mu_x$ is positive/negative is unconnected with whether $y_i - \mu_y$ is positive/negative, so there are equal numbers of positive and negative terms in the covariance sum. If the two variables are dependent, then very often, $\mathrm{Cov} \neq 0$. However, unfortunately this isn't always the case; it is sometimes possible for dependent variables to have $\mathrm{Cov} = 0$, so covariance is not a safe test of dependence.

However *correlated* variables always have $\mathrm{Cov} \neq 0$. If large values of x tend to go with large values of y, then when $x_i - \mu_x$ is positive/negative, $y_i - \mu_y$ will also tend to be positive/negative. If the variables are anti-correlated, then when $x_i - \mu_x$ is positive/negative, $y_i - \mu_y$ will tend to be negative/positive.

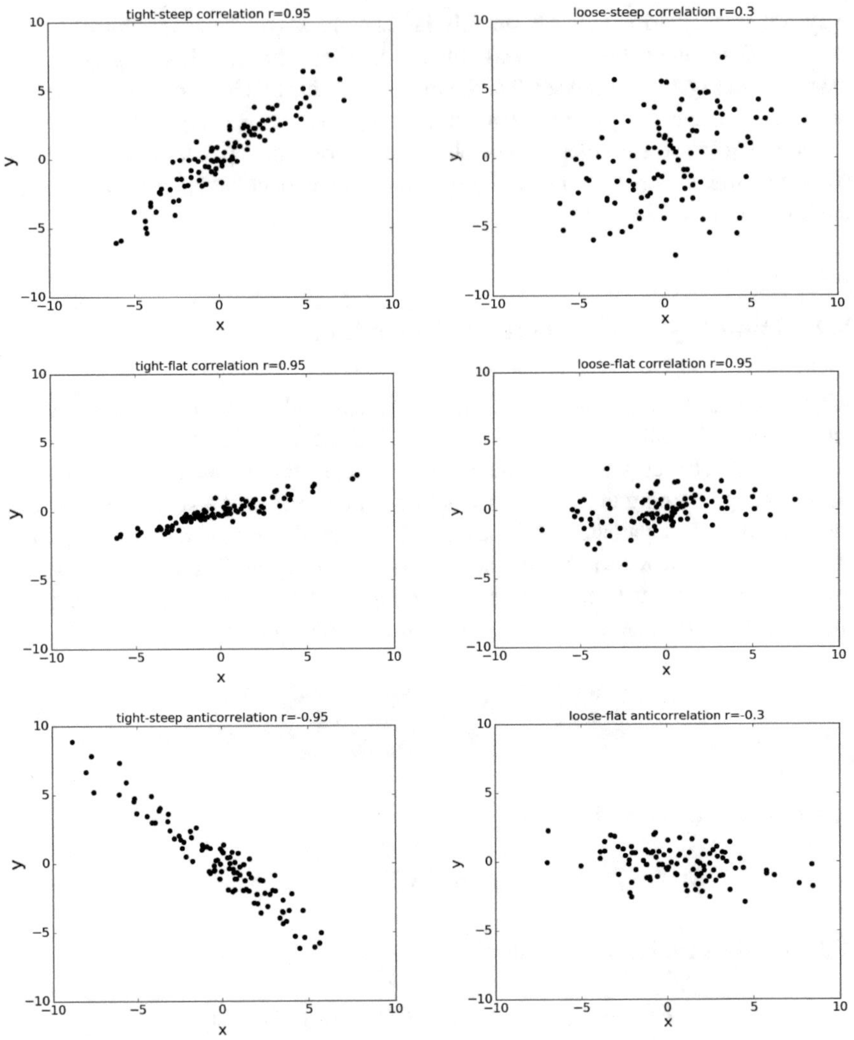

Fig. 9.3 Illustrating the difference between the tightness of correlation, and the steepness/flatness of correlation. The correlation coefficient measures the tightness, not the strength of effect of one variable on the other. Each sample is drawn at random from a bivariate Gaussian distribution with different values of σ_x, σ_y, and σ_{xy}. The slope is determined by the relative values of σ_x and σ_y, whereas the scatter in y at a given x is determined by σ_{xy}

9.4.1 Correlation Coefficient

The value of Cov seems to be a good choice to quantify the degree of correlation of two variables. However, a snag is that covariance carries the units of the measurements in question, and its absolute size can vary a lot; these two things make it difficult to compare one situation with another. For correlation testing we therefore *normalise* covariance to the variances of the two variables, to give us the *correlation coefficient*. As usual we can define sample and population versions:

$$r = \frac{s_{xy}}{s_x s_y} \quad \text{and} \quad \rho = \frac{\sigma_{xy}}{\sigma_x \sigma_y}. \tag{9.2}$$

A handy formulation can be constructed if we write $\mathrm{d}x_i$ for the ith deviation from the mean, i.e. $\mathrm{d}x_i = x_i - \bar{x}$. Then the $1/N$ factors cancel out and the correlation coefficient can be computed as

$$r = \frac{\sum \mathrm{d}x_i \mathrm{d}y_i}{\sqrt{\sum \mathrm{d}x_i^2 \sum \mathrm{d}y_i^2}}.$$

The correlation coefficient is dimensionless, and easy to compare from one situation to another. A value of 1.0 is perfect correlation, and 0 is no correlation; intermediate values show degrees of correlation. Negative values show anti-correlation.

9.4.2 Tightness Versus Sensitivity of Correlation

It is tempting to think of the correlation coefficient as describing the "strength of correlation", but unfortunately we could potentially mean two different things by this term, as illustrated in Fig. 9.3. By "strength" we might mean how sensitive y is to x—i.e. how big an effect in y is caused by a given change in x. This might be better described as the *sensitivity* of the correlation, and has nothing to do with covariance—it is essentially the *slope* of the relationship. The correlation coefficient quantifies how tight the connection between y and x is—the variance in y at a given x. For some bivariate PDFs, the slope and the tightness can be connected, but in general they will not be. Unfortunately the scientific literature often contains references to the "strength" of a correlation without being clear whether the authors mean sensitivity or tightness—so read carefully!

9.5 Testing for Correlation

Consider the two datasets shown in Fig. 9.4. The right-hand sample is drawn at random from a correlated population (parent) distribution, whereas the left-hand sample is drawn from an uncorrelated population distribution. The population distributions are shown in grey-scale. However, if these were not shown, you would not be confident of which sample was which. The data points on the right-hand side could

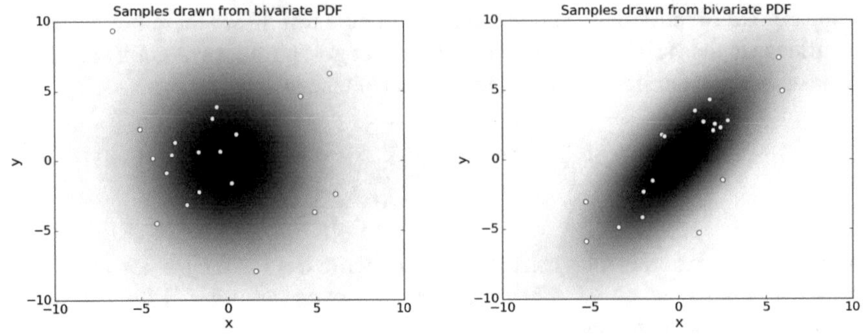

Fig. 9.4 Random samples of 20 points drawn from two bivariate PDFs. The parent distribution is uncorrelated in the left hand case, but in the right hand case has correlation coefficient $\rho = 0.75$

perhaps just be a lucky drawing from the population distribution of the left-hand side. But in that case, how likely would we be to get something convincing?

We can approach this question using the sample correlation coefficient r as a *test statistic*. If two variables are uncorrelated then the true underlying population correlation coefficient ρ will be zero. The sample correlation coefficient r is the *estimator* for ρ. This will on average be zero, but the calculated value for real specific datasets will not generally be zero of course. If we want to test the significance of an apparent correlation, what we need is the *sampling distribution* of the estimator r. We can calculate this on the basis of some null hypothesis. The simplest such hypothesis is that the two variables are drawn from an uncorrelated bivariate normal distribution, with $\rho = 0$, like the left hand side of Fig. 9.4. As discussed in Chap. 5, Sect. 5.8.1, this has a PDF given by

$$p(x, y) = \frac{1}{2\pi\sigma_x\sigma_y} \exp\left(-\frac{1}{2}\left[\frac{\mathrm{d}x^2}{\sigma_x^2} + \frac{\mathrm{d}y^2}{\sigma_y^2}\right]\right),$$

where dx and dy are the mean-centered variables. Note that when we estimate ρ with the sample correlation coefficient r, we do not know the true values of σ_x, σ_y and σ_{xy}, but have to estimate them from the data using the sample variances. This situation is just like that of Sect. 8.34, when we saw how to calculate confidence intervals for an estimated mean, in the case where we do not know the population variance in advance, but have to calculate it from the data, using the t-statistic. The analogous statistic in the 2D case is

$$t = r\sqrt{\frac{N-2}{1-r^2}}. \tag{9.3}$$

For uncorrelated variables with a bivariate normal distribution, this can be shown to follow the standard t-distribution with $\nu = N - 2$ degrees of freedom. Given values of r and N and hence t we can therefore look up the probability of getting a value of t that large or larger. Usually a two tailed test is appropriate, unless for example we know in advance that an anti-correlation is not possible. Note that for uncorrelated variables we expect $r = 0$; for perfectly correlated variables we expect $r = 1$, and for perfectly anti-correlated variables we expect $r = -1$.

9.5.1 Correlation Test with Fisher's z

What if we want to test the hypothesis of some non-zero value of r, and/or construct a confidence interval for r? If ρ is not zero, the PDF that the points follow will still be a bivariate normal, but it will have a slope, like the right hand side of Fig. 9.4. However, given a true value ρ and N data points, Fisher showed that the transformed variate

$$z = \frac{1}{2}\ln\left[(1+r)/(1-r)\right] \tag{9.4}$$

is approximately normally distributed with mean $\eta = \frac{1}{2}\ln\left[(1+\rho)/(1-\rho)\right]$ and variance $1/(N-3)$. This gives us a way to test values of r by using standard normal distribution tables.

9.5.2 Bayesian Correlation Testing

Is there a Bayesian equivalent of correlation testing? No and yes. The spirit of Bayesian methods is not to give yes/no answers, such as "is there a correlation", but rather to assess the relative probabilities of a set of hypotheses. However, if we can calculate the relative probability of different ρ values, we can use this to make a Bayesian equivalent of the traditional correlation test. To do this, we follow the methods of Chap. 8, and treat ρ as a parameter. First, we propose a prior distribution $\pi(\rho)$, for example simply taking it as flat between -1.0 and 1.0. Next, we need the joint likelihood of the observed set of data points D, given a specific value of ρ. This is the tricky part, but for now lets assume we have a way to calculate $L(D|\rho)$. Finally we repeat for different ρ values to get the likelihood distribution $L(\rho)$, and multiply by the prior $\pi(\rho)$ to get the posterior probability $P(\rho)$. Finally, we can then ask whether $\rho = 0$ is inside or outside a chosen probability region, e.g. a symmetrical 90% region centred on the mean of $P(\rho)$,

So how do we calculate the likelihood of the data, $L(D|\rho)$? The simplest thing is to assume a bivariate normal distribution, with the appropriate degree of correlation. Because we are assuming a non-zero ρ, we need to get the likelihood of each point x_i, y_i from the full blown bivariate PDF, as given in Chap. 5, Sect. 5.8.3

$$L_i(x_i, y_i) = \frac{1}{2\pi\sigma_x\sigma_y\sqrt{1-\rho^2}} \exp\left(-\frac{1}{2(1-\rho^2)}\left[\frac{\mathrm{d}x_i^2}{\sigma_x^2} + \frac{\mathrm{d}y_i^2}{\sigma_y^2} - 2\rho\frac{\mathrm{d}x_i}{\sigma_x}\frac{\mathrm{d}y_i}{\sigma_y}\right]\right).$$

where as usual $\mathrm{d}x_i, \mathrm{d}y_i$ are the deviations of the ith data point from the means. We then multiply all the L_i values together to get the total joint likelihood for the whole data set. In principle, the advantage of the Bayesian method is that we could make a variety of other assumptions about the underlying bivariate PDF. Conversely, one has to bear in mind that the probabilities we obtain only make sense in the context of the underlying data distribution we are assuming.

9.5.3 Rank Correlation Tests

The standard correlation coefficient method relies on the variables x and y being normally distributed; it also assumes that each point is drawn from the same bivariate parent distribution, i.e. all data points have the same σ_x and σ_y. In principle we could develop methods for other PDFs, and using data points with individual σ_i values, but there is a more general method of dealing with such problems. We take all the x_i values and place them in numerical order, and replace the x_i values with their ranks, $X_i = 1, 2, 3 \ldots N$. We then do the same with y; and then look to see if the *ranks* are correlated, i.e. if the top point in X is also near the top in Y etc. We will not pursue here the theory of one how calculates the distribution functions for statistics developed this way, but just quote some results. The most popular statistic is the **Spearman rank correlation coefficient**

$$r_s = 1 - 6\frac{\sum_i^N (X_i - Y_i)^2}{N^3 - N}. \tag{9.5}$$

This statistic has a range from 0 to 1. Just as with r, there exist tables which show how r_s is distributed for the null hypothesis of $r_s = 0$, so that we can perform significance tests, but we can also calculate a transformed version which follows a standard t-distribution:

$$t_s = r_s\sqrt{\frac{N-2}{1-r_s^2}}. \tag{9.6}$$

This follows a t-distribution with $\nu = N - 2$ degrees of freedom, and as usual with the t-statistic, is close to the Gaussian z-test for a reasonably large number of data points, $N \gtrsim 30$.

Another popular choice is the *Kendall rank correlation coefficient*, and another useful technique is the *permutation test*; we calculate almost any statistic we like, and then perform many random permutations (which would destroy any correlation) and see empirically how often our anomalous statistic value comes up.

9.6 Worked Example

For our real world example in Fig. 9.1 it seems fairly clear that the correlation is real, even though the large scatter makes us uncertain what the true relation is between the two variables. Let us look at a simple artificial dataset where it is not so intuitively obvious whether there is a correlation. Consider these five (x_i, y_i) data points: (2.1, 9 .6), (2.9, 5.4), (4.1, 17.5), (5.1, 9.8), (5.9, 14.5). They are plotted in the left hand side of Fig. 9.5, together with the x and y means. The points certainly seem to line up bottom left to top right. A good rough test is to see how many points lie in each of the four quadrants defined by the mean-lines. However, with such a small number of points, this arrangement could perhaps be a fluke?

The legend in Fig. 9.5 shows the sample variances calculated from the five points, and the resulting correlation coefficient, $r = 0.5$, which seems like a moderately good correlation. To assess the significance, we calculate the t statistic from equation (9.3), and then look up in a t-table to see the probability of getting that value of t or larger for $\nu = N - 3 = 3$ degrees of freedom. The result is $P(>t) = 19.6\%$. So the result is not significant at all, for the assumed null hypothesis that the points are actually drawn from a bivariate Gaussian PDF with a correlation coefficient of $\rho = 0$. With the traditional method, that is all that it is sensible to say; we would then ignore that dataset and move on.

Trying a Bayesian correlation test gives a somewhat more nuanced insight. We start by assuming that the data points are drawn from a bivariate Gaussian PDF, with some value of the correlation coefficient ρ. Of course that assumption may be wrong, but within this context, we can look at the relative performance of different values of ρ. Next, we assume a uniform prior for ρ. Then, for each assumed value of ρ, we calculate the likelihood of each data point, and multiply them together to get the joint likelihood for the dataset; then finally we multiply the prior by the likelihood to get the posterior. Note that we do not know the variances σ_x^2 and σ_y^2, so we have only relative likelihoods, and likewise have only the relative posterior P' (see Chap. 8, Sect. 8.5). However because ρ is defined over a finite interval, it is easy to numerically integrate P' and normalise. The resulting posterior probability density is shown in the right hand side of Fig. 9.5. The result is roughly consistent with the t-based significance test, in that $\rho = 0$ sits well within any sensibly chosen 90% integrated probability region, and the mean of $P(\rho)$ is exactly the same as the sample correlation coefficient r.

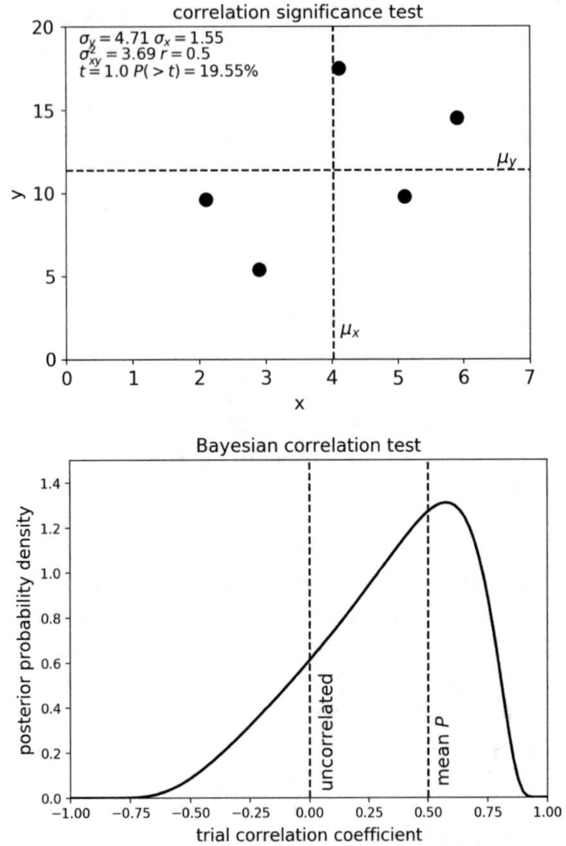

Fig. 9.5 **Upper**: Five test data points, as described in the text. The dashed lines mark the *x* and *y* means. The legend shows the variances calculated from these points, together with the correlation coefficient, *t* statistic, and the significance of the correlation. **Lower**: Bayesian correlation test for the same five data points. The posterior probability density function for ρ has been calculated, on the assumption of the points being drawn from a bivariate Gaussian PDF, and assuming a uniform prior for ρ between -1.0 and 1.0

On the other hand, you can see that the most probable value of ρ is larger than the mean value, and it is twice as likely as $\rho = 0$. The curve confirms one's instinct that the five points most likely are correlated, but you wouldn't bet your shirt on it. It is also clear that $P(\rho)$ is asymmetric; the data points are considerably more likely to be correlated than anti-correlated. Overall, $P(\rho)$ gives us much more information.

9.7 Correlation Test Pitfalls

(i) Correlation does not necessarily imply causation. The old chestnut is the claim that during the twentieth century the size of feet in China was correlated with the price of fish in Billingsgate market. These two quantities are not causally connected, but they are linked by two separate connections with a common variable — time. Finding such spurious and humorous correlations is a popular internet sport these days, but there is a serious issue too. Returning to our initial example in Fig. 9.1, is there a causal link between black hole size and galaxy size? Or have they just grown in the same environment? This is a contentious research question.

(ii) The r statistic is most sensitive to simple *linear* relationships, or more generally monotonic ones. A peaked relationship may show little or no correlation, because there is no overall tendency for large x to go with large y. An example can be seen in Fig. 9.2 in the bottom row. There does seem to be a genuine connection between the two variables, but it has $r = 0$.

(iii) As we discussed in Sect. 9.3 we need to carefully distinguish the tightness of the correlation from its sensitivity or slope. The value of r tells us the *tightness* of the correlation. It tells us nothing about whether about the relationship is a strong or weak one, just how statistically noisy it is. To measure the slope, we need the line fitting techniques that we discuss next.

9.8 Line Fitting

Suppose we have a model that describes how one variable depends on the other, so that $y = f(x)$. For example theory may tell us that the luminosity of a star should be a non-linear function of its mass, such that $L = A.M^{\alpha}$, but we don't know what the values of A and α are—they are the *parameters* of the model. In general our model $f(x)$ will be characterised by a set of parameters $\bar{\theta} = \theta_1, \theta_2, \ldots$. If we measure x, y values for a sample, can we estimate "best" values for $\bar{\theta}$? There are many possible ways to do this, most of which involve deriving a test statistic from the data set. To simplify the problem, we can start by assuming that we are in "Situation B", so that x is an independent variable, and y is the dependent variable, with predicted values $y_i = f(x_i)$. Finding the best parameters is then known, for historical reasons, as the "regression of y on x". The simplest and most popular method is *least squares fitting*, which we look at next.

9.8.1 *The Method of Least Squares*

The method of least squares fitting is a generalisation of the χ^2 minimisation technique that we used for parameter estimation. For a *single variable*, in equation (7.10) we defined

$$\chi^2 = \sum \left(\frac{x_i - x_t}{\sigma_i} \right)^2,$$

where x_t is the "true" value of x and x_i are the data values, each of which arises from an independent measurement estimating x_t, with error σ. We can see the x_i as repeated drawings from a parent distribution with mean x_t and variance σ^2.

Now, for our *two variables*, we are assuming that x is an independent variable, and that we have chosen specific values x_i at which we measure the corresponding dependent value y_i. However we also have a model $y = f(x)$ which predicts the value of y for each x_i. What we want to do is to compare the prediction with the measurement at each point. If the error on each y_i value is σ_i, what we want to do is to minimise the quantity

$$\chi^2 = \sum \left(\frac{y_i - f(x_i)}{\sigma_i} \right)^2. \tag{9.7}$$

9.9 Least Squares Fit to a Straight Line

Let us start with the simplest example, where $f(x)$ is a linear function, so that $y = a + bx$. Let us further assume that each point y_i is drawn from its own normally distributed parent distribution, where each distribution has the same standard deviation σ as all the others, but a different mean value $\mu_y = a + bx_i$. Often we simply don't have estimates of the σ_i values, and the best we can do is to assume they are all the same. In the χ^2 expression above, the constant σ then comes out of the sum and we simply need to find the minimum of the quantity

$$S^2 = \sum_{i=1}^{N} (y_i - a - bx_i)^2.$$

Minimising with respect to each of a and b we get

$$\frac{\partial S^2}{\partial a} = -2 \sum (y_i - a - bx_i) = -2 \sum y_i + 2Na + 2b \sum x_i = 0$$

$$\frac{\partial S^2}{\partial b} = -2 \sum x_i (y_i - a - bx_i) = -2 \sum x_i y_i + 2a \sum x_i + 2b \sum x_i^2 = 0.$$

So we have a pair of simultaneous equations for the unknowns a and b in terms of the data points x_i, y_i. Taking the first equation and dividing through by N, the solution for a in terms of b is

$$a = \bar{y} - b\bar{x}. \tag{9.8}$$

which tells us that the solution must pass through the point $(\bar{x}, \bar{y})$. Substituting this expression into the second equation we find

$$b = \frac{\sum x_i y_i - N\bar{x}\bar{y}}{\sum x_i^2 - N\bar{x}^2} = \frac{\sum (x_i - \bar{x})(y_i - \bar{y})}{\sum (x_i - \bar{x})^2}. \tag{9.9}$$

and of course we can substitute this expression for b back into our expression for a above. We thus have solutions for a in b in terms of sums of the data points. For historical reasons this process is sometimes known as *linear regression* and the fit obtained is known as a *regression line*.

9.9.1 Least Squares for Data Points with Individual Errors

If the each of the data points has a known and different error σ_i, then we minimise

$$\chi^2 = \sum \left[\frac{y_i - a - bx_i}{\sigma_i} \right]^2$$

with respect to a and b which gives us

$$\sum \frac{y_i}{\sigma_i^2} - a \sum \frac{1}{\sigma_i^2} - b \sum \frac{x_i}{\sigma_i^2} = 0$$
$$\sum \frac{x_i y_i}{\sigma_i^2} - a \sum \frac{x_i}{\sigma_i^2} - b \sum \frac{x_i^2}{\sigma_i^2} = 0.$$

We could proceed as we did previously, solving for for a in terms of b and then for b, but it is somewhat cumbersome, and also it is useful now to proceed a little more formally because it sets us up for more complicated examples later. Its useful to define the various sums as follows

$$S_1 = \sum \frac{y_i}{\sigma_i^2} \quad S_2 = \sum \frac{1}{\sigma_i^2} \quad S_3 = \sum \frac{x_i}{\sigma_i^2} \quad S_4 = \sum \frac{x_i y_i}{\sigma_i^2} \quad S_5 = \sum \frac{x_i^2}{\sigma_i^2}.$$

Each of the terms $S_1, S_2 \ldots$ is simply a number made of a sum over the data values and their errors. The equations now read

$$S_1 = S_2.a + S_3.b,$$
$$S_4 = S_3.a + S_5.b.$$

The standard way to solve such a pair of linear equations is by the method of determinants. The solution is

$$\Delta = \begin{vmatrix} S_2 & S_3 \\ S_3 & S_5 \end{vmatrix} = S_2 S_5 - S_3^2 \tag{9.10}$$
$$a = \frac{1}{\Delta} \begin{vmatrix} S_1 & S_3 \\ S_4 & S_5 \end{vmatrix} = \frac{1}{\Delta} (S_1 S_5 - S_4 S_3)$$
$$b = \frac{1}{\Delta} \begin{vmatrix} S_2 & S_1 \\ S_3 & S_4 \end{vmatrix} = \frac{1}{\Delta} (S_2 S_4 - S_1 S_3)$$

The procedure then is that first we combine the data values to form the sums $S_1, S_2, \ldots$, then we combine those values with the equations above, and this gives us the "least squares" solution for the parameters a and b.

9.10 Testing the Fit Using χ^2

The procedure above gives the best fit, but is the fit statistically acceptable? To test the absolute quality of the fit, we need to have values for the errors on each data point, σ_i, so that we can calculate χ^2, as in equation (9.7). Because we have estimated two parameters from the data, the remaining degrees of freedom is $\nu = N - 2$. We can assign a probability to the fit, $P(> \chi^2)$, based on the χ^2 distribution for that value of ν, and then possibly reject the fit if $P < 5\%$ or whatever, depending on our chosen confidence level. As with our previous use of χ^2, it is standard for χ^2 fit tests to consider only a one-tailed test—a bad fit will always give a large χ^2. If the model is wrong, this will not make the data points cluster nearer to the regression line than they ought to.

A χ^2 value that is unexpectedly large may occur (i) simply by chance, (ii) because the model is wrong, or (iii) because the errors have been underestimated. The third possibility should always be considered!

A χ^2 value that is unexpectedly small may occur (i) simply by chance, (ii) because the errors have been overestimated. Although the more common failing is *under*estimating your errors, *over*estimating errors does happen more often than you might think...

9.10.1 *Errors on Fitted Parameters*

We can compute the errors on our best fit parameter estimates using the propagation of errors formula. Because we are treating x as the independent variable, with no errors, then in this case $\sigma_x = \sigma_{xy} = 0$. Above we solved for a and b as in terms of x_i, y_i, σ_i, involving sums over all N values of i. All the σ_i values contribute to the error in a. The net error in a is

$$\sigma_a^2 = \sum \left[\sigma_i^2 \left(\frac{\partial a}{\partial y_i} \right)^2 \right]$$

Plugging into $a = \frac{1}{\Delta}(S_1 S_5 - S_4 S_3)$ from above, the algebra is straightforward but tedious...Sparing you the details, we end up with the pleasantly simple result

$$\sigma_a^2 = \frac{1}{\Delta} \sum \frac{x_i^2}{\sigma_i^2}, \qquad \sigma_b^2 = \frac{1}{\Delta} \sum \frac{1}{\sigma_i^2}. \tag{9.11}$$

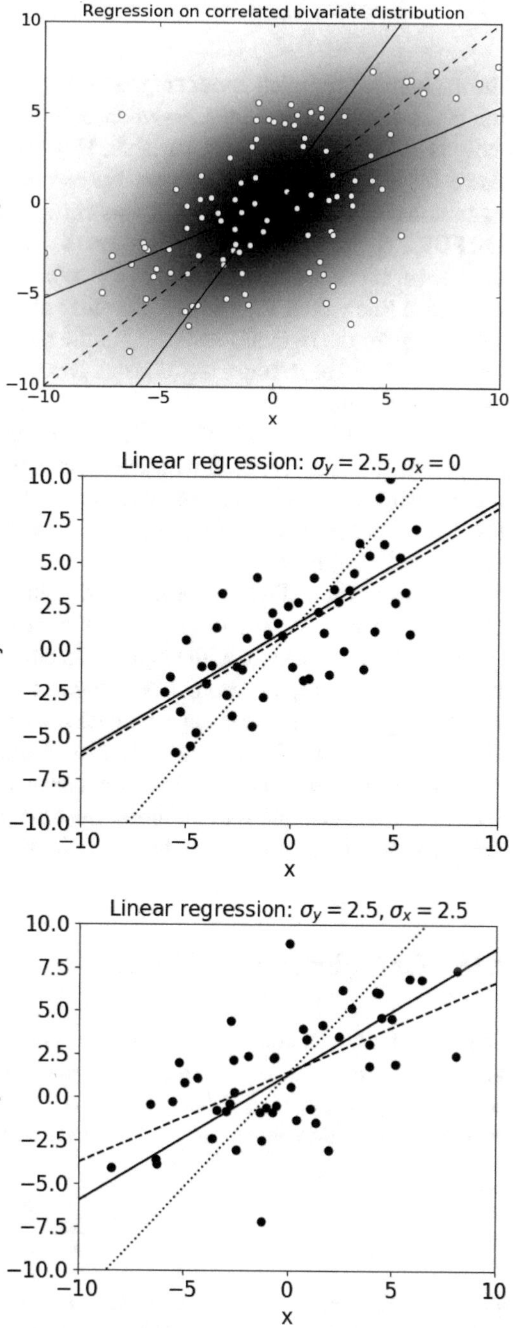

Fig. 9.6 Regression analysis on three simulated datasets. In each case the middle line shows the true input relationship, and the other two lines show the result from regressing y on x and x on y respectively. See text for further details

9.10.2 Regression Model Consistency Checking

The mathematics applied in our "regression analysis" assumes that we are in "situation B" as described in Sect. 9.2—where x is an independent, not random, variable, and y is a dependent, random, variable. However, line fitting is sometimes applied where this is not a correct assumption. For example, the upper plot in Fig. 9.6 shows a regression analysis on a sample of points drawn from a bivariate but correlated Gaussian PDF, i.e. corresponding to "Situation A". When we regress y on x, the resulting line does not follow the major axis of the underlying PDF. If we try regressing x on y we get a line which is wrong in the other direction. In fact, the bisector of these two lines gives the right answer. The middle plot shows a simulation of Situation B. First, a fixed grid of x values was chosen. Next, predicted y values were chosen with a straight line function $y = ax + b$. Finally, the predicted y values were randomly perturbed with a Gaussian PDF. The line resulting from a regression analysis of y on x is shown, and it agrees well with the input straight line relationship. A regression of x on y however gives an incorrect slope. It is hard by eye to distinguish the point distribution in the upper and middle plots, which emphasises how important it can be to perform this check.

The lower plot of Fig. 9.6 shows a third situation, where there is a straight line relationship between the two variables, but there is a random scatter on *both* variables. This shouldn't be confused with Situation A, where the points share a common mean; in the right hand plot, each point has its own underlying mean. This third situation is in fact a very common one in experimental science, with errors on both quantities. In this case, like in the common-mean case, regressing y on x gives the wrong answer, as also does regressing x on y. The correct answer is given by the bisector of the two lines. Another technique is to minimise the othogonal distance from each data point to the line under test, rather than the vertical Δy_i distance.

9.11 Curvilinear Line Fitting

The next step up in complication from a straight line is to consider a polynomial $y(x) = a_1 + a_2 x + a_3 x^2 + \cdots a_m x^{m-1}$. This is a curved line, but it is linear in the parameters $a_1, a_2, a_3, \ldots$, and so may be referred to as "curvilinear". In fact, more generally, we could consider any set of m functions $f_k(x)$ such that

$$y(x) = \sum_{k=1}^{m} a_k f_k(x),$$

as long as the the parameters a_k do not appear inside the functions $f_k(x)$ and the combination of the functions f_k is linear in the parameters a_k. So for example, the $f_k(x)$ could be $f_1 = 1$, $f_2 = x$, $f_3 = x^2$ etc; or could be $f_1 = \sin x$, $f_2 = \sin 2x$, etc. We can always form

$$\chi^2 = \sum_{i=1}^{N} \frac{1}{\sigma_i^2} (y_i - y(x_i))^2 .$$

Note that if we use the data to estimate the m parameters a_k, the resulting χ^2 will be distributed with degrees of freedom equal to $\nu = N - m$. To find those parameters, we require that $\partial\chi^2/\partial a_k = 0$ for each a_k, and so get m coupled linear equations. As with the straight line, there is an analytic solution to these equations, but as we increase m the solution gets much more complicated. Lets look at the simplest step beyond a linear fit, the quadratic with $y(x) = a_1 + a_2 x + a_3 x^2$. If we manually work through the solution, it turns out to be:

$$a_1 = \frac{1}{\Delta} \begin{vmatrix} \sum y_i \frac{1}{\sigma_i^2} & \sum \frac{x_i}{\sigma_i^2} & \sum \frac{x_i^2}{\sigma_i^2} \\ \sum y_i \frac{x_i}{\sigma_i^2} & \sum \frac{x_i^2}{\sigma_i^2} & \sum \frac{x_i^3}{\sigma_i^2} \\ \sum y_i \frac{x_i^2}{\sigma_i^2} & \sum \frac{x_i^3}{\sigma_i^2} & \sum \frac{x_i^4}{\sigma_i^2} \end{vmatrix}$$

$$a_2 = \frac{1}{\Delta} \begin{vmatrix} \sum \frac{1}{\sigma_i^2} & \sum y_i \frac{1}{\sigma_i^2} & \sum \frac{x_i^2}{\sigma_i^2} \\ \sum \frac{x_i}{\sigma_i^2} & \sum y_i \frac{x_i}{\sigma_i^2} & \sum \frac{x_i^3}{\sigma_i^2} \\ \sum \frac{x_i^2}{\sigma_i^2} & \sum y_i \frac{x_i^2}{\sigma_i^2} & \sum \frac{x_i^4}{\sigma_i^2} \end{vmatrix}$$

$$a_3 = \frac{1}{\Delta} \begin{vmatrix} \sum \frac{1}{\sigma_i^2} & \sum \frac{x_i}{\sigma_i^2} & \sum y_i \frac{1}{\sigma_i^2} \\ \sum \frac{x_i}{\sigma_i^2} & \sum \frac{x_i^2}{\sigma_i^2} & \sum y_i \frac{x_i}{\sigma_i^2} \\ \sum \frac{x_i^2}{\sigma_i^2} & \sum \frac{x_i^3}{\sigma_i^2} & \sum y_i \frac{x_i^2}{\sigma_i^2} \end{vmatrix}$$

$$\Delta = \begin{vmatrix} \sum \frac{1}{\sigma_i^2} & \sum \frac{x_i}{\sigma_i^2} & \sum \frac{x_i^2}{\sigma_i^2} \\ \sum \frac{x_i}{\sigma_i^2} & \sum \frac{x_i^2}{\sigma_i^2} & \sum \frac{x_i^3}{\sigma_i^2} \\ \sum \frac{x_i^2}{\sigma_i^2} & \sum \frac{x_i^3}{\sigma_i^2} & \sum \frac{x_i^4}{\sigma_i^2} \end{vmatrix}$$

This is probably enough to convince anybody not to try the cubic solution by hand! Programming routines exist to solve coupled equations of more or less arbitrary size, but at this point, most people would simply resort to numerical methods.

9.12 Arbitrary Function Fitting

What if the function $f(x)$ that we wish to fit is not curvilinear in the sense defined above? The function will still be characterised by a set of parameters a_k, so that the problem is to find the values of a_k which mimimise χ^2. However, in general, there will not be an analytic solution to the best-fit parameters, and we need to resort to the general-purpose model fitting techniques we discuss in Chap. 10.

Sometimes however a non-linear problem can be transformed into a linear one by a suitable transformation of variables. The most common example is wishing to fit a dataset with an exponential form, e.g. $y = ae^{-bx}$. If we take logs then $\ln y = \ln a - bx$. If we apply the same transformation to our y_i values, then we can fit a straight line to those transformed data, i.e. we calculate

$$\chi^2 = \sum \left[\frac{1}{\sigma_i'^2} [\ln y_i + \ln a - bx_i]^2 , \right]$$

where we need to use also the transformed errors

$$\sigma_i' = \frac{\mathrm{d}(\ln yi)}{\mathrm{d}y} \sigma_i = \frac{1}{y_i} \sigma_i .$$

9.13 Key Concepts

Some of the key concepts from this chapter are:

- The subtle distinction between dependence and correlation
- The idea of covariance as a generalisation of variance
- The links between covariance, correlation, and dependence
- The definition of correlation coefficient r as as normalised covariance
- The use of the t-transform of r to test the null hypothesis of $r = 0$
- The use of the Fisher's z-transform of r to test non-zero values of r
- The idea of using ranks rather than values for a robust test of correlation
- The idea of predicting y as a function of x, and minimising the scatter (i.e. χ^2 between observed and predicted y values)
- How to find a best fit line by minimising χ^2 with respect to several parameters simultaneously
- The idea of a curvilinear function as one linear in its parameters but not necessarily linear in x
- The idea of linearisation to turn a hard problem into an easier one.

The key formulae from this chapter are: a formula calculating the sample covariance (9.1); two formulae defining population and sample correlation coefficients (9.2); the t-transform (9.3) and z-transform (9.4) of the correlation coefficient; the Spearman rank correlation coefficient (9.5) and its t-transform (9.6); the definition of χ^2 for function fits (9.7); the solutions for the parameters of the best fit straight line where all errors on the points are the same ((9.8) and (9.9)), and where the errors are all different (9.10) and formulae for the errors on fitted parameters (9.11).

9.14 Further Reading

Standard correlation testing, and least-squares line fitting is covered in many standard statistics textbooks (see the Chap. 1 references). A more in-depth look at line-fitting can be found in Bevington (2002), which is a very good guide to data analysis in general. Assessing the significance of the correlation coefficient relies on the assumption that the variables have a bi-variate Gaussian distribution. If you want more detail on multi-variate Gaussians, a good place to start is the relevant wikipedia page. As usual, it may be a bit more maths-y than you want, but there are good links.

The Bayesian approach to correlation-testing is not in normal textbooks. There is a quite a nice blog post on this topic by Rasmus Bååth.

In this chapter, we treated the non-parametric approach to correlation testing rather briefly. Actually, this is an important topic for physical scientists, as often the assumption of Gaussian distributions is a poor one, and indeed sometimes it is very hard to assign meaningful errors at all. In these circumstances, assigning ranks, and looking at the sampling distributions of statistics calculated from those ranks, is about all one can do. As well as correlation testing, scientists often want to test whether two sample distributions are consistent with each other. Non-parametric tests such as the Mann-Whitney U-test and the Kolmogorov-Smirnov test are very useful for this task. For many many years the bible in this area has been Siegel and Castellan (1988), which in turn was derived from a 1956 text by Siegel. It is still a very clear, simple, and practical book. A more modern book covering similar material is Corder and Foreman (2014), and a more advanced comprehensive survey is given by Wasserman (2007).

The concepts and techniques of correlation and "regression" testing were first developed by Francis Galton in the late 19th century, and then improved and refined by Karl Pearson and Ronald Fisher (but see the earlier references given in Stanton (2001)). All three of these of giants of statistics were keenly interested in eugenics—the idea that you can and should improve the genetic quality of a population—and Galton actually coined the term. Eugenics was controversial at the time, but is even more so now, with its subsequent connections with Nazism and racism more generally. This somewhat unsavoury history is also the reason for the odd term "regression" to describe looking for statistical connections between variables. It arises from the fact that these early workers were keenly interested in heritability, and "regression towards the mean". The idea they were testing was that, for example, tall people tend to have tall children, so you see a correlation if you plot heights of mothers versus daughters, but that—unless you control breeding—over successive generations the connection will weaken.

9.15 Exercises

9.1 In the notes, it was stated that a perfect linear correlation should give $r = 1$. Can you prove this? Assume that $y = a + bx$ is exactly true for all points x_i, y_i, and then consider the definition of r.

9.2 A small group of children take exams in maths, physics, and art. Their scores are given in the table below. Is there any correlation between (a) maths and physics, and (b) maths and art?

Student	Maths mark	Physics mark	Art mark
A	41	36	38
B	37	20	44
C	38	31	35
D	29	24	49
E	49	37	35
F	47	35	29
G	42	42	42
H	34	26	36
I	36	27	32
J	48	29	29
K	29	23	22

9.3 Using the maths and physics exam data above, calculate the best fit linear regression line between maths and physics, (i) taking maths as the independent parameter, and (ii) taking physics as the independent parameter. Assume that all points have the same errors. Are these slopes the same?

9.4 Experienced markers say that the standard deviation on a mark is 5. Assuming this is right, calculate the error on the slope for the case of regressing physics on maths and comment. Is the difference seen between the slopes in the previous exercise significant?

9.5 Suppose we have a dataset x_i, y_i. Fitting the regression line of y on x, i.e. assuming x is the independent parameter, gives a solution $y = a + bx$; regressing x on y gives a solution $x = c + dy$ which corresponds to $y = a' + b'x$. How do the two slopes b and b' relate to the correlation coefficient?

9.6 Children from two different catchment areas are tested on maths and history and the marks tested for correlation. A sample of 75 children from Richville gives

$r_1 = 0.67$; a sample of 63 children from Poortown gives $r_2 = 0.42$. Is there any evidence that these correlations are different?

9.7 Problem 9.2 asked if there was any evidence of correlation between Maths and Arts scores. You probably examined this by using the standard correlation coefficient. However, this relies on the data points having Gaussian errors, which may not be the case here. Is the result any different if we try using Spearman's rank correlation coefficient? (On the other hand, if you already used the Spearman coefficient, do it again now with the standard correlation coefficient!)

References

Bevington, P., Robinson, D.K.: Data Reduction and Error Analysis for the Physical Sciences, 3rd edn. McGraw-Hill Education, New York (2002)

Corder, G.W., Foreman, D.I.: Nonparametric Statistics: A Step-by-Step Approach, 2nd edn. Wiley, Hoboken (2014)

Magorrian, J., et al.: The demography of massive dark objects in galaxy centers. Astron. J. **115**, 2285–2305 (1998)

Siegel, S., Castellan, N.J.: Nonparametric Statistics for the Behavioral Sciences, Revised 2nd edn. McGraw-Hill Publishing Company, New York (1988)

Stanton, J.M.: Galton, Pearson, and the Peas: a brief history of linear regression for statistics instructors. J. Stat. Educ. **9**(3) (2001)

Wasserman, L.: All of Nonparametric Statistics, Revised Corrected 3 printing edn. McGraw-Hill Publishing Company, New York (2007)

Websites (all accessed March 2019):

Blog post by Rasmus Bååth on Bayesian correlation testing. http://www.sumsar.net/blog/2013/08/bayesian-estimation-of-correlation/

Amusing list of spurious correlations by Tyler Vyglen. http://tylervigen.com/spurious-correlations

Wikipedia page on multi-variate Gaussians. https://en.wikipedia.org/wiki/Multivariate_normal_distribution

Chapter 10
Model Fitting

10.1 Outline of Content

- Concepts of model fitting
- Numerical grid searching
- Constructing parameter intervals in multiple dimensions
- Goodness of fit
- Comparing models
- Machine Learning techniques

Fitting models to data is the daily work of modern physics. Is the Standard Model of particle physics consistent with the data from the several instruments installed at the LHC, and if so, can it tell us the mass of the Higgs Boson? How do we go about answering a question as complex as this? We have laid the groundwork in earlier chapters by seeing how we can test a hypothesis, estimate a parameter, and construct an interval expressing our uncertainty; and how to fit a line where two variables are connected. We now need to generalise to multiple parameters and complex datasets. This generalisation brings in some interesting complications. Some of these complications are about numerical problems and computational technique, especially how to locate the "best" parameter set efficiently. We will deal with those technical issues briefly. However, moving up to two or more dimensions in parameter space also produces some new conceptual issues—how to locate contours containing a required amount of probability, or of confidence, in multi-dimensional space; the difference between interesting and uninteresting parameters; and how to account for correlations between parameters. We will start by spelling out the logical steps in model fitting.

A. Lawrence, *Probability in Physics*, Undergraduate Lecture Notes in Physics,
https://doi.org/10.1007/978-3-030-04544-9_10

10.2 The Logic of Model Fitting

What is a model? Its a kind of sausage machine that accepts input data values and predicts output data values, which can be compared to the measured values. When we make that comparison, the measured value is usually a random variable, so we have to know its probability distribution. The logic of the model fitting process is illustrated in Fig. 10.1. The discussion that follows is a little abstract, but hopefully all will become clearer in the worked example in the following section.

10.2.1 Input and Output Data

In earlier chapters we considered first a single data point x, and then a compound dataset of N data points x_i, which in the case of significance testing required us to invent the idea of a test statistic computed from the data. Next we looked at bivariate datasets (x_i, y_i), usually with the x_i being the independent variable, and y_i being the dependent variable, and looked at how to find our best estimate of functional dependence between them, $y = f(x)$. This problem is only analytically solvable for a limited range of cases. More generally, we might have multiple independent variables and multiple dependent variables—for example we might set our equipment to a variety of settings for both voltage and time delay, and at each pair of settings measure both particle mass and particle velocity. We can represent the collection of input data values symbolically as a vector $\bar{x}_{\text{in}}$. We then make a series of experimental measurements that produces a set of observed output data values which we can likewise represent as $\bar{y}_{\text{obs}}$.

10.2.2 The Model

The model itself is a mathematical machine for predicting output values $\bar{y}_{\text{pred}}$ given input values $\bar{x}_{\text{in}}$. This may be a single equation, or a set of steps with multiple

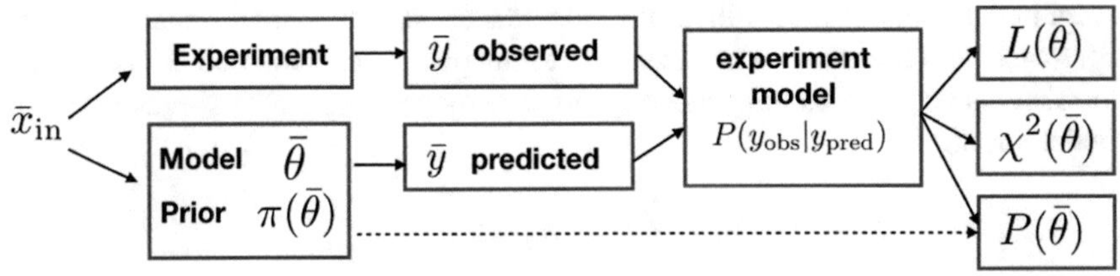

Fig. 10.1 Illustrating the steps in the process of model fitting

equations. All that matters is that it is an algorithmic and quantitative process which will produce definite predictions.

It is possible for a model to be a fixed process, but it is more normal for it to be defined by a set of tuneable parameters. If there are m parameters, the set of values $(\bar{\theta} = \theta_1, \theta_2, ...\theta_k, ..., \theta_m)$ can be seen as a point in an m-dimensional space, or as a *parameter vector*. Some of the parameters may be crucial to the physical theory we are trying to test, whereas others may be uninteresting or incidental parameters, that we nonetheless have to estimate—for example parameters characterising the behaviour of our detector. As we shall see below, the number of parameters, and how many of these are interesting/uninteresting, makes a big difference to how we assess the errors on our parameter estimates.

If we take a Bayesian approach, then our model needs not just a list of parameters $\bar{\theta}$, but a *prior probability distribution* for each parameter, $\pi_k(\theta_k)$. From these we can make the *joint prior distribution* $\pi(\bar{\theta})$. Note that this is a multi-variate probability density, such that $\pi(\bar{\theta})d\bar{\theta}$ is the amount of probability in the box $\mathrm{d}\bar{\theta} = \mathrm{d}\theta_1 \mathrm{d}\theta_2...$

For simplicity, in most of what follows, we will drop the "bar" notation, and just write θ as shorthand for the complete set of parameters $\bar{\theta}$.

10.2.3 The Experiment Model

In order to test our model, and obtain parameter estimates and errors, it is not enough to predict each output value y. We also need to know the probability distribution of a given measurement, given the prediction, $p(y_{\mathrm{obs}}|y_{\mathrm{pred}})$—or in other words the likelihood of the data. To know this, we have to have a clear understanding of our experimental process. Most typically, this means associating an *error* σ_y with each data value, which we take to be the variance of a Gaussian distribution. We can then calculate a joint likelihood $L(\bar{y})$ for the complete dataset.

Deriving the data errors can be a non-trivial process, for two reasons. Firstly, the data values we use are often heavily processed, rather than the "raw" measurements. We need to know the mathematical processing that was undertaken to arrive at our data values, and use the appropriate transmission-of-errors formulae from Chap. 2, Sect. 2.6. Secondly, we need to understand how our equipment works. If we are using an NMR machine, or a confocal microscope, or a large reflecting telescope, this may be non-trivial, and requires an *experiment model*, with its own parameters that may be imperfectly known. For example, we know we should allow for the reflectivity of our mirror, but believe that this has been degrading with time. What was the reflectivity on the day the measurement was made?

Finally, we should be aware that the errors will not *always* be Gaussian. The probability of a given data value might for example be subject to the Poisson distribution, or the Lorentz distribution. Bayesian methods can cope with other distributions, at the cost of numerical complexity, but note that the χ^2 method assumes data points with Gaussian errors.

10.2.4 Testing the Model

The general idea is to roam over parameter space, testing how the model compares to the data—a multi-dimensional version of the process we stepped through in Chap. 8. Each point in our parameter space θ corresponds to a testable hypothesis. At that point, we will have predicted values to compare to the observed values, and a prior probability for that set of parameter values. We can then assign a test-value to that parameter-space point in several different ways—we could compute the χ^2 value, the likelihood, the posterior probability, or potentially other quantities, such as the "negative log likelihood". These things are all closely connected. Posterior is (normalised) likelihood times prior, and for normally distributed data points, χ^2 is closely related to the log of likelihood (see equation (8.13)). Sometimes the quantity $e^{-\chi^2/2}$ is used, which is proportional to likelihood.

We then repeat over a grid of parameter space points and make a *map* of $P(\theta)$ (or $\chi^2(\theta)$ or $L(\theta)$), where as usual θ is shorthand for the complete set of parameters θ_1, θ_2.... The problem then is to take the multi-dimensional surface $P(\theta)$ etc and find the point in parameter space that is "best" in some sense. This will normally mean locating maximum posterior, maximum likelihood, or minimum χ^2. Following this, we want to construct a *region* in parameter space that expresses our uncertainty on where the best point is—for example that encloses an integrated posterior probability of 90%. Finally, we will want to know not just where the "best" point is in the context of the model, but whether the model itself is any good at all. All of these questions have both conceptual and numerical complications, which we will address in the coming sections. But first, let's look at a concrete example.

10.3 Mapping Out Parameter Space: Worked Example

Figure 10.2 shows a real world example, taken from work by my own research group. Working through how we deal with this will illustrate the main features of the process, but will also bring out some common problems in model fitting.

10.3.1 The Data and Its Errors

The middle plot in Fig. 10.2 shows the light-curve of a quasar, i.e. how it's brightness changes with time. A quasar is the centre of a distant galaxy that has a very luminous nucleus, thought to be powered by the accretion of matter onto a supermassive black hole. All such quasars tend to be variable—their brightness fluctuates erratically up and down by a few tens of percent. However the event shown in Fig. 10.2 is unusual in that it shows an increase in brightness by a large factor over several years. Each data point comes from an observation with one of several different telescopes on

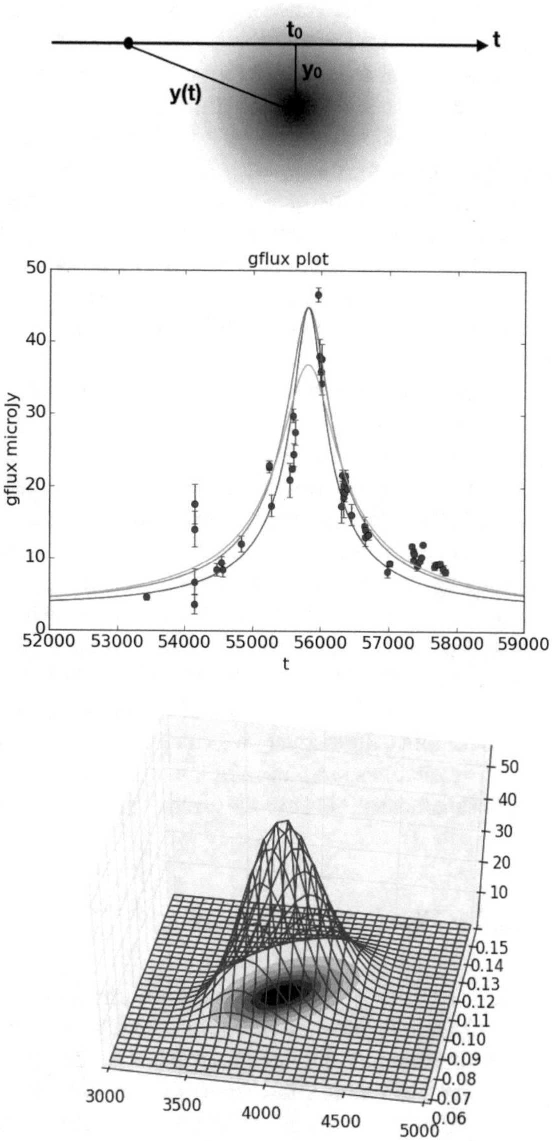

Fig. 10.2 Fitting a micro-lensing model to an outburst in a quasar. The **upper plot** illustrates the physical nature of the model. The grey-scale blob indicates the lensing effect caused by an intervening star, decreasing with angular distance from the star in the plane of the sky. The horizontal line indicates the track of the quasar relative to the star. The **middle plot** shows the observed brightness of the quasar versus time, with errors estimated during the data reduction process. The axis is in days (Modified Julian Date). The curves show brightness versus time predicted by the lensing model, for three different choices of the parameters of the model. The **lower plot** shows the calculated (relative) posterior probability versus a grid of values for the two key parameters—the Einstein timescale t_E, and the impact parameter y_0

a specific night. The time of the observation t_i can be taken as the independent variable, and the measured flux F_i as the dependent variable. (Actually, the quasar was measured at several different wavelengths each night, but we will simplify by looking at one specific wavelength). The raw data is a CCD image with some number of electron-counts in each pixel. We first sum up neighbouring pixels over a small region surrounding our quasar. The error on this total count is the Poisson error. To find the flux of the quasar in SI units, we repeat the process for other stars in the image that have already been measured by a previous survey, and get a scaling factor. The Poisson variance is scaled by the same factor. The counts for the quasar are large enough that we can treat the Poisson variance as a Gaussian error, although this approximation becomes a little poor for some of the points with the largest error bars.

The curves drawn through the data points relate to the model we will discuss below. You can see that this seems to be a fairly good overall description of what is going on in the observations. However, the points are scattered about the smooth curve more than you would expect given the estimated error bars. Sometimes this is a sign that our experimental errors have been underestimated. However, in this case, it is likely that it is because there are two separate reasons why the quasar is changing in brightness; something is causing a long-slow outburst over years, but on top of this we have smaller, month-to-month erratic variability that has a different physical cause. The long-slow outburst is the phenomenon we are trying to understand. Ideally our model would include *both* effects, but the problem is that we don't really understand the month-to-month erratic variability, so we don't know how to do this. Alternatively we can treat the erratic variability as an extra "cosmic noise", estimate its size, and add it in quadrature to our experimental errors. This is also unsatisfactory, because the erratic variability may not be a Gaussian process.

10.3.2 The Physical Model

There are a variety of possible explanations for the long-slow quasar outburst, but we would like to test a specific idea. The theory is that a second galaxy is in the line of sight between us and the quasar. Most of the time this has little effect, but as the stars in the intervening galaxy move around, occasionally the quasar will be in almost perfect alignment with one specific star. The gravity of the star can then bend the light from the background quasar, and produce a focusing effect known as gravitational lensing, which has the effect of magnifying the light from the quasar. The upper plot of Fig. 10.2 illustrates the strength of the magnifying effect as a function of the distance from the foreground star (in angular terms on the plane of the sky) as a greyscale. If, as we expect, the quasar and star are in relative motion, we can see how moving along the track indicated would cause an apparent brightening and then fading with time.

The scale of the magnifying region, as seen on the plane of the sky, is characterised by an angular scale known as the *Einstein angle* θ_E which depends on the geometry of the situation. We don't know how fast the star and quasar are moving relative to each other, but we can characterise this by the *Einstein timescale* t_E which is the time taken to cross the Einstein angle. We can then set up the situation in terms of the time relative to the Einstein timescale, t/t_E, and the angular distance between the star and the quasar in units of the Einstein angle, $y = \theta/\theta_E$. The physics of gravitational lensing then says that the magnification factor at position y will be

$$\mu(y) = \frac{y^2 + 2}{\sqrt{y(y^2 + 4)}}.$$

From Fig. 10.2 you can see however that y will be changing with time following

$$y(t) = \sqrt{y_0^2 + (t - t_0)/t_E},$$

where y_0 is the *impact parameter*, or distance of closest approach. Finally, the brightness (flux) of the quasar will be given by

$$F(t) = F_0\mu(t),$$

where F_0 is the baseline flux, i.e. the flux of the quasar before the event. The model then predicts the flux of the quasar as a function of time, through the three equations above, with four adjustable parameters—F_0, t_0, y_0, and t_E. If we determine the values of those four parameters which give a model that best describes the data, we can then see what they imply in terms of other astrophysical parameters, such as the mass of the star, the distance to the star, and the distance to the quasar.

10.3.3 The Calculation

The middle plot shows three curves defined by different example values of the four parameters above. You can see that they are all roughly plausible, but the red curve looks better than the green curve. For any one of those curves we can calculate the joint likelihood of all 55 data points. Given that we are assuming the errors are Gaussian, we could alternatively calculate χ^2. Or we could decide priors for the parameters and calculate the posterior. We can then in principle calculate our chosen metric over a grid of values for each of the parameters F_0, t_0, y_0, and t_E, and examine its behaviour. Of course it is hard to visualise a multi-dimensional surface. In our case the parameters y_0 and t_E are getting at the essence of the physical problem, whereas t_0 and F_0 seem rather accidental and uninteresting. In the lower plot of Fig. 10.2 we have fixed the values of t_0 and F_0 at reasonable values estimated by inspection, and show the variation of posterior probability against the other two parameters, $P(y_0, t_E)$. For

t_E we took a uniform prior between minimum and maximum values; for y_0 we used a prior weighted linearly towards larger values of y_0, with a maximum value. There is a lot of data, so that the shape of the surface was in fact not very sensitive to the priors.

The resulting $P(\theta)$ is smooth and fairly close to a bivariate Gaussian. The position of the peak of P seems fairly clear. In the next section we will discuss briefly techniques for locating peaks objectively, along with the uncertainties. Note that the parameters y_0 and t_E are strongly correlated—the 2D shape is tilted in the parameter space. This makes a big difference to estimating our errors, as we will see later. In principle we should calculate a four dimensional surface for all the parameters, but of course this is quite hard to visualise.

You will note that the z-axis is in units of relative P value, rather than normalised probability density. This is because of two difficulties, both of which are fairly typical in model fitting. The first difficulty is knowing the correct probability distribution for the data values. As discussed in Sect. 10.3.1 the data points are scattered more than you would expect from the estimated errors, which could be because of an additional variability not included in our model. If we calculate χ^2 rather than P, we find $\chi^2 \sim 180$ rather than the $\sim$55 we might expect for a good model. We can artificially increase σ_y to account for this, using $\sigma^2_{\mathrm{net}} = \sigma^2_y + \sigma^2_{\mathrm{vblty}}$. However, because we don't have a model for the underlying erratic variability, it is hard to pick an objective value for σ_{vblty}. The choice of σ_{vblty} changes the absolute values of P but changes the shape of the surface very little.

The second reason for using relative values of P is sheer numerical difficulty. To get the normalising factor E we need the multidimensional version of equation (8.8)

$$E = \int L(\theta)\pi(\theta)\mathrm{d}\theta.$$

In Chap. 8, Sect. 8.5.3, we already noted that calculating E even in one dimension can be problematic. When we move into multiple dimensions, and large numbers of data points, it can be close to impossible. Also, for large numbers of data points the likelihood will be a very small number, so programs usually work in log space or re-scale as they go along.

10.4 Techniques for Finding the Best Parameter Set

The "best" value of the parameter vector θ can be taken to be where we find the maximum of P, or the minimum of χ^2. How do we locate this point? We will look briefly at types of technique. We will talk in terms of maximum P, but the same logic applies to maximum L or minimum χ^2.

10.4.1 Grid Search Methods

The simplest thing is to map a complete grid on a reasonably fine scale, and record the largest value. For nearly all practical problems, this is far too slow, even on big computers. One way to speed up is to start with a few coarse cells, locate the cell with largest P, then subdivide this cell into smaller cells and repeat, gradually zooming in to an accurate location of the maximum. A danger in this approach is that the absolute maximum might be located within a broader minimum. A better method is the **zig-zag search**, which for χ^2 methods is usually known as a ravine search. Consider being on the side of a hill but not yet at the very top. If you move along one parameter while holding the others constant you will cross a kind of local maximum. The trick then is to locate the highest point in your neighbourhood, then pick another parameter in an orthogonal direction and repeat. The result will be a kind of zig-zag towards the overall maximum, and the number of calculations will be much smaller than for a complete grid-search.

10.4.2 Gradient Searches and Related Methods

The zig-zag method follows the hill up towards the overall maximum, but in a zig-zagging manner. The idea of the gradient search method is find the direction of steepest ascent in a small region surrounding the current location, and then take the next step in that direction. The components of the gradient are calculated from $\partial P/\partial\theta_j$ along each parameter direction, but scaled to make each parameter step dimensionless. Near the top the gradient approaches zero, so it is best to overshoot and then use the Gaussian/parabolic approximation to locate the maximum, as described in the next sub-section. There are clever variations on the gradient search method which don't need explicit calculation of the derivatives. One of these is the "amoeba" method. For m parameters, this calculates P at a cluster of $m + 1$ points. The point with maximum P is found, and then the centroid of the remaining points. The direction moved is the line joining these two. Another very popular method is the "Monte Carlo Markov Chain (MCMC)" method. This again starts with a cluster of $m + 1$ points, calculates the covariance matrix of recent points (see below) and takes a random step based on that covariance matrix. The idea of a "Markov Chain" will be explored more generally in Chap. 12.

10.4.3 Sub-grid Maximum Location

In any one-dimensional slice, we can locate the local maximum using the Gaussian/parabolic method of Chap. 8, Sect. 8.5.2. We refer to this as "sub-grid location" because the solution can be calculated to greater accuracy than the spacing of the grid points. Within the context of the grid search/gradient search methods described above, we can take one parameter at a time, use three data points, solve for the local maximum, and iterate. We could also in principle construct a solution for the parameters of a paraboloid in m dimensions and solve in one go.

10.5 Constructing Uncertainty Intervals

To simplify the discussion of constructing parameter intervals, we will discuss the case of $m = 2$ parameters, which we will label $\theta_1 = a$ and $\theta_2 = b$. The two parameter case brings out all the key features, and it is not too hard to generalise to more parameters. Likewise, working with posterior probability is somewhat easier to understand than working with χ^2, so we will look at that first.

Having found the point (a_0, b_0) where $P(a, b)$ is a maximum, what we want to do next is is to construct a contour which encloses the required amount of integrated probability. There may in general be many such contours, and they may not even be closed loops. Traditionally what we do is to find the ellipse centred on (a_0, b_0) which approximates a contour enclosing the right amount of probability. However, as discussed in Sect. 10.3.3, very often in practice we are not able to calculate the marginal likelihood E in order to get a properly normalised probability density P, and so need to work with $P'(a, b) = L(a, b)\pi(a, b)$. If we are not sure how to normalise our priors, we may be working with a quantity which in turn is only proportional to P'. Just as in the $1D$ case, we can get round this lack of normalisation by assuming that P' is Gaussian.

10.5.1 The Gaussian Approximation

We assume that we already have the peak (a_0, b_0), but now we want the variances[1] which describe the bivariate Gaussian which approximates the surface $P(a, b)$. Generalising the "curvature" method of Sect. 8.5.2, we can find the variance in each axis by

$$\sigma_a^2 = \left(\frac{\partial^2 \log P}{\partial a^2}\right)^{-1} \qquad \sigma_b^2 = \left(\frac{\partial^2 \log P}{\partial b^2}\right)^{-1}. \tag{10.1}$$

The covariance is given by

$$\sigma_{ab} = \left(\frac{\partial^2 \log P}{\partial a \partial b}\right)^{-1}. \tag{10.2}$$

[1]Not to be confused with the variances of the data points!

Note that $\sigma_{ab} = \sigma_{ba}$, and the complete set of quantities is known as the "covariance matrix"

$$\begin{pmatrix} \sigma_a^2 & \sigma_{ab} \\ \sigma_{ab} & \sigma_b^2 \end{pmatrix}.$$

Following the discussion of Sect. 5.8.3, the iso-density contours of the bivariate Gaussian are given by ellipses described by

$$\frac{a^2}{\sigma_a^2} + \frac{b^2}{\sigma_b^2} - 2\frac{\rho ab}{\sigma_a \sigma_b} = k, \quad \text{with} \quad \rho = \frac{\sigma_{ab}}{\sigma_a \sigma_b}. \tag{10.3}$$

where $k = 1$ describes the 1σ contour etc. How we then use the variances/covariances to construct uncertainty intervals needs a little more careful thought. We will divide the discussion into three cases.

10.5.2 The Case of One Interesting Parameter

Suppose we consider b to be an accidental or uninteresting parameter, in the sense that we don't care what it's value is. (An example might be the time of maximum, t_0, in our earlier worked example). We could fix the value of b at $b = b_0$ and construct the conditional distribution of a, i.e. the slice through $P(a, b)$ at fixed $b = b_0$. The standard deviation of the conditional distribution will be given by σ_a as in the analysis above, and we can construct any required interval using standard Gaussian areas. Somewhat better, we may ask "what is the spread of a, regardless of the value of b?" This would be given by the distribution of a marginalised over b. For the bivariate Gaussian, this will give the same answer as the conditional distribution, but for a more general shape this will not be the case.

10.5.3 The Case of Two Interesting But Uncorrelated Parameters

If both parameters are interesting, we have to consider their *joint error* rather than marginalising over one of the parameters. In Sect. 5.8.1 we considered the problem of integrating the bivariate Gaussian to a radial distance $r = (x^2 + y^2)^{1/2}$, where the one dimensional variances are given by the same σ^2 in each of x and y. To achieve integrated probability of 68.3%, equivalent to $\pm 1\sigma$ for the 1D Gaussian, we need to extend out to $r = 1.52\sigma$. If the variances are unequal, then we trace out the ellipse given by equation (10.3), with $\sigma_{ab} = 0$ and $k = 1.52$. This means that the 68.3% probability interval for a is given by $a_0 \pm 1.52\sigma_a$, and for b by $b_0 \pm 1.52\sigma_b$.

For other probability intervals, to include integrated probability P_{int} we use $k = \sqrt{-2\ln(1 - P_{\mathrm{int}})}$.

10.5.4 The Case of Two Interesting and Correlated Parameters

If the parameters are correlated, the iso-density ellipse is tilted in the a, b plane. The most appropriate thing to is to show the ellipse, or to quote both the variance and the covariance; but if one wants to quote a range for a single parameter, a sensible thing to do is to quote the limits of the bounding box enclosing the tilted ellipse, as shown in Fig. 10.3. In Sect. 5.8.3 we discussed how to calculate the rotation angle and major axis of the rotated ellipse, and a little more algebra shows how to derive the bounding box limits. Alternatively, one can use standard matrix manipulation routines to find the eigenvectors of the covariance matrix, which gives the principal axes of the ellipse.

10.5.5 Finding χ^2 Confidence Intervals

There are two ways to find a suitable interval if one has calculated $\chi^2(a, b)$. The first method is convert χ^2 to (relative) likelihood using $L \propto \text{const} + e^{-\chi^2}$. The likelihood surface can then be dealt with in exactly the same way as $P(a, b)$. The variances and covariances of L can be calculated using

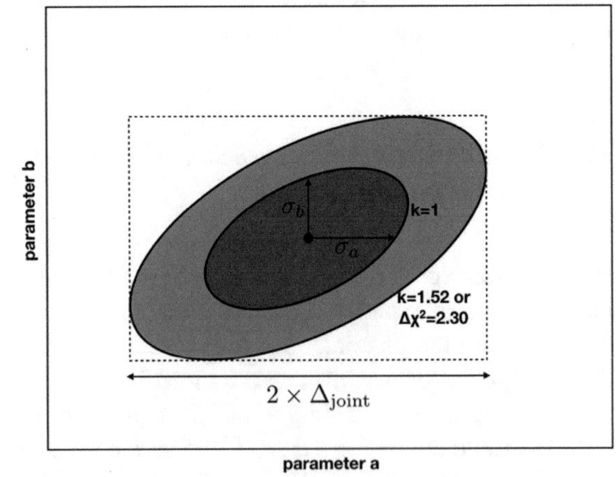

Fig. 10.3 Constructing posterior probability / χ^2 confidence intervals for two correlated parameters. Two effects need to be taken care of. The contour that encloses 68.3% joint probability is further out than the 1σ contour; and to enclose the range of that contour, we need to construct the bounding box enclosing it

$$\sigma_a^2 = \left(\frac{\partial^2 \log L}{\partial a^2}\right)^{-1} \quad = \quad -\frac{1}{2}\left(\frac{\partial^2 \chi^2}{\partial a^2}\right)^{-1},$$

and similarly for σ_b and σ_{ab}. Alternatively, one can generalise the $\Delta\chi^2$ method of Sect. 8.6.4. First we find the minimum value of $\chi^2 = \chi^2_{\min}$, at the best fit (a_0, b_0). When we examine values of χ^2 at other values of a, b, the test statistic $\Delta\chi^2 = \chi^2 - \chi^2_{\min}$ will be distributed as χ^2 with $\nu = 2$ degrees of freedom. We then find for example that 68.3 and 90% confidence are located at $\Delta\chi^2 = 2.30, 4.61$ respectively. For two interesting correlated parameters, the technique is then to locate the required *critical* χ^2 *contour*, and either show this graphically, or construct the ellipse that approximates the contour and find its bounding box.

10.5.6 Modelling with Many Parameters

The techniques described above generalise fairly straightforwardly in principle to larger numbers of parameters, $\bar{\theta} = \theta_1, \theta_2, ...\theta_m$. One can marginalise over any subset of uninteresting parameters, but look at full joint errors for the remaining parameters. Of course it is hard to visualise multi-dimensional surfaces. A common technique is to take two parameters at a time, marginalise over all the others, and examine the contours for the two chosen parameters. (See the example in Fig. 10.4.) Taking every possible pair, one can spot which parameters are correlated with which others, and decide how to quote overall joint errors. When using the Gaussian approximation, we can calculate the covariance matrix by computing every term

$$\sigma_{ij}^2 = \left(\frac{\partial^2 \log P}{\partial\theta_i \partial\theta_j}.\right)^{-1}$$

We can then use standard matrix inversion techniques to find the eigenvectors of the covariance matrix and so the principal axes of the multi-dimensional ellipsoid. How do we find the integrated probability within some region of the general m-dimensional Gaussian ellipsoid? We can define a radial distance in m-dimensional space by $r^2 = \theta_1^2 + \theta_2^2 + \cdots \theta_m^2$. By suitable co-ordinate transformations (rotation and stretching) we can always transform to co-ordinates where each of the σ_θ are the same, and ask, to what value of $k = r/\sigma$ do I need to reach out to include 68.3% probability, or 90% probability etc. This problem is not generally analytically solvable, but has well known numerical solutions. Because of the correspondence between log likelihood and χ^2 noted in equation (8.13), this problem is in general the same as the problem of integrating χ^2 for the appropriate degrees of freedom. So what we do is to look up tables of the integral of χ^2 for $\nu = m$ degrees of freedom, which will tell us the $\Delta\chi^2$ we need. The corresponding $k = r/\sigma$ is given by $k = \sqrt{\Delta\chi^2}$. Table 10.1 summarises the required values of k and $\Delta\chi^2$ for various numbers of parameters and probability/confidence levels.

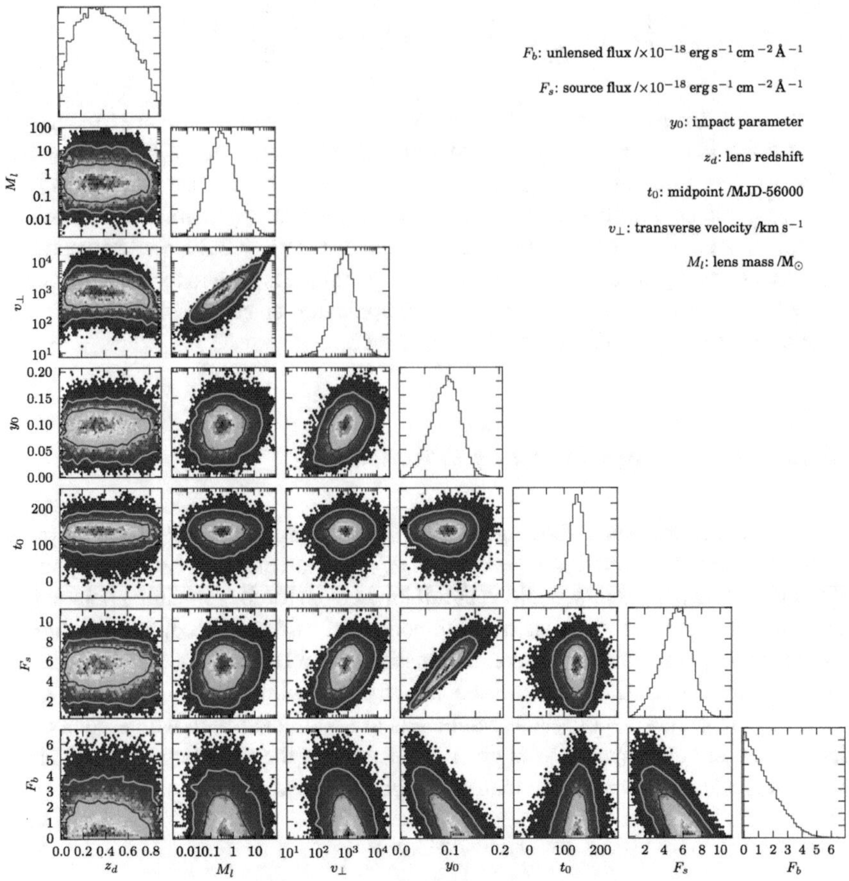

Fig. 10.4 A "corner plot" taken from Bruce et al. (2017). This originates from fitting a microlensing model to a quasar light curve, as outlined in Sect. 10.3. There are seven parameters in the model, leading to a seven-dimensional posterior surface $P(\theta)$. Each parameter is represented in both a row and a column. The 2D colour plot at the interaction of every pair of parameters is a contour plot representing the $P(\theta)$ surface marginalised over all the remaining parameters—so at position (2, 5) for example we can see that our estimate of M_l (the mass of the lens) and $v_\perp$ (the relative velocity of the lens and the source) are highly correlated. At the end of each row/column we see for each parameter in turn, the one-dimensional PDF resulting from marginalising over all the other parameters

10.5.7 Limits of the Gaussian Approximation

The Gaussian approximation for either the likelihood surface or the posterior surface is good at small multiples of σ_θ, and as long as there are large numbers of data points. As well as making the surface Gaussian, having large numbers of data points means that we will not be very sensitive to the choice of prior, and so the likelihood and

Table 10.1 Gaussian/$\Delta\chi^2$ regions for multi-parameter fits

	68.3%	90%	95%	99%
$m = 1$	$k = 1.0$ $\Delta\chi^2 = 1.0$	$k = 1.64$ $\Delta\chi^2 = 2.71$	$k = 1.96$ $\Delta\chi^2 = 3.84$	$k = 2.58$ $\Delta\chi^2 = 6.63$
$m = 2$	$k = 1.52$ $\Delta\chi^2 = 2.30$	$k = 2.15$ $\Delta\chi^2 = 4.61$	$k = 2.45$ $\Delta\chi^2 = 5.99$	$k = 3.03$ $\Delta\chi^2 = 9.21$
$m = 3$	$k = 1.88$ $\Delta\chi^2 = 3.53$	$k = 2.50$ $\Delta\chi^2 = 6.25$	$k = 2.95$ $\Delta\chi^2 = 7.81$	$k = 3.37$ $\Delta\chi^2 = 11.34$
$m = 5$	$k = 2.43$ $\Delta\chi^2 = 5.89$	$k = 3.04$ $\Delta\chi^2 = 9.24$	$k = 3.27$ $\Delta\chi^2 = 11.07$	$k = 3.85$ $\Delta\chi^2 = 15.09$
$m = 10$	$k = 3.40$ $\Delta\chi^2 = 11.54$	$k = 4.00$ $\Delta\chi^2 = 15.99$	$k = 4.28$ $\Delta\chi^2 = 18.31$	$k = 4.82$ $\Delta\chi^2 = 23.21$
$m = 20$	$k = 4.74$ $\Delta\chi^2 = 22.44$	$k = 5.33$ $\Delta\chi^2 = 28.41$	$k = 5.60$ $\Delta\chi^2 = 31.41$	$k = 6.13$ $\Delta\chi^2 = 37.57$

posterior surfaces will be very similar. If we have small numbers of data points, the Gaussian approximation may be poor, and we may get quite different answers for the χ^2 and Bayesian methods.

For any number of data points, the Gaussian approximation will start to break down at larger multiples of σ_θ. In Sect. 8.5.3 when considering 1D parameter fitting, we already noted the rule of thumb that 68% or 90% regions may be safe, but a 99% region may be badly underestimated. For larger numbers of parameters, this problem sets in much earlier, because one has to stretch out to large multiples of σ_θ even for modest amounts of integrated probability, as we can see in Table 10.1. Even for a modest $m = 3$ parameters, for 90% probability we need to reach a 2.5σ contour. Another way to look at this, is that for a trivariate Gaussian, the 1σ contour contains only 20% of the integrated probability.

Rather than set guidelines for what is safe and what is not, the thing to do is to examine the calculated surface. Do the slices/marginalised profiles look Gaussian? Do the contours look symmetrical, or are they banana-shaped? Issues like this are common in model fitting. It is always healthy to show the actual contours. In Bayesian fitting, one should ideally follow the contours far enough out to get a reasonable estimate of the marginal likelihood E in order to get an empirical normalisation of posterior probability. For large numbers of parameters however, this becomes computationally very expensive.

10.6 Goodness of Fit

Once the best fit set of parameter values θ_0 has been found, can we make a statement about whether the model is a good description of reality? If we have used χ^2 we can certainly do this. The set of best-fit parameter values constitutes a hypothesis that we can test in the usual way, using the minimum value χ^2_{min}, which will be distributed as χ^2 with $\nu = N - m$ degrees of freedom, where N is the number of data points and m is the number of parameters fitted. Note that to make an absolute test based on χ^2, we need to assume that the data errors are Gaussian, and we have to be confident that we know the errors accurately. Over-estimating errors may lead to false confidence in a model, and under-estimating errors can lead a false rejection of a model.

With the Bayesian method, we do not have a measure of absolute goodness of fit for a single model. Bayesian methods always concern relative degrees of belief. Within the context of model fitting, this means *assuming* the model, and calculating the relatively believability of different values of the parameters. A growingly common procedure is to calculate $P(\theta)$ to get the best model, but then to calculate χ^2 at the position of the best-fit model, to quantify the model's overall believability.

It is tempting to take the marginal likelihood $E = \int \pi(\theta)L(D|\theta)$ as a measure of the overall agreement of the data with the model, integrated over all parameter values. It is after all referred to as the "Bayesian Evidence". However, there is no absolute scale for E. It is sensitive to the units of the data, the number of data points, and so on. On the other hand, if those things are fixed, the relative values of E can be used to compare two models, as we shall describe shortly.

10.7 Model Comparison

Suppose we have two models M_1 and M_2, and have calculated P or χ^2 for each of these. Even if we can't reject one model or the other, can we say which we prefer? There are several ways to do this.

10.7.1 Model Comparison Using $\Delta\chi^2$

Suppose our two models M_1 and M_2 have degrees of freedom ν_1 and ν_2, and that when compared to our data they give best fit χ^2 values of χ_1^2 and χ_2^2. If $\chi_1^2 > \chi_2^2$ we might want to prefer M_2, but how often does this happen by chance? The trick is to form a test statistic using the χ^2 and ν values for which we know the sampling distribution. This is a solvable problem if the two models are "nested", i.e. one is a subset of the other. For example we might first try fitting a quadratic polynomial, and then try a cubic polynomial. Note that a model with more parameters will usually produce a better fit than one with fewer parameters; if we add a parameter to the same model, this is almost always true. But is it a *significantly* better fit?

The simplest test statistic to use is $\Delta\chi^2 = \chi_1^2 - \chi_2^2$, which we have met before in the context of parameter interval estimation. For two equally good nested models, this quantity will be distributed as χ_ν^2 with $\nu = \nu_1 - \nu_2$. You can then answer the question "if I assume these two models are in fact equally good, how often would I get a $\Delta\chi^2$ value this large or larger?". If you reject the hypothesis that the hypotheses are equally good, then you can (tentatively) prefer one model over the other.

This technique is often used as an **extra parameter test**, i.e. in order to ask the question "am I justified in adding an extra parameter to my model?". If there is no significant difference, then $\Delta\chi^2$ will be distributed like χ_1^2; if we see a large value of $\Delta\chi^2$, we can reject the null hypothesis and conclude that the extra parameter does genuinely help.

10.7.2 Model Comparison Using the F Test

Another way to compare the χ^2 values is to take their ratio. We use the F statistic

$$F_{12} = \frac{\chi_1^2/\nu_1}{\chi_2^2/\nu_2}. \tag{10.4}$$

This follows the rather ghastly distribution

$$p_{\nu_1,\nu_2}(F) = \frac{\Gamma\left(\frac{\nu_1+\nu_2}{2}\right)}{\Gamma\left(\frac{\nu_1}{2}\right)\Gamma\left(\frac{\nu_2}{2}\right)}\left(\frac{\nu_1}{\nu_2}\right)^{\frac{\nu_1}{2}} F^{\frac{\nu_1}{2}-1}\left(1+\frac{\nu_1}{\nu_2}F\right)^{-\frac{1}{2}(\nu_1+\nu_2)}. \tag{10.5}$$

This is a two parameter family of curves, with mean and variance

$$\mu_F = \frac{\nu_2}{\nu_2-2}, \qquad \sigma_F^2 = \frac{2\nu_2^2(\nu_1+\nu_2-2)}{\nu_1(\nu_2-2)^2(\nu_2-4)}.$$

Luckily of course one rarely one has to compute this oneself, but just look up values of F in tables. However, be careful about whether you have calculated F_{12} or F_{21}. For reasonably large ν_1 and ν_2 we can use a Gaussian approximation, and $\nu_1 \approx \nu_2 \approx \nu$ and we get the much simpler version

$$\mu_F \sim 1, \qquad \sigma_F^2 \sim \frac{4}{\nu}. \tag{10.6}$$

10.7.3 Likelihood Ratio Test

The core of any hypothesis testing is the likelihood of the data given the model. Once we have found our best parameter values for both models M_1 and M_2, why not directly compare the likelihoods? We could compute the statistic

$$R = \frac{L_1}{L_2}.$$

The issue then is to find the sampling distribution for our test statistic R. This can in principle be done whatever the probability distribution the data points follow. However, in practice this method is useful for Gaussian distributed data points. From equation (8.13) we have

$$\log L = -\frac{1}{2} N \log 2\pi\sigma^2 - \frac{\chi^2}{2}.$$

If we compare two models, χ^2 changes but N and σ stay the same, so

$$2 \log R = \Delta\chi^2,$$

and the likelihood ratio test boils down to being the same thing as the $\Delta\chi^2$ test.

10.7.4 Bayesian Model Comparison: The Bayes Factor

Although we can't use Bayesian methods to assess a single isolated model, we can use them to compare rival models. Suppose first of all we compare the models at the best fit parameter values. Then this is like the simple case of comparing two fixed hypotheses A and B, as we considered in Chap. 7 and we have

$$P_1 = \frac{\pi_1 L_1}{E}, \quad P_2 = \frac{\pi_2 L_2}{E} \quad \text{where} \quad \pi_1 + \pi_2 = 1 \quad \text{and} \quad E = \pi_1 L_1 + \pi_2 L_2.$$

Note that here the subscripts 1 and 2 refer to the fixed hypotheses corresponding to the best-fit versions of models M_1 and M_2. Then as we noted in Chap. 7, Sect. 7.5, if we take the posterior ratio the marginal likelihood E cancels out and we have

$$\frac{P_1}{P_2} = \frac{\pi_1}{\pi_2} \cdot \frac{L_1}{L_2}.$$

The ratio of likelihoods $B_{12} = L_1/L_2$ is known as the *Bayes Factor* and quantifies the relative agreement of the data with the two hypotheses.

Now suppose we wish to compare two complete models, each with their own parameter vectors θ_1 and θ_2. Within the context of each model, the parameters will have their own normalised prior probability distributions $\pi_1(\theta_1|M_1)$, $\pi_2(\theta_2|M_2)$. When fitting the data, we will get likelihood functions $L_1(\theta_1) = p(D|\theta_1, M_1)$ and $L_2(\theta_2) = p(D|\theta_2, M_2)$. We have written the dependencies more carefully than normal here. The likelihood L_1 is a function of the parameter vector θ_1, but always assuming the context of model M_1. The overall likelihood for each model is the likelihood marginalised over all parameter values, but always in the context of the model concerned. We can write

$$E_1 = \int \pi_1(\theta_1|M_1)L_1(D|\theta_1, M_1)\mathrm{d}\theta_1, \qquad E_2 = \int \pi_2(\theta_2|M_2)L_2(D|\theta_2, M_2)\mathrm{d}\theta_2.$$

Remember that θ_1 and θ_2 are actually vectors, so that these are multi-dimensional integrals. The next step is to assign a prior probability for each model overall $\pi(M_1)$ and $\pi(M_2)$—i.e. our relative degree of belief in each model before we get any data. It is important to distinguish these from the prior probability distributions for the parameters, within the context of each model. Then we can quantify our relative degree of belief in each model overall after getting the data:

$$\frac{P(M_1)}{P(M_2)} = B_{12}\,\frac{\pi(M_1)}{\pi(M_2)}, \tag{10.7}$$

where B_{12} is the Bayes factor

$$B_{12} = \frac{E_1}{E_2} = \frac{\int \pi_1(\theta_1|M_1)L_1(D|\theta_1, M_1)\mathrm{d}\theta_1}{\int \pi_2(\theta_2|M_2)L_2(D|\theta_2, M_2)\mathrm{d}\theta_2}. \tag{10.8}$$

In practice when considering whole models it is normal to treat them equally and so assume $\pi(M_1)/\pi(M_2) = 1$, in which case our relative degree of belief depends only on the Bayes factor. How do we interpret this number? Following the discussion in Sect. 7.5, we can see B as the betting odds on M_1 versus M_2. So if $B = 8.0$ that is 8:1 on, and if $B = 0.2$ it is 5:1 against. Starting with Jeffreys, there have been various attempts to turn the value of B into qualitative words, so that for example, $B = 1 - 3$ is "not worth more than a mention"; $B = 3 - 10$ is "substantial"; $B = 10 - 100$ is "strong"; and $B > 100$ is "decisive".

10.8 Artificial Intelligence Techniques

The growing field of Artificial Intelligence (AI), and especially the subset of AI known as Machine Learning (ML), has some similarities to the topic of probabilistic model fitting which we have considered in this chapter, but also some key differences. We will take a very brief look at these similarities and differences.

In *model fitting*, we have an algorithm that performs calculations on input data in order to predict output values, which are then compared to observed values. The algorithm has tuneable parameters, which are varied to see how the comparison of predicted and observed output values changes; and we have some kind of performance metric—likelihood, χ^2 etc—which is used to judge the performance of the algorithm for different parameter values. We then find the "best" parameter values by maximising/minimising our chosen performance metric. The ultimate point of the exercise is **understanding**. The algorithm is a mathematical representation of a theoretical idea, and the parameters have physical meaning within that context—the parameters are things like the mass of a particle, or the scaling constant that connects luminosity and temperature. We want to know the values of those parameters in order to understand Nature.

In *artificial intelligence*, the aim is not normally understanding, but rather **decision** or **action**. We once again have an algorithm acting on input data to predict output data, but the point is to do something with the output data. For example, the input data may be a stream of images, and the output data one of a set of categories, such as "dog", "cat" or "fish". We run each image through the algorithm and decide what sort of animal we have. Alternatively the output may be a number; for example each input dataset may be a stock market time series, and the output data value is a predicted share value in a week's time. Before using such methods to make decisions, the algorithm has to be *trained* on a set of inputs for which the correct answer is known—for example a set of images for each of which we know the answer to whether it is a dog or a cat etc. This is where the process is similar to model fitting. The algorithm will have a set of parameters, which the software will adjust to try to get the correct answers. The metric for success will often be simply the fraction of correct answers. However rather than a handful of parameters, there may be hundreds or thousands or even millions of parameters—for example the node values in a multi-layer neural network. The algorithm and its parameter values are not interpretable in any simple sense—they just adjust themselves until the system works. This is what we mean by *machine learning*. We do not use a physical model of the situation to impose a logical structure on the parameters—we just let software do its own thing and see where it gets to. The idea is that this is analogous to what your brain is doing—hence the description of such methods as Artificial Intelligence.

Beyond these general ideas, there are many more subtleties, such as whether the algorithm is supervised or unsupervised, and whether the code itself is adaptive, that we don't have space here to go into. An issue that is very relevant to the theme of this chapter however, is how we might assess the uncertainty on the results we obtain. In model fitting, because the parameters are physically meaningful, we want not just their best values, but a probability distribution for the parameters, based on our understanding of the success-metric that we are using. In machine learning, the values of the parameters are of no interest. However, a common situation is that we try several different algorithms—neural networks, random forests, etc—and want to know which one works best. We might find, say, that on our training set the random forest gives the right answer 87% of the time, and the neural network gives the right answer 84% of the time. Does this mean that the random forest algorithm is definitely the one to use? Or did we by chance pick a training set that happened to work better for the random forest? Typically, there is no a priori way to calculate the probability distribution for our success metric, but we can achieve this empirically. One method is choose a separate *validation set*. Another way is to pick many random subsets of our training set, calculate the success metric for each of these, and examine the distribution of resulting values. A recurring issue is that in practice, the performance of any machine learning algorithm is only as good as the training set it is given.

10.9 Key Concepts

Some of the key concepts from this chapter are:

- The idea of a model generating predicted values to compare to observed values
- The need for an experiment model as well as a physical model
- The idea of multiple parameters as defining a multi-dimensional space to explore
- The variety of numerical techniques for finding maximum posterior or likelihood or minimum χ^2, including gradient search, sub-grid location, and the Gaussian approximation
- Using the variances/covariances to construct 2D parameter intervals
- The distinction between interesting and uninteresting parameters
- The distinctions between conditional errors, marginalised errors, and joint errors
- Constructing a bounding box to allow for correlated parameters
- Testing *absolute* goodness of fit with χ^2 values
- Comparing models using $\Delta\chi^2$, the F-test, and the Bayes factor
- The general ideas of Machine Learning algorithms.

The opening of this chapter was mostly conceptual and numerical, but from Sect. 10.5 onwards there were several key formulae: the formulae for obtaining parameter variances, assuming the Gaussian approximation ((10.1) and (10.2)); the equation describing the ellipse approximating the $k - \sigma$ contour (10.3); the definition of the F-ratio statistic (10.4), its PDF (10.5), and its expected mean and variance for large ν (10.6); and finally, the Bayes factor and its use in comparing models ((10.7) and (10.8)).

10.10 Further Reading

The techniques of model fitting and multi-dimensional model fitting are covered in many books. A book that many have relied on for many years is Bevington (2002), which we have referred to in several other chapters. Probably the definitive practical explanation, with detailed code listings, is "Numerical Recipes" by Press (2007). This is a deep well of information for anything to do with Scientific Computation. There is also a good website, and associated online e-book which is free as long as you read a limited amount each month. Another very practical book with code examples, which again is much broader than model fitting, is Ivezic et al. (2014). This is aimed at astronomers, but in fact is very useful for a wide range of scientists.

The differences between joint, marginal, and conditional errors for multi-parameter fits, how they depend on whether the parameters are correlated, and the importance of the distinction between "interesting" and "uninteresting" parameters, has been a notoriously confusing area for many years, for both students and working scientists. Within my own (astronomical) career, the first research paper to really stress these points clearly was Avni (1976), and the definitive statements are as usual in Press et al. (Numerical Recipes). Most researchers do now get these things right. I think part of the historic difficulty was because arriving at the right answer using χ^2 and significance-based thinking was a little on the brain-bending side; once you think in terms of a multi-variate posterior probability distribution, it all seems fairly obvious.

10.11 Exercises

10.1 A model involves two parameters p and q. Priors have been assigned, experimental data obtained, and the posterior probability surface calculated. The peak of the posterior probability surface is found to be at $p = 19.7$, $q = 23.4$. A Gaussian approximation to the surface gives the covariance matrix

$$\begin{pmatrix} 9.7 & 0.0 \\ 0.0 & 16.3 \end{pmatrix}$$

Are the parameters correlated? What is the 90% conditional error range on p? What is the 90% joint error range on p?

10.2 A model involves two parameters a and b. Analysis of experimental measurements gives a posterior probability surface, for which the $1 - \sigma$ contour is shown in Fig. 10.5. By eye, estimate the full joint error range on a.

10.3 For a four-parameter fit, what multiple of the conditional error gives the full joint error? What $\Delta\chi^2$ does this correspond to?

10.4 In Problems 9.2 and 9.3 we examined the correlation between maths, physics, and art test marks obtained by a sample of schoolchildren. (The table of data can be found in the Chap. 9 exercises.) When regressing the physics marks on the maths mark, a fit was found of the form $y = a + bx$ with $a = 4.276$ and $b = 0.658$. Taking the maths mark as the independent variable, and the physics mark as the dependent variable, and taking the typical error on the physics marks to be $\sigma = 5$, calculate the χ^2 for this fit. Suppose instead $\sigma = 3.5$. Re-calculate the χ^2. What is the quality of the fit in these two cases? What general lesson does this hold?

10.5 An astronomer measures the X-ray spectrum of a quasar in 9 energy channels and fits the data with a simple power law model, with just the power-law slope as a free parameter, and gets a fit quality of $\chi^2 = 12.6$. A rival astronomer fits the

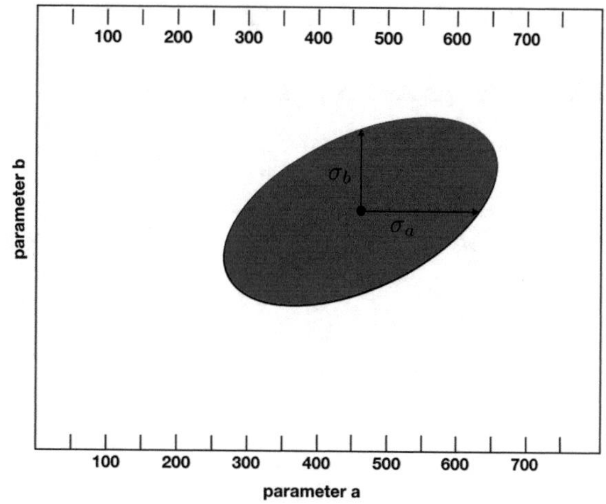

Fig. 10.5 One sigma contour of posterior probability for Problem 10.2

same data with a more sophisticated model that involves fitting the temperature, the density, and a Comptonisation parameter, and gets $\chi^2 = 4.8$. Using the $\Delta\chi^2$ test, can the second astronomer definitely conclude they have a better model?

10.6 Try comparing the two models from the problem above using the F-test. Do we get the same answer?

10.7 A scientist fits a dataset with a model that has 15 parameters, using the χ^2 method. The resulting χ^2 surface can be approximated as a multi-dimensional Gaussian, where in each dimension, after suitable re-scaling, the width of each 1D Gaussian is given by σ. If we want to reach 95% confidence, what multiple of σ do we need to go out to?

References

Avni, Y.: Energy spectra of X-ray clusters of galaxies. Ap. J. **210**, 642–646 (1976)

Bevington, P., Robinson, D.K.: Data Reduction and Error Analysis for the Physical Sciences, 3rd edn. McGraw-Hill Education, New York (2002)

Bruce, A., et al.: Spectral analysis of four 'hypervariable' AGN: a micro-needle in the haystack? MNRAS **467**, 1259–1280 (2017)

Ivezic, Z., Connolly, A.J., VanderPlas, J.T., Gray, A.: Statistics, Data Mining, and Machine Learning in Astronomy: A Practical Python Guide for the Analysis of Survey Data. Princeton University Press, Princeton (2014)

Press, W.H., Teukolsky, S.A., Vetterling, W.T., Flannery, B.P.: Numerical Recipes: The Art of Scientific Computing, 3rd edn. Cambridge University Press, Cambridge (2007)

Websites (all accessed March 2019):

Numerical Recipes Home Page: http://numerical.recipes

Part IV
Selected Topics

Probabilistic reasoning is at the core of physics, and indeed is central to modern science in general. We have seen examples from many areas of physics and astronomy as we stepped through the basics of probability, examined how probability distributions arise in the natural world, and developed the techniques of statistical inference. To reduce, analyse, and interpret our data, the concepts of probability are crucial. This is true for every area of physics. However, there are certain topics where probability is particularly important, or controversial, or leads to deep insights. We pick four such areas to look at in the closing part of this book.

The first such topic—Information Theory—isn't really physics as such, but is really a different angle on the concepts of probability. However, the Information Theory way of looking at things is very important for many physics problems, and also for data analysis. Every well-educated physicist should have a basic grasp. The second topic—erratic change with time—has long been interesting to physicists, and has become particularly fashionable recently in my own field of quasar variability, and somewhat notorious because of its mixed success in predicting the behaviour of the stock market. The third and fourth topics—probability in quantum physics, and the concept and interpretation of entropy—are areas of enduring controversy and confusion. They bring us face to face with some of the deepest issues in natural philosophy—is the world really inescapably unpredictable, and where does the arrow of time come from? We won't solve these difficult problems in this book, but looking carefully at how probability works in quantum physics and in statistical physics, I hope some bright students will be armed to solve them for us some time soon!

Chapter 11
Information, Uncertainty, and Surprise

11.1 Outline of Content

- Measuring uncertainty/information
- Binary search games
- Uncertainty associated with probability distributions
- Messages and alphabets
- The Maximum Entropy Principle

In writings about current affairs, culture and politics, we often hear that we live in the "Information Age". This phrase seems to be about computers, the Internet, connectedness, and the ease of exchanging information. On the other hand, if you open a textbook on *Information Theory*, it is full of dry stuff about channels, encoding, bits, and efficiency. More puzzling to a physicist, information theory seems to talk constantly about "entropy", whereas you thought this had something to do with thermodynamics, and the dissipation of energy. How do these apparently disparate things tie together? At its heart, information theory is essentially a mathematical re-casting of the ideas of probability, but done in a way that gives us new insights. Just as important, while "information" has a perfectly good but rather broad English meaning, the definition of information in information theory is a very narrow technical one. This narrowing of concept is analogous to what happened to the word "energy" in the development of physics.

We begin this chapter with a fresh look at the concept of probability, and how it relates to ideas of uncertainty and information. We will start by looking at single events, then broaden out to consider uncertainty attached to probability distributions. This will lead us to an understanding of how information theory can be usefully applied in various places, but especially in the transmission of messages, and in statistical inference.

A. Lawrence, *Probability in Physics*, Undergraduate Lecture Notes in Physics,
https://doi.org/10.1007/978-3-030-04544-9_11

11.2 Probability, Uncertainty, Surprise, and Information

Imagine yourself about to roll a die; we can say that the probability of rolling a 4 is $p = 1/6$. When we make this statement, we are using probability as a measure of the "amount of likeliness". Another way of looking at the same situation is that we are *uncertain* what the number rolled will be. In general, if some possible event has a probability p of occurring, then the lower the value of p, the more uncertain we are. Now suppose that the event in question has in fact just occurred. If the event in question had a low value of p, we will be *surprised* to see it happen; the larger the uncertainty before the event, the greater the "surprisal" after the event. Finally, we can see this surprise as *information*—when an event that is expected to be rare has happened, we feel that this tells us a lot, whereas a more likely event is unsurprising and tells us little. Returning to the before-the-event view, we could say that we are *lacking information* that we would need to have in order to remove our uncertainty. Suppose the die has already been rolled but our friend is covering it up with their hand. If the friend simply tells us "its a three" then that information would remove all the uncertainty. If the friend says "its less than four", then our uncertainty would be reduced, but not completely removed; and we would agree that we have been given less information than in the previous case.

Uncertainty and information are then two ways of looking at the same thing—information is what we would need to supply an observer in order to remove the uncertainty. Both concepts are clearly closely related to probability. However, uncertainty and information, while useful English words, are rather fuzzy. Can we firm up their meanings, and get a mathematical relationship with probability? Let us concentrate first on uncertainty. Once we have a mathematical definition, then our technical meaning of information should be quantitatively the same, but with a different interpretation.

11.2.1 A Measure of Uncertainty/Information

Can we come up with a formal definition of "amount of uncertainty" which corresponds reasonably to the normal English meaning, but which is mathematically useful? We need a definition for which uncertainty increases as probability decreases. The trouble is that there are many potential ways of doing this—we could use $1/p$ or $1 - p$ for example. In fact the standard definition, first proposed by Shannon in 1948, is to use

$$h = \log_2(1/p) \quad = \quad -\log_2 p. \tag{11.1}$$

Why is this a good definition to use? To understand this, we need to switch to the information view of the same quantity. The quantity $h = -\log_2 p$ tells us the *number of questions* you need to ask to solve any binary search problem. Lets look at this.

11.2.2 Quantifying Information: Binary Search Games

Suppose we have 8 boxes. A friend has placed a coin in one of them, and our task is to find out which box contains the coin, by asking yes/no questions. We might laboriously ask "is it in box 1?" and then "is it in box 2?" and so on. Sometimes we would get lucky and find the coin on the first question, and sometimes we would need 8 questions. Given that the probability of the coin being in any one box is 1/8, its not hard to see that if we played the game many times, on average we would need $n_Q = (1 + 2 + \cdots + 8)/8 = 4.5$ questions. However, there is a smarter method. We draw a dividing line between box 4 and box 5, and ask "is it in the lower half?". Whatever the answer, we are left with four boxes. We then divide again, to get two boxes, and then we need only one more question, so that we will always find the coin after $n_Q = 3$ questions.

In general if we have $n = 2^k$ boxes, this "binary search" method will succeed after k questions. For a number n that is not a power of two, there will always be a value of k such that 2^k is less than n but 2^{k+1} is greater than n. In this case, sometimes we will need k questions, and sometimes $k + 1$ questions. For large values of n, the difference between k and $k + 1$ is negligible, so to a good approximation, for a n-box problem, the number of questions we need is $n_Q \approx \log_2 n$. Now, the probability of the coin being in any one box is $p = 1/n$. We therefore find

$$n_Q = \log_2 n = \log_2 (1/p) = -\log_2 p \equiv h.$$

We have arrived at our earlier definition of h. For the n-box problem therefore, we can interpret our uncertainty measure h as the minimum number of yes/no questions you would need to ask on average to remove the uncertainty. A large number of problems boil down to, or approximate, the n-box problem, especially in problems related to computer coding and the transmission of messages. For other problems which don't necessarily map onto the n-box problem, $h = -\log_2 p$ is still a perfectly good measure of uncertainty to use, even though it's interpretation as "information" may be less clear. The quantity h has therefore become the standard definition of information and uncertainty.

11.2.3 The Units of Uncertainty/Information

Note that we are so far dealing with discrete probabilities, rather than probability densities. (We will look at how to deal with the continuous case in Sect. 11.3.7).

The probability p is just a number, so h likewise is just a number. It does not have physical units. However, within the yes/no question picture, we can attach units of a sort. The answer to each question in our binary search can be represented by a 1 or 0, and so the whole answer sequence can be encoded as a binary string—left-left-right-left becomes 1101 and so on. The quantity $h = n_Q$ can therefore be thought of the *number of bits* needed to supply the information. To avoid confusion with the actual bits (binary digits) of storage in a computer system, some workers use "Shannons" as the units of h. It also common to define variants of h using logarithms to different bases. If we define $h_e = -\ln p$, then h_e is said to be in units of "nats"; if we use $h_{10} = \log_{10} p$, then h_{10} is said to be units of "bans". (See "Further reading" for the explanation of this strange term.)

11.2.4 Combining Uncertainties

Suppose now we have two events A and B, with probabilities $p(A)$ and $p(B)$. As we discussed in Chap. 1, to get the probability of both events occurring, we need to know $p(B|A)$, the conditional probability of B happening, given that A has already happened. Then we find that $p(A, B) = p(A) \times p(B|A)$. Now, we can attach an uncertainty to the event of both A and B occurring, just as we can for event A or B alone, so that $h(A, B) = -\log_2 p(A, B)$. Then it is easy to see that

$$h(A, B) = h(A) + h(B|A), \tag{11.2}$$

where $h(B|A)$ is the *conditional uncertainty* (or conditional information, depending on point of view). Of course if the events are independent, $h(A, B) = h(A) + h(B)$. The additivity of h is one of the things that encouraged Shannon to pick the logarithmic definition.

11.2.5 Information Theory Terminology

We have been using a technical definition of our quantity h. It is clearly related to the broader English sense of "information", in that unexpected events tell us things that we otherwise didn't know, and upon which we might act. However, this general natural language sense of information is receiver-dependent. For example, a data packet which contains a section of HTML can be understood by a web browser, but if you sent the same packet to an image analysis programme, it would be meaningless. Likewise, a page of Shakespeare and a page of the telephone directory, which take the same number of bits to express, may be seen by different people to contain quite different amounts of information. Our technical definition of h on the other hand, gives a fixed value. For these reasons, I prefer using the term "uncertainty" most of the time, so that one can keep the broader sense of "information" available in discussion.

In information theory literature, h is sometimes referred to as "entropy", but there is an argument to be made to avoid this usage, to stay out of the long standing controversial debate about the interpretation of entropy (see Chap. 14). On the other hand there clearly is a simple connection between the Boltzmann/Gibbs view of entropy, and the Shannon view of uncertainty/information, as we will discuss in Chap. 14, and this connection becomes clearer when we talk about probability distributions, rather than single events. Our solution to these delicate matters will be to use the terms "uncertainty" and "information" fairly freely, but to avoid the unqualified use of the term "entropy", preferring the term "Shannon Entropy" for the uncertainty connected with a probability distribution, which is what we look at in the next section.

11.3 Uncertainty Inherent in Probability Distributions

Now let us consider a discrete random variable X which has n possible outcomes X_i, and a probability distribution $P(X)$. The probability of outcome X_i is P_i, and it has an associated uncertainty $h(X_i) = \log_2(1/P_i)$. Over many trials, with various outcomes, each with a different associated uncertainty, what will be the average uncertainty associated with $P(X)$? What we want is the *expectation value* of $h(X)$:

$$H(X) = E[h(X)] = -\sum_i P_i \log_2 P_i. \tag{11.3}$$

Note the capital H, to distinguish this concept from the uncertainty h of a single event. The quantity H can be referred to as "average uncertainty" or the "expected uncertainty", or the "Shannon entropy". It can be seen as the average number of yes/no questions you would need to ask to find out which of the n outcomes has actually occurred, using a kind of probability-weighted binary search. Imagine a thought experiment where your friend knows which value of i is the actual outcome, and you are asking questions to find out this value. Rather than simply dividing at $i \sim n/2$, you would use your knowledge of $P(X)$ to find the value of i which divides the summed amount of probability in half, and then ask whether the outcome is above or below that. Then you take the successful half, and once again work out which value of i divides the remaining probability in half, and so on until you have the actual outcome.

Let us at now look at H in a variety of useful cases.

11.3.1 Two-Box Experiments: Bernoulli Trials

As described at the beginning of Chap. 4, a Bernoulli trial is an experiment which has a yes/no, either/or outcome, and a Bernoulli process is an ongoing sequence of such trials. Examples might be coin tosses, or recording whether a radiation counter detects a particle or not in each one-second time window. We can conceive of any such experiment as having two boxes, each of which may have a differing probability, e.g. because the coin is weighted, or because the boxes are of different sizes. If box-1 has probability $P_1 = p$ for each trial, then box-2 has probability $P_2 = 1 - p$, and we have defined a very simple two-state probability distribution. Using equation (11.3) we then find that the average uncertainty for the two boxes is

$$H(p) = -\left(p \log_2 p + (1-p) \log_2 (1-p)\right).$$

The function $H(p)$ is plotted in Fig. 11.1. Note that H becomes small for both large and small values of p. What value of p gives the maximum? Differentiating and setting $\mathrm{d}H/\mathrm{d}p = 0$ we find that we require $p = 1 - p$, i.e. $p = 1/2$, to give us the maximum uncertainty. Note that at this p-value, the two boxes have the same probability and the maximum value is $H_{\max} = -2p \log_2 p = 1.0$.

11.3.2 Multi-box Experiments

Now let us consider n boxes, with an arbitrary set of P_i values. What distribution gives us the maximum uncertainty? Imagine first a case where all the boxes except one are the same: $n - 1$ boxes have probability p, but the final box has probability

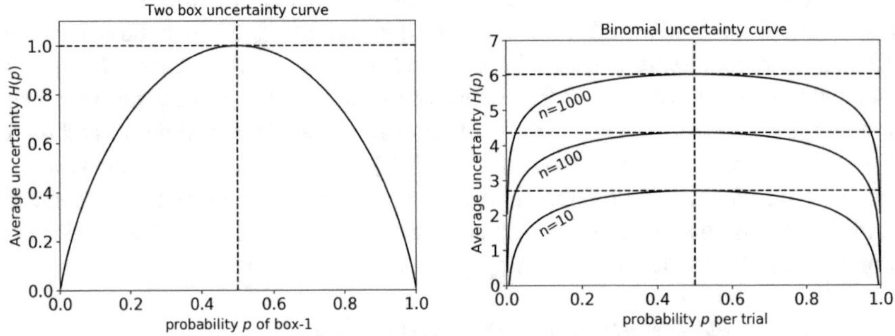

Fig. 11.1 **Left**: Average uncertainty $H(p)$ for a two-box system, where the probabilities are p and $1 - p$. The maximum uncertainty is $H_{\max} = 1.0$ at $p = 0.5$ **Right**: Average uncertainty $H(p)$ for the binomial distribution, for various values of of n. The maximum occurs at $p = 0.5$ in all cases, and $H_{\max}$ varies logarithmically with n. The expected maximum given by the formula shown in Sect. 11.3.4 is shown in each case

p_2. Of course the probabilities have to add up to one, so $p_2 = 1 - (n-1)p$. Then we have

$$H = -\left((n-1)p\log_2 p + [1-(n-1)p]\log_2[1-(n-1)p]\right).$$

Differentiating and setting $dH/dp = 0$ you find that

$$(n-1)\log_2 p = (n-1)\log_2[1-(n-1)p].$$

The expressions inside the two logs must therefore be equal and so

$$p = 1-(n-1)p = 1 - np + p \quad \Longrightarrow 1 = np \quad \Longrightarrow p = 1/n.$$

If $p = 1/n$ then we find that $p_2 = 1-(n-1)/n = 1/n$ so in fact all the probabilities are forced to be the same. We can make the same argument whichever P_i we choose to make the anomalous one. It is clear therefore that to maximise uncertainty we must have a *uniform distribution*. With n boxes each having probability $p = 1/n$ a little more algebra shows that the maximum value is

$$H_{\max} = \log_2 n \qquad \text{for uniform } n\text{-box distribution.} \tag{11.4}$$

You may note that the argument we have made here is essentially the same as we made in Chap. 3, Sect. 3.6, when considering partitions. We looked at the number of microstates per macrostate—the multiplicity—and found that the uniform distribution is what maximised the multiplicity. It makes sense that we get the same answer. The more microstates there are that correspond to the same macrostate, the less certain we will be concerning which microstate the system is actually in.

Any non-uniform distribution will give a smaller uncertainty than the uniform case. Suppose we take a distribution with four possible outcomes. For a uniform distribution, with $P_i = 0.25$, we find $H_{\max} = \log 4 = 2$ bits. Now suppose $P_1 = 0.5$, $P_2 = 0.25$, and $P_3 = P_4 = 0.125$. Then we get

$$H = 0.5 \times \log 2 + 0.25 \times \log 4 + 0.125 \times \log 8 \quad = 1.375 \text{ bits}.$$

Another good example might be the experiment of rolling two six-sided dice, and looking at the total rolled, as discussed in Chap. 1, Sect. 1.4.1. There are 11 possible outcomes ($T = 2$ through 12). If the 11 outcomes were equally likely, the maximum Shannon entropy would be $H_{\max} = \log_2(11) = 4.46$ bits. Using the actual P_i values for a two-dice experiment (see Fig. 1.1), we find $H = 3.27$.

11.3.3 Binomial Distribution

We can compute H for any distribution defined by a mathematical formula. For example, consider the binomial distribution with number of trials n_t, and probability of success per trial p:

$$P_i = \binom{n_t}{r} p^r (1-p)^{n_t - r}, \quad \text{where} \quad r = i - 1 \quad \text{and} \quad n = n_t + 1.$$

Note that following our earlier convention we have n different outcomes running from $i = 1$ to n, but when we have n_t trials, the number of successes can be from $r = 0$ to $r = n_t$. For $n_t = 3, n = 4$, a uniform distribution would give $H = 2.0$, whereas with $p = 1/2$ the binomial distribution gives $= 1.81$. Likewise for $n_t = 7, n = 8$, a uniform distribution would give $H = 3.0$, and the binomial gives $H = 2.45$. A good approximation is given by

$$H_{n_t}(p) = \frac{1}{2}\log_2(2\pi e n_t p(1-p)) + O(1/n_t) \quad \text{Binomial uncertainty} \tag{11.5}$$

where the last term indicates terms of order $1/n_t$, which of course become negligible as n_t increases. The function $H(p)$ is plotted in Fig. 11.1 for various values of n_t. You can see that $p = 1/2$ gives the maximum H for any n_t, in which case

$$H_{\max} = \frac{1}{2}\log_2\left(\frac{n_t \pi e}{2}\right).$$

There are $n = n_t + 1$ possible outcomes. Larger values of n_t make more concentrated distributions in relative terms, while being broader in absolute terms, as discussed in Chap. 4, Sect. 4.3, so it makes sense that the the uncertainty increases with n_t but becomes smaller compared to the corresponding uniform distribution.

11.3.4 Poisson Distribution

For a Poisson distribution with mean value μ, it can be shown that

$$H = \frac{1}{2}\log(2\pi e\mu) - \frac{1}{12\mu} + \cdots \quad \text{Poisson uncertainty} \tag{11.6}$$

where the ellipsis indicates higher order terms. We could re-express the binomial formula in terms of $\mu = np$, and if we let $p \to 0$ we can see that the Binomial and the Poisson agree in the limit, as we would expect. Notice that for both binomial and Poisson distributions it is approximately true that

$$H \simeq \frac{1}{2}\log(2\pi e\sigma^2).$$

11.3.5 Bivariate Distributions

If we have two discrete variables X and Y with n and m possible outcomes respectively, and joint probability distribution P_{ij}, then the joint uncertainty of the two variables is

$$H(X, Y) = -\sum_{i=1}^{n}\sum_{j=1}^{m} P_{ij}\log_2 P_{ij}.$$

We can illustrate the meaning of this quantity as follows. Imagine an $n \times m$ grid of boxes. A ball is thrown into one of the $\{i, j\}$ boxes at random, but with probability given by P_{ij}. Then H is the average number of yes/no questions you would need to ask to discover exactly which box contains the ball.

What if we are interested only in the uncertainty in Y? How do we extract that from the joint distribution P_{ij}? Just as with our discussion in Chap. 2, Sect. 2.3, there is more than one way we can ask that question. Suppose we already know the X value X_i. Then what we want is to examine the conditional distribution $g(Y) = g_i(j)$ as in equation (2.1), and extract the uncertainty from that distribution:

$$H_i = H(Y|X_i) = \sum_{j=1}^{m} g_i(j)\log_2 g_i(j), \quad \text{where} \quad g_i(j) = \frac{P_{ij}}{\sum_{j=1}^{m} P_{ij}}. \tag{11.7}$$

We will refer to this as the *specific conditional uncertainty*. Now imagine running this experiment many times, each time with a different X_i. What is the probability of a given X_i? This will be given by the marginal distribution for X, as in equation (2.2). Then the *average conditional uncertainty* for Y is the expectation value of H_i over the probability distribution $f(X)$:

$$H(Y|X) = \sum_{i=1}^{n} H_i fi, \quad \text{where} \quad f(i) = \sum_{j=1}^{m} P_{ij}. \tag{11.8}$$

Finally, the *marginal uncertainty* for Y can be found by calculating the uncertainty attached to the the marginal distribution $g(Y)$:

$$H(Y) = -\sum_{j=1}^{m} g(j)\log_2 g(j), \quad \text{where} \quad g(j) = \sum_{i=1}^{n} P_{ij}. \tag{11.9}$$

We can illustrate the meaning of these quantities by extending our thought experiment of throwing a ball into the $n \times m$ grid at random. $H(X, Y)$ is the average number of questions you need to ask to pin down which grid-box the ball is in, if you have no other information. The marginal uncertainty $H(Y)$ is the number of questions you have to ask to pin down which Y-row the ball is in, if you don't care which X-column it is in. Next, imagine your friend taking a peek and telling you "the ball is somewhere in column $i = 6$". Then the specific conditional uncertainty $H(Y|X_i)$ is the average number of questions you would need to ask to pin down the Y position, given that specific piece of information. Finally, if you imagine running the experiment many times, with your friend telling you each time what the i position is, then the average conditional uncertainty $H(Y|X)$ is the number of questions you need to ask to pin down the Y position, averaged over the many runs of the experiment.

11.3.6 Mutual Information: Testing Dependence

If our two variables X and Y are independent, then their probability distribution can be written $P_{ij} = f_i g_j$ where f and g are the marginal distributions for X and Y. Then the joint uncertainty can be written as

$$\begin{aligned} H(X, Y) &= -\sum_{i=1}^{n}\sum_{j=1}^{m} f_i g_j \left(\log_2 f_i + \log_2 g_j\right) \\ &= -\sum_{i}\sum_{j} f_i g_j \log_2 f_i - \sum_{i}\sum_{j} f_i g_j \log_2 g_j. \end{aligned}$$

Looking at the definition of marginal uncertainty in equation (11.9), if we have $P_{ij} = f_i g_j$, we can see that the two terms above are identical to $H(X)$ and $H(Y)$ respectively. So for independent variables, the joint uncertainty is equal to the sum of the two marginal uncertainties. More generally, this will not be the case. This provides us with a simple numerical test of independence. We can define the *Mutual Uncertainty* or more usually in the literature, the *Mutual Information*:

$$I(X, Y) = \left[H(X) + H(Y)\right] - H(X, Y). \tag{11.10}$$

For independent variables, $I = 0$; for dependent variables, $I > 0$. Noting that $H(X, Y) = H(X) + H(Y) - I(X, Y)$, we can interpret the situation as follows: our uncertainty in the X, Y position of our imaginary ball is determined partly by the uncertainty in X, i.e. $H(X)$, and partly by the uncertainty in Y, i.e. $H(Y)$. However, if the uncertainty in Y is connected to the uncertainty in X, then if we add $H(X) + H(Y)$ we have overestimated the total uncertainty in X, Y, by the amount $I(X, Y)$. It's not too hard to show that it is also the case that $I(X, Y) = H(Y) - H(Y|X)$—i.e. it is the remaining uncertainty in Y after you have taken account of the *average* uncertainty caused by X. Likewise, of course, $I(X, Y) = H(X) - H(X|Y)$.

11.3.7 Continuous Distributions

So far we have considered discrete random variables. When we look at the uncertainty associated with continuous random variables, we hit a problem. Suppose we consider a small range Δx of a continuous random variable x, and find an amount of probability Δp within the range. if we decrease Δx, then Δp also decreases, but the ratio $p(x) = \Delta p/\Delta x$ converges, which enables us to define a probability density function. Can we likewise define an uncertainty density function $h(x)$? We could attempt this by looking at the amount of uncertainty Δh associated with the range Δx, and taking a limit:

$$\Delta h = \log_2\left(\frac{1}{\Delta p}\right) = \log_2\left(\frac{1}{p(x)\Delta x}\right), \qquad h(x) = \lim_{\Delta x \to 0} \frac{\Delta h}{\Delta x}.$$

However unfortunately, as Δx gets smaller, Δh gets larger, so the uncertainty diverges. Suppose however we ignore this and plunge ahead trying to calculate an expectation value for the uncertainty associated with the whole PDF $p(x)$. The technique is to take a discretised version and then take the limit. We therefore consider N samples p_i, separated by distance Δx, so that the amount of probability in each range is $\Delta p_i = p_i \Delta x$. Then the uncertainty connected with each range is

$$\Delta h_i = \log_2(1/\Delta p_i) = -\log_2(p_i \Delta x) = \log_2 p_i - \log_2 \Delta x.$$

Then the discretised estimate of the average uncertainty H is

$$\hat{H} = \sum_{i=1}^{N} -p_i \log_2 p_i - N \log_2 \Delta x,$$

and finally, taking limits, we have

$$H = \lim_{\Delta x \to 0} \hat{H} \quad = \quad -\int p(x) \log_2 p(x)\, \mathrm{d}x \quad + \quad \infty. \tag{11.11}$$

We now simply ignore the second term (infinity), and identify the first term as the continuous equivalent of the average uncertainty, often known as the *Differential*

Entropy. Interestingly, it is of course exactly what you might have naively assumed to be the expectation value of uncertainty. The trick of producing an infinity and then subtracting it is a favourite trick of physicists; in quantum mechanics it is known as “re-normalisation”.

One can now apply this formula to our favourite continuous PDF, the Gaussian. The result found is that

$$H = \frac{1}{2}\log(2\pi e\sigma^2) \quad \text{Gaussian uncertainty} \tag{11.12}$$

Comparing this with Sect. 11.3.4 we can see that this result is the same as the Binomial and Poisson in the limit of large μ, which at least somewhat vindicates the dubious practice of subtracting an infinity.

11.3.8 Comparing Distributions

It can often be useful to compare the H values given by two different distributions—for example when looking at the efficiency of a message encoding system, as we will examine in the next section. For the purpose of discussion, let us assume we are considering a k-box system, with different probability distributions P_i and Q_i, with the special case of a uniform distribution labelled as $U_i = 1/k$. The corresponding uncertainties are H_P, H_Q and H_U. As we have seen $H_U = \log_2 k$, and H_P and H_Q will always be smaller.

The simplest comparison between P and Q would be to take the ratio H_P/H_Q. If we compare to the uniform case, then $R = H_P/H_U = H_P/\log_2 k$ could be seen as the “efficiency” of P, i.e. how close it comes to the uniform case, so that R is a number between 0 and 1. Alternatively, as H is a logarithmic quantity, it may seem more natural to take the difference, $H_P - H_Q$, which has units of bits. Then comparing to the uniform case $\varepsilon = H_P - \log_2 k$ can be seen as measuring how many bits P is from the uniform case. However, the standard method for defining the “distance” between two distributions is a little more subtle. First, we define the *cross entropy* between P and Q:

$$H(P, Q) = \sum P_i \log Q_i.$$

This is the expectation value for Q, but using the probability distribution for P. You could see it as the entropy you get if you use an “alien” probability distribution Q rather than the “true” distribution P. We can then characterise how far we are from the true H_P by defining what is known as the *Kullback–Leibler distance* of Q from P:

$$D_{KL}(P||Q) = H(P, Q) - H(P) = \sum P_i \log \frac{P_i}{Q_i}. \tag{11.13}$$

Sometimes this quantity is known as the "KL divergence" rather than the KL distance. Note the asymmetry in the definition—what we have is the divergence of Q from the ideal distribution P. If we compare to the uniform distribution then you can see that

$$D_{KL}(U||Q) = k - \sum P_i \log Q_i.$$

11.4 Messages, Alphabets, and Data Compression

Shannon constructed his definition of information/uncertainty/entropy as part of his development of communications theory. How do we pass messages most efficiently? To answer this question, we need to understand how much information a message contains. Rather than thinking about specific individual messages, it helps to think about classes of message, and what the structure is for that class of message (Fig. 11.2).

In the Shannon picture, transmitting a message consists of several standard stages. (i) We split the message into a sequence of *symbols*. (ii) We *encode* each symbol as a binary string. (iii) We transmit the codewords through a *channel* with some fixed *capacity* in bits per codeword. (iv) During transmission, *noise* may be added, corrupting some of the codewords. (v) The arriving codewords are then *decoded* to reveal the message at the other end.

We can now look at each of these concepts more carefully. For the purpose of brevity however, we are going to ignore noise, assuming perfect transmission through the channel.

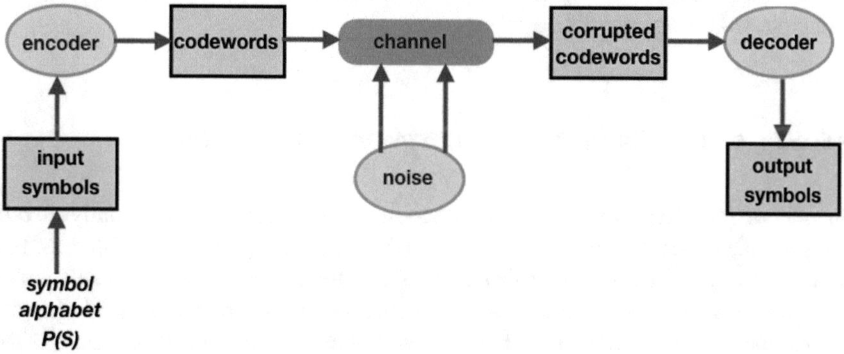

Fig. 11.2 Standards steps in the process of encoding, transmitting, and decoding messages. See text for details

11.4.1 Messages and Alphabets

Any message, or more generally any data structure, can be seen as a sequence of symbols S, each of which is chosen from an *alphabet* of k possible symbols, A_i. The message might for example be a piece of text in the English language, with each symbol being one of the 26 letters of the Western alphabet (or a few more symbols, including spaces, commas, and so on). Alternatively, we might be transmitting an image made up integer pixel values. If we run the rows together one after the other, we can think of this as a long message, where each symbol represents a pixel value as a number between, say, 0 and 256. The symbols will in general not appear with equal frequency in our message; we could describe their relative frequency for a particular class of message by a probability distribution for that alphabet, P_i. That alphabet will then have an associated average uncertainty or Shannon entropy

$$H = -\sum_{i=1}^{k} P_i \log_2 P_i .$$

So, if we were transmitting a sequence of numbers representing the rolls of two dice, we would have $H = 3.27$ as discussed in Sect. 11.3.2. For the case of transmitting a message in English, the distribution P_i is given by the empirically found frequency distribution of letters in English text. (See for example the table in Stone, page 60). We then find that you expect $H = 4.11$. For an image, the frequency with which different grey-levels occur may vary quite a lot depending on what sort of image we are transmitting. If all the numerical values between 0 and 256 occurred equally often, we would have $H = 8.0$, but in practice the observed H will be much less. These various H values tell us the amount of *information per symbol* we would in principle gain from decoding a message.

11.4.2 Codes, Channels, Compression and Efficiency

In order to transmit the message, or the image or whatever, we would normally *encode* the symbols in some binary fashion, to turn them into a sequence of *codewords*. For example, we could use the Unicode standard to express each text character as a binary number. Allowing for space and say 7 punctuation characters, we have 34 options, so that we would ideally need 6 bits per symbol in our codewords. (Or more if we double up for capitals). Likewise, we could express our image pixel values as 8 bit numbers, so needing 8 bits per symbol.

The term "channel" sounds like a physical pipe, but in practice refers to the fixed practical arrangements for transmitting the codewords. For example, we may take a decision to put our data into a sequence of computer words that are 1 byte long, i.e. 8 bits. The word size of 8 bits is then the *capacity* of our channel—$C = 8.0$ bits. But if ideally we only need, say, 6 bits per codeword to express all the symbol options, then we are wasting some space in our fixed-size channel codewords. Suppose for example our message is made up of only four symbol options, that occur with equal frequency—A, B, C, D. Then with just 2 bits, we can encode these as $A = 00$, $B = 01$, $C = 10$, $D = 11$. However suppose instead we have five options—A, B, C, D, E. Two bits is not enough to express these, so we will need to use three bits. A codeword of that size can express eight possibilities, so three of the possible codewords will never be used. In principle our messages have Shannon entropy/information content per symbol of $H_{\text{sym}} = \log_2(5) = 2.32$, whereas our channel capacity is $C = 3.0$. Our encoding can be said to have an *efficiency* of $R = H_{\text{sym}}/C = 2.32/3.0 = 0.77$, or to be wasting $\varepsilon = C - H_{\text{sym}} = 0.68$ bits/symbol.

However, our encoding doesn't have to be a one-to-one mapping from symbols to codewords. For example, we could take the symbols three at a time, so that the possibilities are AAA, AAB, AAC..., with 125 options for these new multi-symbols. Note that if the original symbols A, B, C, D, E occur with equal frequency, then our 125 multi-symbols will also occur with equal frequency. The Shannon entropy for the new alphabet of multi-symbols will therefore have $H_{\text{sym}} = \log_2(125) = 6.92$. We can encode the multi-symbols into 7 bit codewords, which allows for 128 possible options. So now the channel capacity is $C = 7.0$ and the entropy of the messages is $H = 6.92$, giving a much improved efficiency $R = 6.92/7.0 = 0.989$. This packing of symbols three at a time is a simple example of *lossless compression*.

Of course most messages will have symbols that come from an alphabet that does not have a flat P_i distribution. Then even if we pack multiple symbols together, the efficiency may be poor. The distribution of English characters is an obvious example. Our 6 bit codewords have channel capacity $C = 6.0$; a flat distribution of 34 symbols would have $H = 5.1$; but as we discussed above, the actual Shannon entropy of English text has roughly $H = 4.1$. Likewise, for an 8-bit image, an image with extreme contrast, where there might large patches of pure black and large patches of pure white, might have H much smaller than the channel capacity $C = 8.0$. However, there are clever ways to encode the symbols. For a large range of problems, the optimal method is *Huffman coding*, which, roughly speaking, involves frequency-sorting the symbols and arranging them into a binary tree for optimal packing. One of the key aims of Shannon's original 1948 paper was to show that in principle some coding always exists which can bring you arbitrarily close to the channel capacity.

11.4.3 Messages with Mutual Information

In all the discussion above, we have assumed that the symbols are independent. In practice this is often not the case. For example, in English text, the probability that a given symbol will be u is strongly dependent on the previous symbol. Mostly, the letter u occurs with quite low frequency, but if the previous letter was q, then getting a u next is very likely indeed. Successive symbols S_1, S_2 have a very high degree of *mutual information* $I = h(S_1) + h(S_2) - h(S_1, S_2)$. Of course the situation is even more complicated, as patterns of letters are very common. Dealing with such mutual information is beyond our goals here, which is simply to understand the general principles of information theory.

11.5 The Maximum Entropy Principle in Statistical Inference

How do we calculate or assign the appropriate values in a probability distribution? The traditional method, which we followed in the first two parts of this book, is to first identify elemental events which we believe to be equally probable, and then to recognise other situations as compound events made up of the elemental events. So for example, we assume that in rolling a single die, each face has a probability of $1/6$, and it is then simple to calculate the probability of rolling two dice giving a total T (see Sect. 1.4.1). In assigning the elemental events equal probability, we are applying what Laplace referred to as the "Principle of Indifference". We employed the same principle when using Bayesian techniques and assigning prior probabilities. For example if we are testing a coin which we think may be biased, and have no reason in advance to suspect which of heads or tails might be biased, we assign them both equal prior credibilities/probabilities. Finally, when considering particle distributions in Chap. 3, we assumed that our system was ergodic, i.e. spending equal amounts of time in each microstate. This is really just a variant of the principle of indifference.

Starting with Jaynes during the 1960s (see Jaynes 2007), a number of scientists have raised the idea of maximising uncertainty, i.e. the Shannon Entropy, as a similar kind of logical principle—either that Nature tends to arrange itself into a form where we have the least knowledge, or that in making logical inferences, we should do so in a way that makes the fewest unjustified assumptions. This idea is known as the *Maximum Entropy Principle*, and techniques using this idea are said to be using the *Maximum Entropy Method*, usually referred to as "MEM" or "MaxEnt". Mathematically, the Maximum Entropy Principle is more or less the same thing as the Principle of Indifference, but it gives things a different philosophical spin.

11.5.1 Shannon Entropy and Macrostate Multiplicity

In Chap. 3, Sect. 3.6, we looked at partitions, taking distributions of particles in space or energy as an example. The way we approached such problems was to calculate the *multiplicity* W of a macrostate, i.e. the number of microstates corresponding to the same macrostate. If we apply the principle of indifference by assuming each microstate to be equally probable, then the most probable macrostate is the one with the largest W. We showed that a uniform distribution maximises W, unless there is some additional constraint, such as the total energy of all particles being constant. Another way to look at the idea of maximising multiplicity is that the macrostate with the largest multiplicity will be the one that gives us the least information about exactly which microstate the system is actually in; maximising W will also maximise uncertainty.

In fact, this idea—that maximising W and maximising H should be closely connected—leads directly to the Shannon expression for entropy. We start by repeating equation (3.5) for convenience:

$$W = \frac{n!}{n_1!n_2!...n_k!}.$$

Here, k is the number of "boxes" in our partition, n is the total number of particles, and n_i is the population of the various partition boxes. Next we note that maximising W will be the same as maximising $\log W$, and we use the Stirling approximation from equation (3.4), $\log n! \sim n \log n - n$. Assuming that n is sufficiently large and dropping the approximation sign, we get

$$\begin{aligned} \log W &= \log n! - \sum \log n_i! \\ &= n \log n - n - \sum (n_i \log n_i - n_i) \\ &= n \log n - \sum n_i \log n_i, \end{aligned}$$

where we have used the fact that $\sum n_i = n$. Now, for a given particle, the probability of being in box i will be $P_i = n_i/n$. Writing $\log W$ in terms of the P_i values, we get

$$\begin{aligned} \log W &= n \log n \sum n P_i \log n P_i \\ &= n \log n - n \sum P_i (\log n + \log P_i) \\ &= n \log n - n \log n \sum P_i - n \sum P_i \log P_i \\ &= -n \sum P_i \log P_i \\ &= -nH. \end{aligned}$$

So apart from a scaling factor (noting that above derivation is in natural logs rather than logs to the base 2), the log of multiplicity is essentially the same thing as the Shannon entropy. This is a strong argument for the Shannon expression for uncertainty, $-\log_2 p$, rather than for example $1 - p$ or $1/p$.

11.5.2 Using Uncertainty/Entropy to Choose Priors

As we discussed several times in Chaps. 7, 8, and 10, Bayesian logic does not tell us how to choose prior probabilities. Personally, I find this a strength of the approach.

You have to make a initial subjective judgement; the use of priors forces you to make your assumptions explicit. When you have lots of data, the choice of priors makes little difference. When you have less data, different priors will give you different answers—but this is a good thing. Again, it brings the inevitable biases and subjectivity out into the open and gives you a way to discuss them. Nonetheless, there is a desire to find a way to make choosing priors rigorous and in some way more objective.

The principle of indifference is the traditional starting point. If we have k different hypotheses that have various prior probabilities π_i, and have no reason to prefer one hypothesis over another, we should set them all to $\pi_i = 1/k$. Maximising the uncertainty/Shannon entropy $H = -\sum \pi_i \log \pi_i$ gives the same answer, as we saw in Sect. 11.3.2. If we have a parameter θ defined over some finite range $\theta_{\rm min}$ to $\theta_{\rm max}$, then we can maximise the "differential entropy" from equation (11.11), which gives a prior probability density of $\pi(\theta) = 1/(\theta_{\rm max} - \theta_{\rm min})$.

Things get more interesting if we can add further constraints. Suppose we have k evenly spaced possible values of $\theta_i = \theta_{\rm min} + i\Delta\theta$, with the corresponding prior probability being π_i. We can then maximise H given the additional constraints using the Lagrange multiplier technique of Sect. 3.8.2, taking each π_i value as a variable. The simplest constraint is the normalisation of the probability distribution, $\sum \pi_i = 1.0$, which we express as $g_1(\pi_1, \pi_2..) = 1 - \sum \pi_i$. This gives the uniform distribution as a solution. However, we might for example know the *mean value* of θ, μ_θ. If we maximise H with the additional constraint $g_2(\pi_i, \pi_2, ..) = \sum \pi_i \theta_i - \mu_\theta$, it can be shown that the solution we get is exponential in θ, i.e. $\pi(\theta) \propto e^{-\theta}$. Of course, this is exactly analogous to the problem we considered in Sect. 3.8.1, where we constrain the total particle energy, or equivalently, the mean particle energy, as the total number of particles is fixed. Finally, suppose we know not just the mean but also the variance of θ, σ_θ^2. The shape which maximises the Shannon entropy under all three constraints of normalisation, mean and variance is the *Gaussian* distribution with the same values of mean and variance.

11.5.3 Image Restoration

A very popular use of MEM in data analysis is the derivation of the "best" image, given noisy and blurred data. The problem is illustrated in Fig. 11.3. Here we simplify the situation by considering a one-dimensional image with just six pixels. The true brightness distribution as a function of x is $f(x)$. This is not necessarily characterisable as a simple mathematical function with a few parameters—it is a completely arbitrary curve, representing for example the brightness in a scene at different spatial positions, or the flux versus wavelength in a spectrum. We then measure the brightness at discrete points, resulting in the data points shown. As well as sampling $f(x)$, these data points will have added measurement errors; furthermore, rather than sampling the true $f(x)$, the measurements may be sampling $h(x) = f(x) * g(x)$, the convolution of $f(x)$ with the instrument resolution $g(x)$.

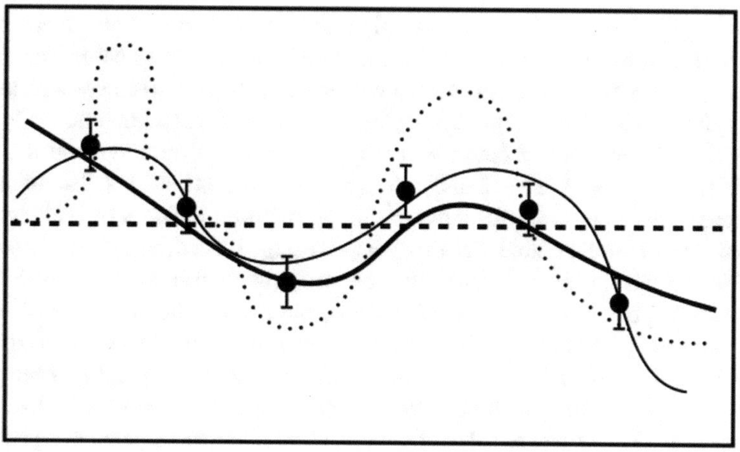

Fig. 11.3 A simple one-dimensional "image" with six data points, and variety of possible true brightness distributions consistent with the data points and their errors. It seems instinctively reasonable that the thick black line is the most sensible solution, because it is the smoothest of the curves consistent with the data

All the curves shown in Fig. 11.3 (except for the horizontal dashed line, which we will come to shortly) are consistent with the data points, in the sense that they would give reasonable χ^2 values. The difficulty is that there are infinitely many such curves; the problem is severely underconstrained. One solution would be to propose some parametric form for the underlying $f(x)$, and fit the parameters, following the methods of Chap. 10. However, there is no good physical basis for choosing such a parameterised form. How then do we choose which curve is best? Instinctively, we feel that the dotted curve is not a sensible solution; we have no evidence for the extreme bumps and wiggles suggested; the thin black line is better, and the thick black looks best. What the eye is doing is preferring the smoothest curve consistent with the data.

MEM formalises these instincts, arguing that we should pick the curve that makes the fewest unjustified assumptions, and that this is best quantified by the average uncertainty or Shannon Entropy. But what do mean by the entropy of an image, as opposed to the entropy of a probability distribution? The trick is to imagine a fine-scaled discretisation. First we divide x into a series of cells at positions x_i, where the true brightness value is $f_i(x_i)$. This discretisation can be finer than than the data pixels, and as fine as we like in order to approximate a continuous curve. Next, we likewise make a fine discretisation of the brightness using "quanta" of size δ, so $f_i = n_i\delta$. Then the n_i can be seen as the populations of the cells. Informally we could think of the quanta as the "photons" that make up the image, although this need not literally be the case.

The total number of quanta is $n = \sum n_i$. Imagine taking these one at at time, and throwing them at random into the k cells. Then the probability that cell i will have number n_i is $P_i = n_i/n$. We then define the entropy of the image in the usual way as $H = -\sum_i P_i \log P_i$. If we then simply maximise H this will give a completely uniform image, as in the dashed line of Fig. 11.3. However, this does not agree with the data points. Given a trial curve $f(x)$ we can quantify the agreement with the data as normal using χ^2. The simplest procedure is to require that a good fit has $\chi^2 = N$ where N is the number of data points. We can then use the Lagrange multiplier technique, with $g_1 = \sum P_i - 1$ and $g_2 = \chi^2 - N$. We then have $k + 2$ simultaneous equations to solve in order to find $f(x)$ to the required fineness of scale we have decided upon. However, note that k may be very large indeed; images may have tens of millions of pixels, and we would often choose to use a discretisation several times finer than the pixel scale. Using MEM is therefore computationally expensive.

As we noted in Sect. 11.5.1, maximising entropy is really about maximising multiplicity. Returning to our n quanta, the set of values n_i could be seen as the "macrostate" of the image, and the specific locations of the n quanta as the various "microstates". Then there are W different ways of achieving a given image f, with W given by (3.5) as usual. Next, we can apply Bayes's theorem:

$$P(f|D) \propto \pi(f)\, L(D|f).$$

The best choice of prior $\pi(f)$ is the one that corresponds to the macrostates with the most microstates. Taking logs, and applying Stirling's approximation again, we find that $\log P \propto H - \chi^2$. We then need to simultaneously maximise both the entropy of the finely gridded $f_i(x_i)$ and the fit with the measured data points.

11.6 Key Concepts

Some of the key concepts from this chapter are:

- The subtle relationships and differences amongst probability, information, uncertainty, and surprise
- The idea of information as the number of binary search questions needed to remove uncertainty
- How to combine uncertainties, and the idea of mutual information
- The idea of the average uncertainty or Shannon entropy associated with a probability distribution
- The idea of messages and alphabets, and the Shannon entropy associated with a message
- The importance of message entropy in data compression
- The connection between maximising entropy and maximising multiplicity
- The idea of using maximum entropy to choose priors
- The use of maximum entropy in image restoration

Key formulae in this chapter include: the definition of the uncertainty connected with a probability (11.1); the formula for combining uncertainties, allowing for conditional uncertainty (11.2); the definition of the average uncertainty/ Shanon entropy connected with a probability distribution (11.3); the formulae for the uncertainty of a uniform n-box distribution (11.4), for a binomial distribution (11.5), and for a Poisson distribution (11.6); for a bivariate distribution, the formulae for specific conditional uncertainty (11.7), average conditional uncertainty (11.8), marginal uncertainty (11.9), and mutual information (11.10); the general formula for differential entropy (11.11), and its value for the Gaussian distribution (11.12); and the definition of the Kullback–Leibler distance between two distributions (11.13).

11.7 Further Reading

There are many good textbooks covering information theory at a variety of levels. Three that I like are Stone (2015), which is very clear and simple; Applebaum (2008), which has a nicely unified approach to probability theory and information; and MacKay (2003), which is very thorough and rigorous. In particular in MacKay you will find a thorough treatment of noise, which I rather skimmed over in this chapter. You will also find clear treatments of various issues in the explanatory introductions to various sections of Press et al. (2007). MacKay also has many amusing touches and historical details. Amongst other things, he explains how the use of the term "ban" to represent factors of ten in information stems from the wartime code-breaking work of Turing and others at Bletchley park, which used specially printed sheets of paper printed in Banbury. The code-breaking task was known as "Banburismus". Although David MacKay has now sadly passed away, his research group maintains his web pages.

The ideas of communication/information theory were swirling around for many years, and had precursors in the work of Gibbs, Nyquist, Tukey, Hartley, von Neumann, and Wiener. However, the famous paper by Shannon (1948) really did crystallise all the key issues. The paper was first published in the Bell Systems Technical Journal, but was very quickly re-published in book form together with a very interesting and readable introduction by Weaver. This book has been re-issued and reprinted many times, e.g. Shannon (1998).

The idea of using maximum entropy in image restoration and related problems started with Frieden (1971), and ballooned quickly from the 1980s onwards. A interesting set of essays can be found in Buck and MacAulay (1991) and an early application to astronomical images in Willingale (1981).

There is an amusing story about how Shannon chose to call his measure "entropy", recounted in Tribus and McIrvine (1971). He originally wanted to call it "uncertainty", but Von Neumann told him "entropy" was a better idea, partly because his uncertainty function was already used in statistical mechanics under that name, but mostly because no one knows what entropy really is, so in a debate he would always have the advantage. As we will discuss in Chap. 14, thermodynamic entropy can in fact be seen as a specialised version of the more general Shannon entropy.

11.8 Exercises

11.1 What is the uncertainty attached to a single roll of a six sided die (a) in Shannons, (b) in bans?

11.2 A card is picked from a standard pack of playing cards. What is the uncertainty attached to (a) drawing a spade? (b) drawing an Ace? (c) drawing the Ace of Spades? Check that $h(c) = h(a) + h(b)$.

11.3 A biased coin has a 60% chance of landing on heads. What is the Shannon-entropy per coin-flip? How badly biased would the coin need to be to have $H = 0.5$?

11.4 Show that the formula for the maximum H for a binomial distribution for a given n agrees with the expression for H for the Gaussian distribution, given the variance of the Gaussian.

11.5 A particle created in interactions at a new particle collider can be described by two quantum numbers Q_1 and Q_2, each of which can be either positive, negative, or zero. The values following a given collision are random, but with some bivariate probability distribution. Three different theoretical models claim to predict these probabilities. The calculations are complicated, but a number of simulations using each model give a distribution of results as in Fig. 11.4. Use these to calculate the corresponding probability distributions, and for each distribution calculate the joint uncertainty, the marginal uncertainties, and the mutual information. By assigning numerical values to positive/negative, also calculate the covariance of each distribution. What do these calculations show about the difference between dependence and correlation?

11.6 Suppose we are repeatedly rolling a six sided die, and transmitting the sequence of results as a coded message. The computer system we are using forces us to to transmit using words which are multiple of 4 bits long. With 4 bit codewords, what is the coding efficiency of our message? How could we go about improving our coding efficiency?

Model A

Q_2 value \ Q_1 value	-	0	+
-	5	10	5
0	9	18	9
+	5	10	5

Model B

Q_2 value \ Q_1 value	-	0	+
-	10	4	2
0	4	19	4
+	2	4	10

Model C

Q_2 value \ Q_1 value	-	0	+
-	10	17	10
0	17	2	17
+	10	17	10

Fig. 11.4 Frequency distributions for the two quantum numbers, calculated for the three different theoretical models. (See Exercise 11.5)

References

Applebaum, D.: Probability and Information: an Integrated Approach, 2nd edn. Cambridge University Press, Cambridge (2008)

Buck, B., MacAulay, V.A.: Maximum Entropy in Action: A Collection of Expository Essays. Oxford University Press, Oxford (1991)

Frieden, B.F.: Restoring with maximum likelihood and maximum entropy. J. Opt. Soc. Am. **62**, 511–518 (1971)

Jaynes, E.T.: Probability Theory: The Logic of Science, sixth printing. Cambridge University Press, Cambridge (2007)

MacKay, D.J.C.: Information Theory, Inference and Learning Algorithms. Cambridge University Press, Cambridge (2003)

Press, W.H., Teukolsky, S.A., Vetterling, W.T., Flannery, B.P.: Numerical Recipes: The Art of Scientific Computing, 3rd edn. Cambridge University Press, Cambridge (2007)

Shannon, C.: A mathematical theory of communication. Bell Syst. Tech. J. **27**, 379–423 (1948)

Shannon, C., Weaver, W.: The Mathematical Theory of Communication. The University of Illinois Press, Champaign (1998)

Stone, J.V.: Information Theory: A Tutorial Introduction. Sebtel Press, Sheffield (2015)

Tribus, M., McIrvine, E.C.: Energy and Information. Sci. Am. **225**(3), 179–190 (1971)

Willingale, R.: Use of the maximum entropy method in X-ray astronomy. MNRAS **194**, 359–364 (1981)

Websites (all accessed March 2019):

MacKay Inference Group web page: http://www.inference.org.uk/mackay/

Chapter 12
Erratic Time Series

12.1 Outline of Content

- Stationary and non-stationary processes
- Practical difficulties characterising observed erratic time series
- Characterising time series using the structure function, autocorrelation, and the periodogram
- Moving Average processes
- Autoregressive processes
- Poisson processes and shot noise
- Conceptual differences in continuous random processes
- Stochastic differential equations and how to solve them
- The Ornstein–Uhlenbeck process and related processes
- Markov chains

A *time series* is a sequence of values of some quantity in time—for example the day by day changes in stock market values, or the flickering brightness level of a quasar, or the "shot noise" seen in electronic components. Such time series $x(t)$ can have a mixture of deterministic and stochastic elements, which leads to a fascinating variety of appearances. Figure 12.1 shows four simulated examples. Top left shows a *secular trend*, an evolution with time that is purely deterministic. The usual method of analysing such a situation is to find a *differential equation* that will reproduce the trend $x(t)$. Top right on the other hand shows a sequence that is *purely random*. Each data value x is drawn independently from some probability distribution $p(x)$. The remaining two examples are *partially random*. Bottom left is an example of a random walk, as discussed in Chap. 6. The key characteristic is that the neighbouring points are strongly correlated, but well separated points are more or less independent. At bottom right we see what is often known as flicker noise. Like the random walk, it seems to be a kind of semi-random drift, but the degree of correlation between neighbouring points is weaker. By eye, it seems intermediate

A. Lawrence, *Probability in Physics*, Undergraduate Lecture Notes in Physics,
https://doi.org/10.1007/978-3-030-04544-9_12

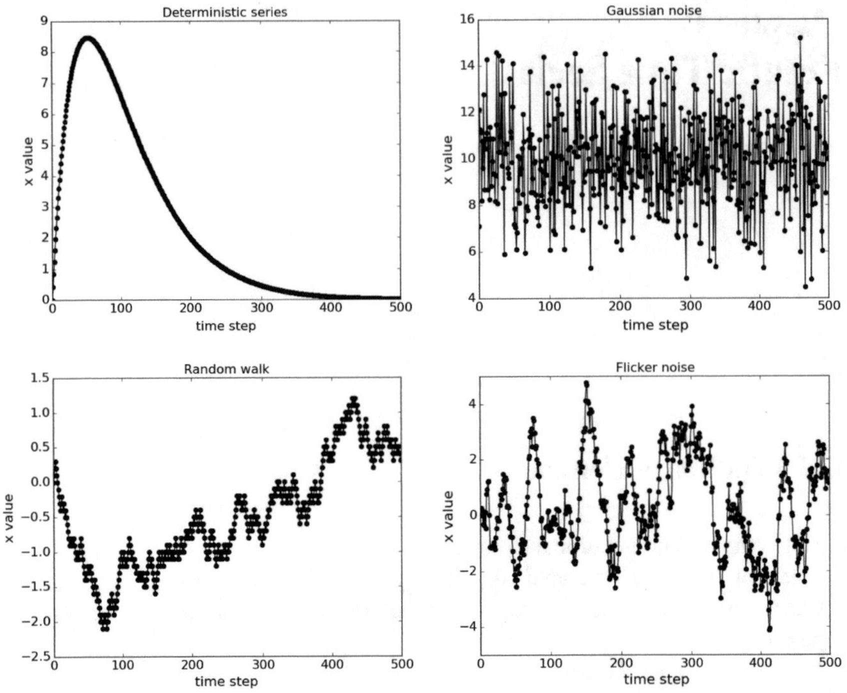

Fig. 12.1 Simulated time series. **Top Right**: Purely deterministic series calculated from a simple equation. **Top Right**: Purely random noise. Each point is independent and is drawn from a Gaussian PDF with $\mu = 10$, $\sigma = 2$. **Bottom Left**: A random walk in one dimension. Generated with a fixed step size in x and a random choice of increasing or decreasing at each time step. **Bottom Right**: Flicker noise. Simulated using an autoregressive process with $\mu = 0$, $\sigma = 0.5$, and $\alpha = 0.95$ (see text for an explanation)

between a purely random process and a random walk. This kind of flicker noise is very common in Nature. How do we explain it? Before we attempt that question, let us consider how to objectively characterise a given time series.

12.2 Characterising Time Series

A time series is essentially an ordered list of sample values. The process itself $x(t)$ may be discrete or continuous in either x or t. We will assume for now that the x values are continuous. In time, the physical process may be intrinsically discrete, producing a sequence of values x_i—for example daily stock prices. Alternatively the process $x(t)$ may be intrinsically continuous, but our measurements sample the values at a set of discrete times t_i. The *sample data series* is therefore always a sequence of values x_i. In much of this chapter, we will write the sequence of values as x_t, understanding that in fact t takes a series of values $t_1, t_2, \ldots$. Note that the sequence of times t_i may not necessarily be uniformly spaced, but for simplicity we will assume that they are. Given the random element involved, no physical model will correctly produce a specific sequence x_t. This situation is much the same as attempting to describe any set of sample values of a random variable x—it's just that the set of sample values x_t are ordered in time, so we have to explain the pairs t_i, x_i. The time ordered plot of sample values x_t as a function of time t_i, such as we see in Fig. 12.1, is often known as a *sample path.*

When considering unordered sample values in Chap. 2, we discussed how to produce descriptive numbers from the probability distribution for x—the mean, the variance, the skewness etc. Those characterising numbers, rather than the values x_t themselves, are then the targets for physical theories of the process to explain. Can we produce similar characterising quantities that capture the time-ordered behaviour of our erratic time series in a statistical sense? Such quantities can be used as a test for theories, but also can be used empirically to describe and separate one data series from another. We will look at three different methods—growth of variance, autocorrelation, and periodogram analysis.

First however, we examine the question of whether a time series has a consistent statistical behaviour at all.

12.2.1 Stationary and Non-stationary Processes

The simplest way to characterise a time series is ignore the time axis and construct the sample distribution of data values, $P(x)$. From this we can derive the usual moments—mean, variance, skewness, kurtosis etc. But do we get a consistent answer? Imagine taking a section of the time series in Fig. 12.1, with some starting value x_0 at time t_0, and continuing for some duration of time. A secular trend such as that in the top left of Fig. 12.1 will look different depending on the starting time of our section. The absolute value of time matters. For the other examples in Fig. 12.1, although the *exact* sequence of values x_i will be different for different starting values x_0, t_0, in a statistical sense, every section would look the same. Absolute time does not matter. A process where any slice of time is statistically the same is known as a *stationary process.*

What do we mean by "in a statistical sense"? Suppose that after deciding our section of time, we construct the histogram of values x_i to get the sample probability distribution $P(x)$. The resulting $P(x)$ would be slightly different each time if we repeat this process many times with different starting values, but will be consistent with always being drawn from the same underlying population/parent distribution. If we were to take longer and longer sections, or equivalently, average together many such sample distributions (making a so-called *ensemble average*) then the result should converge on a consistent answer. We could test the hypothesis that the sample distributions from two such sections are drawn from the same parent distribution, using the statistical inference techniques of Part III. On the other hand, a *non-stationary* process would not produce a consistent convergent result. The same logic could be applied to the other statistical quantities that we will discuss in the following subsections. Generally speaking, processes with a secular component are not statistically stationary. For example, a process might be made of a secular trend with added noise, in which case the process as a whole would not be stationary. A purely random process could also be non-stationary. For example, a process might consistent of purely random values, but with the dispersion of the random values decreasing with time. Quite often, processes in the real world are not intrinsically stationary, but an idealised stationary process is a good model. For the rest of this chapter, we assume that we are dealing with stationary processes.

The flicker noises that we consider in this chapter are *asympotically stationary*—i.e. they will pass the kind of tests discussed above if our time-section is long enough. In the next section we will see that "long enough" may be longer than you think.

12.2.2 Growth of Variance and the Structure Function

The characteristic feature of erratic time series is that they tend to drift with time. Consider Fig. 6.1, where we showed a large number of simulations of a random walk, starting from the same initial value. The right hand side of the figure shows the frequency distribution of resulting values at various different times. These values are reasonably well centred on a mean of zero, but the spread is increasing as the square root of time. By contrast, the purely random process will show the same spread of values regardless of the time we wait following the start. Visually, the flicker noise in the bottom right of Fig. 12.1 seems intermediate in this respect—drifting on short timescales, and looking much the same on long timescales. This suggests that we could use the *growth of variance with time* as a characteristic of a time series.

However, we have to be rather careful how we define the quantity that we are after. There are three ways we could do this.

Ideally, to understand the process behind our time series, what we want is the **path-to-path variance**. Imagine making simulations where we pick a starting value x_0 at time t_0, and generate many different sample paths of length t that all finish at some distant final time $t_f = t_0 + t$. Each path produces a different finishing value x_f. From many paths we look at the distribution $P(x_f)$, and get the variance $\sigma(t)$. This procedure would give us the variance of our process as a function of path-length. For a purely random process it will be constant; for a random walk $\sigma(t) \propto t^{1/2}$, and the process is not strictly stationary. Shortly we shall see what a typical flicker noise does.

Of course, in the real world, we do not usually have many paths starting from the same place, but only one observed path sampling our process. Furthermore, the process is sampled at a set of discrete time steps. If we take a section of length k time steps, the simplest procedure would be to calculate the **within-section variance**—i.e. starting from some initial value x_0, we take all the values x_i from $i = 0$ to $i = k$. We could then see how the variance σ_k^2 grows with section length k. However, this mixes together points close in time that will typically be close together with those further apart in time that will have a larger spread.

The third method of characterising growth of variance using a single sample path is as follows. We pick a starting value x_i at time t_i, and compare it to the value k time steps later, forming $\Delta x = x_{i+k} - x_i$. The value of k is usually known as the *lag*. We then repeat for many different starting times t_i, getting many different Δx values. For a stationary process, the mean value of Δx will be zero, but the variance will depend on k; our calculation of how this changes with lag k gives us what is known as the **structure function**, $\sigma^2_{\Delta x}(k)$. This method is particularly useful for time series with irregular sampling—any pair of data points provides an estimate at some lag k, and we can bin the samples in k.

Figure 12.2 compares the three methods—path-to-path variance, within-section variance, and structure function, for simulations of a flicker noise, using the autoregressive process described in Sect. 12.3.4. For this process, we know the theoretically expected process variance σ_x^2 in the limit of large k, and we know that variance should approach this limit following $\sigma_x^2(1 - e^{-k/k_{\text{ch}}})$ where in, the units used for our simulation, the characteristic scale length $k_{\text{ch}} = 20$ time steps. The path-to-path dispersion converges on the correct value very much as expected. The within-section dispersion for a single simulation on the other hand is very poorly behaved—the variations between three different realisations shows that the behaviour is very noisy, and that it takes many multiples of the scale length to converge. The structure function is better behaved, but is still rather noisy; also the asymptotic value of dispersion is $\sqrt{2}$ larger than σ_x, because of course Δx is the difference of two random variables. Finally, the noise can get *worse* at long lag values k, because the lag is becoming comparable to the whole path length, so that only a few samples of Δx are possible.

The key lessons are that quantities characterising erratic time series from single path realisations can in general be very noisy, and that to estimate a quantity for a lag k, the length of the dataset needs to be many multiples of k—both because behaviour converges slowly, and because one needs many independent samples of length k.

12.2.3 Autocorrelation

Neighbouring points in the random walk and flicker-noise examples of Fig. 12.1 seem to be quite close to each other in x-value, whereas data points with a large separation in time are, on average, further apart. By contrast, for the purely random sequence top right, the difference in x seems unconnected with how far apart in time the data points are. We can quantify this effect. Figure 12.3 shows the values of each point x_t plotted against the point x_{t+k} for two different values of k, for the flicker-noise sequence in Fig. 12.1. Clearly the points with $k = 1$ are more tightly correlated than the points with $k = 10$.

We can calculate the covariance for the points in such a graph, in just the same way we did when testing for the correlation between two random variables in Chap. 9, using the sample covariance s_{xy} (see equation (9.1)). Now however, the second vari-

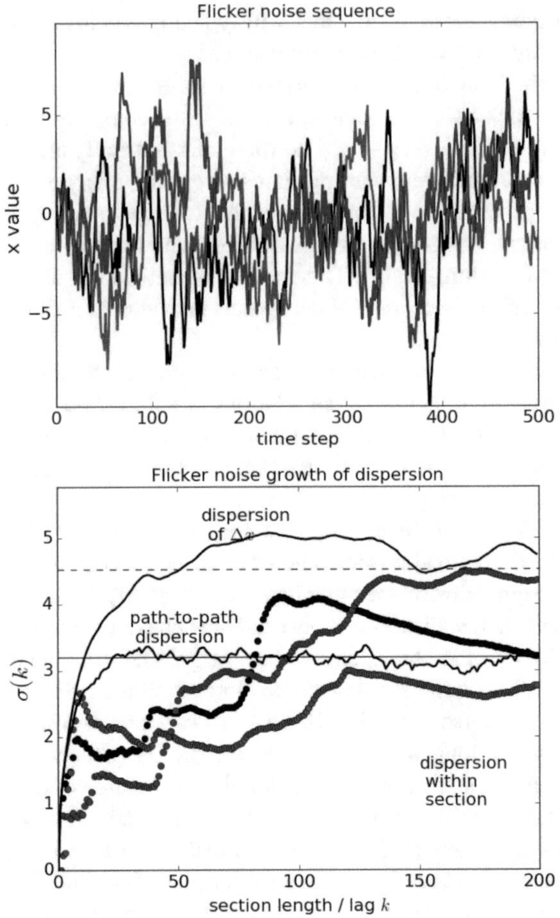

Fig. 12.2 Illustrating growth of dispersion in flicker noise. **Top**: Three sample paths generated using an autoregressive process with $\sigma_z = 1.0$ and $\alpha = 0.95$ (see text for explanation). **Bottom**: Growth of dispersion calculated three different ways. The coloured symbols take the sample paths from the upper panel, and for each one calculates the dispersion between of all the values between the start time at $t = 1$ and finish time at $t = k$. The lower black line is the ensemble dispersion—500 paths were generated, and for each k the dispersion is calculated from one sample path to another. For the upper line, a single path is used, of overall length $t_{\max} = 1000$. Then for a given value of k, and a given starting value x_i, we find the difference with the value a distance k away, i.e. $\Delta x_k = x_{i+k} - x_i$. Finally, for given k, we find the dispersion in Δx_k, using the many different starting values. The horizontal lines indicate the theoretically predicted values for the autoregressive process

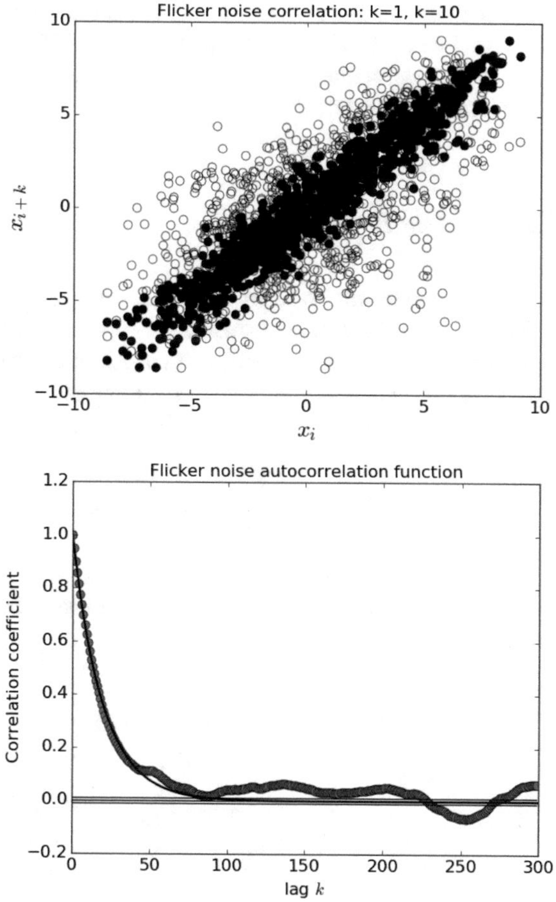

Fig. 12.3 Illustrating autocorrelation in flicker noise, using the sequence in Fig. 12.1. **Top**: Scatter plot of value x_i versus value x_{i+k} for two different values of k. **Bottom**: Correlation coefficient versus lag k. The solid line shows the expected behaviour for the autoregressive process used to make the simulation

able is also x, but shifted by a lag k, and so correlated with itself. We can then define the *sample autocovariance* as a function of lag k:

$$\gamma(k) = \left[\frac{1}{N} \sum_i (x_i - \bar{x})(x_{i+k} - \bar{x}) \right]. \tag{12.1}$$

It is usual to normalise the autocovariance by dividing by its value at $k = 0$, which is of course just the variance s_x^2, yielding the *sample autocorrelation function* $r(k) = \gamma(k)/s_x^2$, sometimes referred to as the ACF. Just like with the standard correlation coefficient, one would normally use $\rho(k)$ to refer to the ACF of a theoretical process, and $r(k)$ for the observed ACF of a specific sample path. Note that $r(k = 0) = \rho(k = 0) = 1.0$ by definition.

This gives us another standard way to characterise the behaviour of a time series. The right hand side of Fig. 12.3 shows how the ACF declines systematically with lag k. As with the structure function, the ACF is well behaved at small lag, but fluctuates considerably at larger values of k.

12.2.4 The Periodogram

Another popular way to characterise the behaviour of a time series is to perform a Fourier analysis. A fundamental mathematical result is that any function can be represented as the sum of a series of *sin* and *cos* terms. (Readers not familiar with Fourier series should consult one of the references given at the end of this chapter.) For a finite series of N values x_t, the Fourier series representation is

$$x_t = a_0 + \sum_{p=1}^{N/2-1} \left[a_p \cos \omega_p t + b_p \sin \omega_p t \right] + a_{N/2} \cos \pi t,$$

where $\omega_p = 2\pi p/N$ is the *angular frequency* and the solutions for the coefficients are

$$\begin{aligned} a_0 &= \bar{x} \\ a_{N/2} &= \sum (-1)^t x_t / N \\ a_p &= 2 \left[\sum x_t \cos \omega_p t / N \right] \\ b_p &= 2 \left[\sum x_t \sin \omega_p t / N \right] \\ p &= 1 \ldots N/2 - 1. \end{aligned}$$

The *amplitude* of a given frequency component p is given by $R_p^2 = a_p^2 + b_p^2$. Some algebra leads to the result

$$\frac{\sum (x_t - \bar{x})^2}{N} = \sum_{p=1}^{N/2-1} \frac{R_p^2}{2} + a_{N/2}^2.$$

The left hand term of the equation above is the variance of x. We can then interpret the terms R_p^2 as partitioning the variance into harmonics. A plot showing the calculated coefficients R_p^2 as a function of the frequency index p is known as the *periodogram*.[1] Such a plot tends to fluctuate rather wildly from point to point, so it is normal to

[1] At a detailed level, different authors define the periodogram in subtly different ways—careful reading required!

smooth the periodogram. In the continuous approximation, we can define the *spectral density* per unit ω

$$I(\omega_p) = \frac{NR_P^2}{4\pi}.$$

One can show that the structure function, the autocorrelation function, and the periodogram are mathematically derivable from each other. They therefore contain the same information and so to some extent choosing which to look at is a matter of taste. However, they can give differing insights. For example, a process that has a sinusoidal component, possibly hidden under noise, will show oscillations in the ACF, but will show a distinct peak in the periodogram at the appropriate ω_p.

12.3 Probabilistic Models for Discrete Time Series

Nature has many ways to make something that looks like flicker noise. We will look at the two simplest classes of mathematical process that produce the desired result—*autoregressive processes*, and *moving average processes*. Although simple, they are frequently used successfully as models of real physical processes. In both cases, we start by assuming that we are dealing with a sequence at discrete times x_t, with $t = t_1, t_2, \ldots$. Both processes have continuous analogues, but these have subtleties that are best left until after we have examined the discrete versions.

12.3.1 Purely Random Processes

In a *purely random process* each value in the sequence x_t is drawn from the probability distribution of a separate, independent random variable Z_t. Normally we assume that the random variables Z_t all have the same probability distribution. Then for a specific sample path, the sequence of values is $x_t = Z_t$ where now we assume that each Z_t is in fact a value independently drawn from the same random variable Z, which has distribution $P(Z)$, with some mean μ_Z and variance σ_Z^2, so that $\mu_x = \mu_Z$ and $\sigma_x = \sigma_Z$. Most often we assume that Z has a Gaussian distribution, but sometimes we might assume for example a Lorentzian distribution or a Poisson distribution. The periodogram of such a random process has equal contributions to variance from all frequencies, so that this type of process is sometimes referred to as "white noise".

Some time series may be modelled as purely random, but the random process is also the base process for our methods of generating flicker noise. Used this way, we usually assume $\mu_Z = 0$.

12.3.2 *Moving Average (MA) Processes*

In a *moving average process of order q*, denoted MA(q), we start with a random process Z_t and then construct x_t as the weighted sum of Z_t plus q previous random seed values:

$$x_t = c + \beta_0 Z_t + \beta_1 Z_{t-1} + ...\beta_q Z_{t-q}. \quad (12.2)$$

Here each Z_i is an independent random variable, all with the same $\mu_z = 0$ and dispersion σ_z. Usually Z is assumed to have a Gaussian distribution. (We will discuss possible variants in Sect. 12.3.7.) The constant c is to allow x to have a non-zero mean. We can get the essence of such a process by taking $c = 0$ and $\beta_0 = 1$, which we will do henceforth. The value of x_t is therefore a weighted average of a sequence of Z values, which we "slide along" for each new x_t value.

We can see the base sequence Z_t as the input, the sequence of weights $1, \beta_1, \beta_2 ... \beta_q$ as a filter, and the result x_t as the output. Viewed this way, a moving average process is an example of a *linear system.* Such a system can be expressed as $b(t) = \int h(u)\, a(t-u)du$, where $a(t)$ is the input, $b(t)$ the output, and $h(u)$ the *impulse response function*, i.e. the output we would get if the input was a delta function spike. The special features of the moving average process are that the sequences are in discrete steps of time, and the input is a sequence generated by a set of independent random variables. Both the input Z and the output x are therefore statistical entities, different at each implementation.

What are the statistical characteristics of the resulting sequence? As x_t is a weighted sum of random variables each with $\mu = 0$ and $\sigma = \sigma_Z$, we can see that

$$\mu_x = 0 \quad \text{and} \quad \sigma_x^2 = \sigma_Z^2 \sum_{k=0}^{q} \beta_k^2.$$

Note that σ_x^2 here is the true process "path-to-path" variance. In other words, if we implemented the MA process many times, and compared the values of x_t for the many implementations at some fixed time, this would be the variance in the distribution of values. What about the autocorrelation function of this process? To do this calculation, note two things about the covariance of two independent variables Z_a and Z_b. First, $\text{Cov}(Z_a, Z_b)$ is zero unless $a = b$. Next, because our variables have zero mean, then $\text{Cov}(Z_a, Z_b)$ is simply $E[Z_a Z_b]$. Now lets look at the simplest case, the MA(1) process with $\beta_1 = \beta$ and all other $\beta_k = 0$. At lag k we have

$$\begin{aligned}
x_t &= Z_t + \beta Z_{t-1} \\
x_{t+k} &= Z_{t+k} + \beta Z_{t+k-1} \\
\gamma(k) &= E[x_t x_{t+k}] \\
&= E[\underbrace{Z_t Z_{t+k}}_{①} + \underbrace{\beta Z_t Z_{t+k-1}}_{②} + \underbrace{\beta Z_{t-1} Z_{t+k}}_{③} + \underbrace{\beta^2 Z_{t-1} Z_{t+k-1}}_{④}]
\end{aligned}$$

The trick now is to look at the four labelled terms in that last expression. For $k = 0$, terms ② and ③ vanish, because the indices are different. Term ① gives $E[Z_t^2] = \sigma_Z^2$, and ④ gives $E[\beta^2 Z_{t-1}^2] = \beta^2\sigma_z^2$. For $k = 1$, only term ② is non-vanishing, giving $E[\beta Z_t^2] = \beta\sigma_z^2$. We then have

$$\gamma(0) = \sigma_Z^2(1+\beta^2) = \sigma_x^2 \qquad \rho(0) = 1$$

$$\gamma(1) = \beta\sigma_Z^2 = \sigma_x^2\frac{\beta}{1+\beta^2} \qquad \rho(1) = \frac{\beta}{1+\beta^2}$$

By similar but more laborious arguments we can derive the autocovariance function of an MA(q) process, cross-multiplying all the terms and keeping only the ones with matching indices for a given k. What we find is

$$\gamma(k) = \sigma_Z^2\sum_{i=0}^{q-k}\beta_i\beta_{i+k}, \qquad \rho(k) = \sum_{i=0}^{q-k}\beta_i\beta_{i+k}\Big/\sum_{i=0}^{q}\beta_i^2. \tag{12.3}$$

12.3.3 *Useful Filters*

In much of social science, and sometimes in physics, the aim of characterising the statistical properties of a time series is to enable *forecasting*—for example, if we have a time series of recent stock prices, can we predict tomorrow's price, or next week's? The typical procedure is then to assume a relatively low order process (MA(1), MA(3) or whatever), then to measure the sample mean, variance, and autocorrelation function of the series, and finally to use these to estimate c, σ_Z, and the β_i values. This will not allow us to predict the future values with certainty, but rather gives us a range of forecast values, which will diverge with time. Having estimated the parameters of the MA process, you can imagine taking todays stock price, or quasar brightness, and then creating many simulated paths starting from that value, which will give a diverging set of paths, like in Fig. 6.1.

In physics we are more interested in the interpretation of the filter, seen as an impulse response function, as this will give us insight into the nature of the physical system. For example, the impulse response function may correspond to how an electronic component responds to impinging electrons, or how the gas surrounding a black hole responds thermally to random fluctuations of energy. If we propose a mathematical form for our filter, what will be the resulting ACF, or the resulting structure function?

A simple example is a *top hat filter* of size q, with all β_k the same. Correlating this filter with itself will produce a triangular $\rho(k)$, reducing to zero by $k = q$. For a *Gaussian filter* centred on $q/2$ with dispersion σ_G, the ACF is also a Gaussian, but with $\sigma = \sqrt{2}\sigma_G$.

Possibly the most important example however is a *sawtooth exponential filter*. A wide range of natural systems respond to fluctuations with a fast rise, followed by an exponential decay on some timescale $k_{\rm ch}$. Note that $k_{\rm ch}$ is the decay time in units of the timestep of our discrete time series. We can then express such a filter as $\beta_k = e^{-k/k_{\rm ch}}$, with $q = \infty$. The variance of the resulting time series will be

$$\sigma_x^2 = \sigma_Z^2 \sum \beta_k^2 = \sigma_Z^2 \left[e^{-0/k_{\rm ch}} + e^{-2/k_{\rm ch}} + e^{-4/k_{\rm ch}} + ... \right]$$

For the autocovariance we need to calculate the sum of terms $\beta_i \beta_{i+k}$. For $i = 0$ we get $e^{-0/k_{\rm ch}} e^{-k/k_{\rm ch}}$. For $i = i$ we get $e^{-1/k_{\rm ch}} e^{-(1+k)/k_{\rm ch}} = e^{-2/k_{\rm ch}} e^{-k/k_{\rm ch}}$, and so on. Putting these terms together we get

$$\begin{aligned} \gamma(k) &= \sigma_Z^2 e^{-k/k_{\rm ch}} \left[e^{-0/k_{\rm ch}} + e^{-2/k_{\rm ch}} + e^{-4/k_{\rm ch}} + ... \right] \\ &= \sigma_x^2 e^{-k/k_{\rm ch}}, \end{aligned}$$

and so for the ACF we find that

$$\rho(k) = e^{-k/k_{ch}}. \tag{12.4}$$

So an exponential filter produces an exponential ACF with the same decay time, a pleasingly simple result. What about the structure function? We have two random variables a distance k apart, x_t and x_{t+k}. Both of these have variance σ_x^2, but they also have a mutual covariance given by $\gamma(k)$. Following the usual transmission of errors formula (see equation (2.9)), if $c = a - b$, then $\sigma_c^2 = \sigma_a^2 + \sigma_b^2 - 2\sigma_{ab}$. So here we have

$$\sigma_{\Delta x}^2(k) = 2\sigma_x^2 \left(1 - e^{-k/k_{ch}}\right). \tag{12.5}$$

where $\sigma_{\Delta x}^2$ is the variance of the differences between pairs of measurements k steps apart. Note that the asymptotic value is twice the "path-to-path" variance, as we would expect.

12.3.4 Autoregressive (AR) Processes

In an *autoregressive process of order p*, denoted AR(p), each data value x_t is a linear combination of past values, plus a random term:

$$x_t = c + \alpha_1 x_{t-1} + \alpha_2 x_{t-2} + ... \alpha_p x_{t-p} + Z_t. \tag{12.6}$$

As with the MA process, each Z_t is an independent random variable with the same σ_Z and $\mu_Z = 0$, not necessarily Gaussian but usually assumed to be so. The simplest

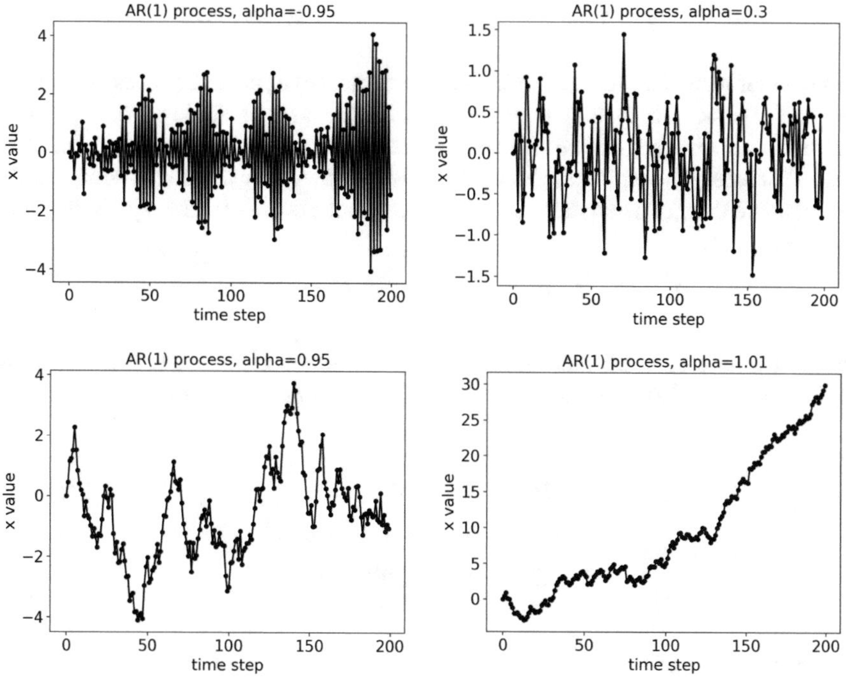

Fig. 12.4 Simulations of the AR(1) process for different values of α. Note that $\alpha = 0$ would produce a purely random sequence, and $\alpha = 1$ would produce a random walk. The value $\alpha = 0.95$ is what was used in simulations of flicker noise earlier in the chapter

process is AR(1) with $c = 0$, in which case

$$x_t = \alpha x_{t-1} + Z_t. \tag{12.7}$$

Some examples are shown in Fig. 12.4. The parameter α tunes the process:

- For $\alpha < 0$ the sequence oscillates.
- For $\alpha = 0$ we have a purely random process, where each value is determined independently purely by the probability distribution $p(Z)$.
- For α in the range 0 to 1 we get flicker noise—a sequence like a random walk, but smoother. This is often known as a *damped random walk*.
- For $\alpha = 1.0$ we get a random walk.
- For $\alpha > 0$ the sequence diverges rapidly.

The sequences we see for $\alpha = 0 - 1$ are reminiscent of the outputs we get from MA processes. This is not a coincidence. Using the AR(1) formula above, we can get Z_{t-1} in terms of Z_{t-2} and so forth; applying this recursively you can see that another way to express x_t is

$$x_t = Z_t + \alpha Z_{t-1} + \alpha^2 Z_{t-2} + +\alpha^3 Z_{t-3} + \cdots .$$

So in other words, an AR(1) process can be expressed as an MA process with $\beta_k = \alpha^k$ and $q = \infty$. Because x_t is made of an infinite sum of random variables, it is immediately obvious why the process blows up for $\alpha > 1$. The identification of an AR(1) process with an infinite MA process also allows us to use our previous results to calculate the characteristics of x. For example, the overall (path-to-path) variance will be:

$$\sigma_x^2 = \sigma_Z^2 \sum \beta_k^2 = \sigma_Z^2 (1 + \alpha^2 + \alpha^4 \cdots) = \frac{\sigma_z^2}{1 - \alpha^2},$$

where we have used the fact that $(1 - x)^{-1} = 1 + x + x^2 + \cdots$. Likewise we can get the autocovariance $\gamma(k) = \sigma_Z^2 \sum_{i=0}^{\infty} \beta_i \beta_k$. The $\beta_i \beta_k$ term gives

$$\begin{array}{lcl} i = 0 & \rightarrow & \alpha^k \\ i = 1 & \rightarrow & \alpha \alpha^{k+1} = \alpha^2 \alpha^k \\ i = 2 & \rightarrow & \alpha^2 \alpha^{k+2} = \alpha^4 \alpha^k \end{array}$$

and so you can see

$$\begin{aligned} \gamma(k) &= \alpha^k (1 + \alpha^2 + \alpha^4 + ...) \\ &= \alpha^k \sigma_x^2 = \frac{\alpha^k \sigma_Z^2}{1 - \alpha^2} \end{aligned}$$

and so the ACF is given by $\rho(k) = \alpha^k$. This power-law filter shape is actually close to exponential. For $\beta_k = \alpha^k$ we have $\log \beta_k = k \log \alpha \sim k(\alpha - 1)$, where in the last step we have used the standard power law approximation for $\log x$. So then we have

$$\beta_k \sim e^{-k/k_{\mathrm{ch}}} \quad \text{with} \quad k_{\mathrm{ch}} = \frac{1}{1 - \alpha}.$$

Following our MA results, we then expect

$$\rho(k) = e^{-k/k_{ch}} \qquad \text{and} \qquad \sigma_{\Delta x}^2(k) = 2\sigma_x^2 (1 - e^{-k/k_{ch}}) \tag{12.8}$$

with

$$k_{ch} = \frac{1}{(1-\alpha)}, \quad \text{and} \quad \sigma_x^2 = \frac{\sigma_Z^2}{1-\alpha^2}. \tag{12.9}$$

12.3.5 The General ARIMA Process

The simple AR and MA processes we have considered can be generalised further. First, they can be combined, so that x_t can be formed from a linear combination of both past x values and past Z values. This forms what is known as an ARMA process:

$$\text{ARMA(p,q)}: \qquad x_t = c + Z_t + \sum_{i=1}^{p} \alpha_i x_{t-1} + \sum_{i=1}^{q} \beta_i Z_{t-1}.$$

Such ARMA models are very popular in the social sciences, but are not so popular in physics, as the physical interpretation of the parameters isn't usually obvious. Real time series often have trends or other deterministic components as well as the random element; a common practice is to take the differences between successive values to try to remove such trends, and then model the remaining difference series with an ARMA process. One can repeat this differencing process multiple times. The whole modelling process is then known as an "autoregressive integrated moving average (ARIMA)" model. It is also possible to model several variables at the same time with a combined AR model, which is then known as a "Vector AR (VAR) model".

12.3.6 Wiener, Cauchy and Poisson Processes: Shot Noise

We have so far assumed that our seed process Z is Gaussian. What if we loosen this assumption? Any process where the Z_t values are independent random variables is known as a *Levy process*. If the distribution of Z is Gaussian, this is known as a *Wiener process*. If the distribution is Cauchy (Lorentzian) then this is known as a *Cauchy process*.

Another possibility of interest is that the seed is a *Poisson process*—i.e. a series of delta-function spikes, with waiting time between events given by the waiting time distribution equation (6.4). The resulting output is known as "shot noise" and is clearly a set of overlapping "events" (see Fig. 12.5). The MA-Wiener process with exponential filter can be seen as a limiting case, where many "shots" occur in each time bin. In general, the ACF will reproduce the input shot shape.

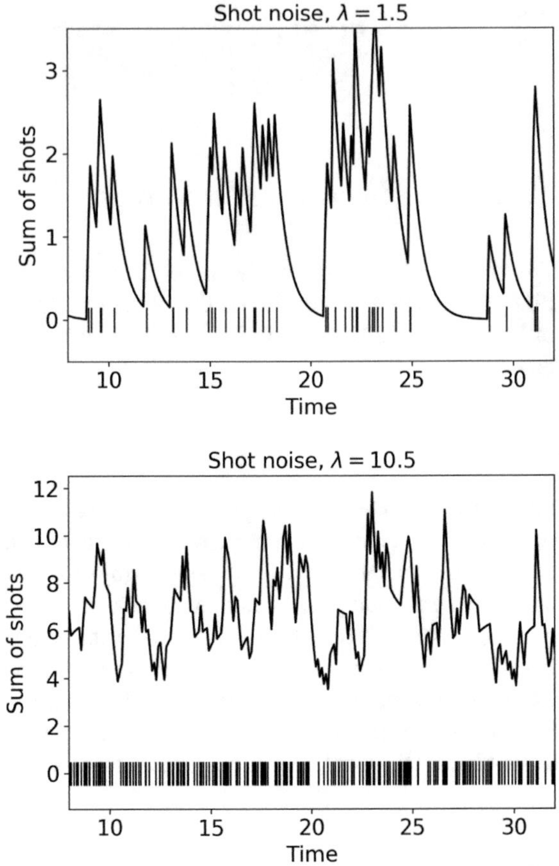

Fig. 12.5 Simulations of shot noise process. In each case, events are produced at random using the waiting time distribution equation (6.46) with the rate per unit time given by $\lambda = 1.5$ (upper panel) and $\lambda = 10.5$ (lower panel). The event times are indicated by the vertical bars. Each event produces an exponential shot with height $h = 1.0$ and decay time $\tau = 1.5$; these are then added together to produce the final time series. In the upper panel, the individual shots can be easily seen; in the lower panel, the shots are heavily overlapping

12.3.7 Other Variations

Potentially there are many more variations on the basic ideas of AR and MA processes. For example, we have assumed that current values are linear combinations of past values, or of past random seed values. What if we allow arbitrary non-linear combinations? We could also relax the assumption that the random seed values are independent of each other. But there is little point adding mathematical complexity for its own sake. (Mathematicians may disagree, because that is their idea of fun...) We should make mathematical implementations of whatever seems to emerge from physical models of real situations when we require them. From a physicist's point of view, the most obvious thing we are missing is how to treat systems in continuous time, which is what we look at in the next section.

12.4 Continuous Stochastic Processes

We have so far modelled erratic processes as discrete time series. This is appropriate for some situations, where measurements are naturally taken at discrete intervals, for example daily stock prices, or seasonal effects. However many physical systems are most naturally modelled in continuous time. Turning physical laws into mathematics generally results in differential or integral equations. Can we recast our time series techniques as continuous processes? For example, we noted that a MA process is equivalent to:

$$x(t) = \int_0^\infty h(u)Z(t-u)\mathrm{d}u,$$

where $Z(t)$ is the input, $h(u)$ is the filter, and $X(t)$ is the output. If $Z(t)$ were a simple deterministic function of time, then this would be a normal linear system, but here we require $Z(t)$ to be a random function of time. In a similar fashion, for an AR process, in principle we should be able to arbitrarily shrink the time step between successive values, and deduce a differential form for the process in terms of a continuous random process $Z(t)$. But what do we *mean* by such a continuous random function of time $Z(t)$? Its a rather problematic concept. Suppose Z is a random variable with some probability distribution $P(Z)$, a mean μ, and a dispersion σ. Then at each possible time t, $Z(t)$ is a distinct and independent such variable. Consider two points in time separated by a small amount of time Δt. The difference ΔZ between the two Z values will have a mean $\mu_{\Delta Z} = 0$ and a dispersion $\sigma_{\Delta Z} = \sqrt{2}\sigma$. If we make Δt smaller and smaller, $\sigma_{\Delta Z}$ will always be the same, and $\Delta Z/\Delta t$ therefore becomes ever larger on average. The function $Z(t)$ is therefore *not differentiable*, and it becomes hard to use normal techniques to solve differential equations. Similarly, if we try to evaluate the integral of $Z(t)$ over some region by summing up using small steps Δt, and then letting $\Delta t \to 0$, we don't get a convergent answer.

Although a pure random continuous input process therefore seems to be pathological, the output erratic sequences themselves are not pathological. Although the x values of our AR and MA processes flicker erratically, the correlation we have introduced by these processes means that neighbouring values are statistically closer together; as you can see from equation (12.5), as $k \to 0$, $\sigma_{\Delta_x} \to 0$. What then is our route forward to using differential and integral equations to model noisy systems?

12.4.1 The Wiener Process as an Approximation

We could consider continuous random noise as a mathematical abstraction which in practice can only be implemented at a specific discretisation Δt. Sometimes $Z(t)$ has no physical meaning; it is just a modelling device to enable us to characterise the observed process $x(t)$. Sometimes however, both $Z(t)$ and $h(u)$ do have physical meaning. The flicker noise we observe may be caused by underlying rapid fluctuations in some system, which are then smoothed out by the physical response. The classic example is *Brownian motion*, the jiggling movements of particles, such as pollen grains, suspended in a fluid, which is caused by the bombardment of those grains by the random motions of the even smaller molecules of the fluid. On short enough timescales, and with enough spatial resolution, the motion of the molecules is deterministic. However, in practice we do not know what the molecules are doing, so the force acting on the grain must be considered random from one moment to the next. On very short timescales, the net force acting on the grain, while random, will be highly correlated over time. However, on slightly longer timescales, the random force will be effectively independent from one time step to the next. The timescale on which this effective independence happens will still be much shorter than the sampling timescale with which we observe the motion of the grain. Finally then we arrive at the idea that although the buffeting of the grains by the molecules is not in truth a pure random continuous process, we can model it with that mathematical abstraction.

The buffeting of grains leads to a random walk in 3D, as first shown by Einstein, and subsequently set out with more mathematical rigour by Norbert Wiener. We will consider the equivalent 1D process for simplicity. Imagine a sequence of values $x(t)$ separated by some small (but not infinitesimal) time Δt. Suppose that, on this timescale, each x value is connected to the next by a random increment that we could write as ΔW. Each increment is an independent random variable with mean zero and dispersion $\sigma_{\Delta t}$. If the probability distribution of each ΔW is Gaussian, then the increments are known as "Wiener noise", and the resulting $x(t)$ sequence as a "Wiener process". (Hence the use of the symbol W.) Now consider the x value N time steps later, at a time $t + T$ where $T = N\Delta t$. The result is given by the sum of N successive increments:

$$x(t+T) = x(t) + \sum_{i=1}^{N} \Delta W_i.$$

As each of the ΔW_i are independent random variables with dispersion $\sigma_{\Delta t}$, then the sum of these random variables will give another random variable, with mean value equal to $x(t)$ and a dispersion given by $\sigma_T = \sqrt{N}\sigma_{\Delta t} \propto \sqrt{T}\sigma_{\Delta t}$. Note the square-root spreading with time, which we already met in Chap. 6 when talking about simple random walks.

When we use Wiener noise to model real processes, the customary procedure is as follows. We define a standardised Wiener noise[2] $\mathrm{d}W_t$ as that which has unit variance on some infinitesimal timescale $\mathrm{d}t$. On this timestep $\mathrm{d}t$ we then apply increments $\Delta W = \sigma \mathrm{d}W_t$ where σ is a free parameter used to model the process. We then calculate the properties of the process on macroscopic timescales, for example finding the total observed dispersion σ_x, which will be some function of the modelling parameter σ. In essence then we don't have to worry what happens to the noise in

[2]There are some subtly different conventions for normalising Wiener noise in the literature, so do read carefully.

the continuum limit. For small time steps, the approximation of independent random fluctuations will fail, but most of the effect on the output is caused by fluctuations on much longer time steps, and σ is just a parameter that allows us to model the process on macroscopic timescales. This will become a little clearer when we discuss an example in Sect. 12.4.3. But first, we must think more carefully about how we solve difference equations.

12.4.2 Solving Difference Equations

How do we use these ideas to solve the behaviour of dynamical systems driven by noise? We will take a very brief look here at a big subject, following but further condensing the treatment given by Jacobs (see the Further reading section). First, let's recap techniques for deterministic differential equations. Sometimes one can integrate directly. For example if we know that $\mathrm{d}x/\mathrm{d}t = a$ it is clear that $x = at + C$. Often we need to separate variables. So for example

$$\frac{\mathrm{d}x}{\mathrm{d}t} = -\gamma x \quad \Rightarrow \quad \frac{\mathrm{d}x}{x} = -\gamma \mathrm{d}t \quad \Rightarrow \quad \ln x = -\gamma t + C,$$

and so the solution is $x(t) = x_0 e^{-\gamma t}$ where $C = \ln x_0$. Let us look at an alternative way of solving this equation, by expressing it as a *difference equation.*

$$\frac{\mathrm{d}x}{\mathrm{d}t} = -\gamma x \quad \Rightarrow \quad x(t + \mathrm{d}t) = x(t) - \gamma x(t)\mathrm{d}t \quad = x(t)\,(1 - \gamma \mathrm{d}t)\,.$$

Next, we note that if a is small, then $e^a \approx 1 + a$. So if $\mathrm{d}t$ is small and we can ignore terms in $\mathrm{d}t^2$, $\mathrm{d}t^3$, ... then the above equation becomes

$$x(t + \mathrm{d}t) \approx x(t)e^{-\gamma \mathrm{d}t}.$$

So moving by step $\mathrm{d}t$ is accomplished by multiplying by $e^{-\gamma \mathrm{d}t}$. We can repeat this process so that $x(t + 2\mathrm{d}t) = x(t)e^{-2\gamma \mathrm{d}t}$ and so on. If we move from time $t = 0$ at $x(t = 0) = x_0$ to time $t + \tau$ where $\tau = N\mathrm{d}t$ then we get

$$x(\tau) = x_0 e^{-N\gamma \mathrm{d}t} \quad = x_0 e^{-\gamma\tau}.$$

which is the same solution we found before. This method can be generalised to more complicated situations. For example if γ is a function of time, you can show that

$$x(t + \tau) = x(t) \exp\left(-\int_t^{t+\tau} \gamma(t)\mathrm{d}t\right).$$

Alternatively if we add a "driving term" $f(t)$ we find that

$$\frac{dx}{dt} = \gamma x + f(t) \quad \Rightarrow \quad x(t) = x_0 e^{-\gamma t} + \int_0^t e^{-\gamma(t-s)} f(s)ds.$$

12.4.3 *The Ornstein–Uhlenbeck Process*

This last variant leads us to the idea that we can think of our random increments as the "driving term" in an equation like the one above. At each time step in our difference equation, $f(t)$ could be an independent random variable, with mean zero and dispersion depending on the time step dt. This leads to a *stochastic difference equation (SDE)*. The simplest such difference equation can be written

$$dx = -\gamma x_t dt + \sigma dW_t, \tag{12.10}$$

where we have used the standard normalisation of the Wiener noise, where dW_t are increments with unit variance, and σ is a parameter characterising the model. Note that the equation is written in difference equation style; we can't write in terms of derivatives because we can't guarantee they exist. However, using dt rather than Δt is meant to imply that we are extrapolating to the continuum limit.

In the mathematical literature, this equation is known as the *Ornstein–Uhlenbeck (OU) equation.* In Physics, it is exactly the same as the *Langevin equation* which models Brownian motion in the presence of drag. In economics it is known as the *Vlasicek* model, and is used for example to model the time history of interest rate changes, and where σ is referred to as the "volatility". The "OU process" resulting from the OU equation is also sometimes known as a *damped random walk.* By comparing with the previous section, you can see that the solution is

$$x(t) = x_0 e^{-\gamma t} + \sigma \int_0^t e^{-\gamma(t-s)} dW(s).$$

Note however the peculiar meaning of "solution" in the world of SDEs. What we have above is a predicted sample path $x(t)$, given a starting point x_0, but also given a specific realisation of the Wiener noise, i.e. a particular set of dW values. The solution also involves a "stochastic integral", which is technically problematic because of the non-differentiability of W. However, if we take a specific discretisation (in which case what we have essentially is an AR(1) process) we can always numerically compute results to the precision we require.

What is often more useful than one sample path is to look at the ensemble of paths from many realisations of W, and then to characterise the probability distribution of $x(t)$—so for example this could give you a range of uncertainty on future predicted interest rate values. The solution for the OU process is that the mean of $x(t)$ is zero, and the variance is

$$\mathrm{Var}\left[x(t)\right] = \frac{\sigma^2}{2\gamma}\left(1 - 2e^{-2\gamma t}\right), \tag{12.11}$$

which you will recognise as essentially the same thing as equation (12.5).

12.4.4 Other SDEs

The simplest variant is the OU process with drift:

$$\mathrm{d}x_t = \gamma(\mu - x_t)\mathrm{d}t + \sigma \mathrm{d}W_t.$$

This process has $E[x] = \mu$, known as the "mean reversion level", with γ being referred to as the "mean reversion rate". Another popular variant is Geometric Brownian motion, which has the increment size proportional to the current x value:

$$\mathrm{d}x_t = \alpha x_t \mathrm{d}t + \sigma x_t \mathrm{d}W_t.$$

The solution to this equation produces tracks which grow systematically with time, and has $E[x_t] = x_0 e^{\alpha t}$ and $\mathrm{Var}(x_t) = x_0^2 e^{2\alpha t}(e^{\sigma^2 t} - 1)$. This equation is the basis of the *Black Scholes model* for predicting the behaviour of the stockmarket. It can be seen as one of a family of Vlasicek models where the increments are $\sigma x_t^\beta \mathrm{d}W_t$.

Many other variations have been considered in the literature. There could be multiple noise sources, or multiple variables. The parameters could be a function of time. An interesting variant on this idea is the "two factor" Vlasicek model, where the mean reversion level μ is not a constant, but is itself produced by its own OU process. Potentially one could add stochastic and damping terms to any set of differential equations. Two interesting examples discussed by Lemons are the damped harmonic oscillator, and stochastic cyclotron motion (see the Further Reading section). Finally, we could relax the assumption that the increments are Gaussian noise; assuming a Cauchy/Lorentzian distribution for the increments produces quite different looking results. Alternatively, the input random fluctuations could be a Poisson process; i.e. at each time step there is a probability that either there is a increment or not. Obviously this is closely related to the shot-noise process we discussed earlier; in the context of SDEs it is known as a *jump process*.

For any proposed process, we can in principle simulate paths using a suitable discretisation, and we can likewise take a numerical approach to finding the distribution of x_t by simulating a large number of paths. Ideally however, we would like to find analytic solutions. This is where the non-differentiability of Wiener noise (let alone Cauchy noise) becomes a problem. Solving regular differential equations often involves guessing a suitable change of variables; however changing the differentials involves using derivatives which we can't do. Likewise, in evaluating stochastic integrals, we can't assume that terms in $\mathrm{d}W^2$ vanish with respect to $\mathrm{d}W$. Alternative rules of *stochastic calculus* have been developed, by Ito and by Stratonovich, but these are beyond the scope of this introductory material.

12.5 Markov Chains

A *Markov process* is one that has no memory; its future state depends only on its current state and some kind of procedure for deciding how it evolves. A random walk is therefore an example of a Markov process, as is the more general AR(1) process we have discussed in this chapter. The MA process on the other hand depends on past history, and a purely random process doesn't even care about the current value.

Markov processes can be in continuous time or discrete time, and the values in the sequence can likewise be continuous or discrete. A *Markov chain* is a Markov process in discrete time steps, with a set of discrete possible x values—usually a finite set, but sometimes a countably infinite set. Most often it refers to a system that has a relatively small set of possible states, and hops between those states in a sequence. At each hop, there is some probability of hopping from the current state to each of the other states. If those probabilities are always the same from one hop to the next, the chain is said to be *homogeneous*.

An example might be a a multi-state quantum system, with transition rules for jumping between the states. Another popular example is the idea that the probability of tomorrow being a rainy day depends on whether today is rainy or sunny (see exercises).

We can express these ideas more formally. The chain is a sequence of random variables $X_1, X_2, X_3, \ldots X_n, \ldots$. The set of all possible states is called the *state space*, and we can label the states $S_1, S_2, \ldots S_i, \ldots$. At any time n, the system is in some specific state $X_n = S_i$. (The states may not be numerical values, but we can take the = sign as indicating that the system is in that state). The probability that at the next step the system will hop from state S_i to state S_j is given by

$$A_{ij} = P(X_{n+1} = S_j | X_n = S_i).$$

This is known as the *transition matrix* $\mathbf{A}$. Note that it includes the probability of staying in the same state, A_{ii}. Given this updating rule, starting with some initial value $X_0 = S_i$, we can generate a sequence of values. Some updating rules might result in the system getting a stuck in a specific state, but more often the process will result in an ever changing sequence of values. However, at any one time step n there will be a probability distribution of being in state i, which we could think of as a *state probability vector* $\mathbf{p}^{(n)}$ where $p_i^{(n)} = P(X_n = S_i)$. We have written the superscript as (n) to emphasise that this indicates the nth iteration, rather than raising to the nth power. Then the probability distribution at the next time step will be given by

$$\mathbf{p}^{(n+1)} = \mathbf{p}^{(n)} \times \mathbf{A},$$

where we follow the usual rules of matrix multiplication. If the system starts in say state S_k, then we can express the initial probability distribution by setting $P_k = 1$ and all other $P_i = 0$. Then after n steps we have $\mathbf{p}^{(n)} = \mathbf{p}^{(0)} \times \mathbf{A}^n$ where $\mathbf{A}^n$ means the matrix $\mathbf{A}$ multiplied by itself n times. Does this probability distribution settle down to a stable result after some period of time? In other words, can we find a state probability vector π such that $\pi \times \mathbf{A} = \pi$? This problem can in principle be solved by diagonalising $\mathbf{A}$, and finding its eigenvalues. In general this could be quite a tough problem. The simplest case is the two state Markov system. This has a wide variety of applications, but including for example studying random transitions between spin-states of particles.

Suppose then we have two states, labelled 1 and 2. Suppose the probability of hopping from state 1 to state 2 is a, and that of hopping from state 2 to state 1 is b. Then the transition matrix is

$$\mathbf{A} = \begin{pmatrix} 1-a & a \\ b & 1-b \end{pmatrix}.$$

Multiplying $\mathbf{A}$ by itself many times gives

$$\mathbf{A}^n = \frac{1}{a+b}\begin{pmatrix} b & a \\ b & a \end{pmatrix} + \frac{(1-a-n)^n}{a+b}\begin{pmatrix} a & -a \\ -b & b \end{pmatrix} \qquad \lim_{n\to\infty}(\mathbf{A}^n) = \frac{1}{a+b}\begin{pmatrix} b & a \\ b & a \end{pmatrix}.$$

Solving the eigenvector problem shows that the steady state solution is

$$\pi = \frac{1}{a+b}(b \quad a). \tag{12.12}$$

If you calculate $\pi \times \mathbf{A}$ you can easily verify that π is reproduced.

Another interesting physics related problem is the *Ehrenfest model of diffusion*. Here, we consider two chambers containing gas molecules, connected by a narrow pipe. There are N molecules in total, with x in chamber-1 and $N - x$ in chamber-2. At each time step we pick one of the N molecules at random and consider moving it; we give it probability p of moving to the other chamber, and probability $1 - p$ of not moving. However, we are of course more likely to pick a particle from the chamber currently containing more particles. Suppose now we watch the sequence of values x_n. We can set up the transition matrix for changing from any value of x to any other value; however, because we decided to move one molecule at a time, most of the matrix elements are zero, and we have only

$$p_{i,i+1} = \frac{p(N-x)}{N}, \quad p_{i,i-1} = \frac{px}{N}, \quad p_{i,i} = 1 - p.$$

It is fairly obvious that the most probable state is with $x = N/2$ and that in equilibrium the probability of other values of x will follow the binomial distribution. The stable solution is what we found of course in Chap. 3, Sect. 3.6.4. The Markov chain we have set up is a kind of "toy model" of how a gas will approach that equilibrium over time. As it does so, its entropy increases, as we will discuss in Chap. 14.

12.6 Key Concepts

Some of the key concepts from this chapter are:

- That many natural processes seem to be partially random.
- The importance of distinguishing stationary from non-stationary processes, and how hard that is in practice.
- How to characterise a time series by its structure function, its autocorrelation function, or its periodogram.
- How to model an erratic time series as a moving average process, or as an autoregressive process.
- The conceptual difficulties of modelling noise-driven processes in continuous time, and how they are addressed.
- The idea of a Markov chain, and the related state vector and transition matrix.

The key formulae from this chapter are as follows: the definition of the autocovariance function (12.1); the definition of a moving average (MA) process as a filtered version of random noise (12.2); the autocovariance function of a general MA process (12.3), and the autocorrelation function (12.4) and structure function (12.5) for a MA process with an exponential filter; the definition of an autoregressive (AR) process of order p as a weighted sum of past values, plus a noise term (12.6), and the simplest AR(1) process (12.7); the autocorrelation function and structure function for an AR(1) process (12.8), and how the timescale and variance depends on the parameter α of the AR(1) process (12.9); the equation defining the Ornstein–Uhlenbeck (OU) process (12.10), and the structure function resulting from it (12.11); and the stable state vector resulting from a two-state Markov chain (12.12).

12.7 Further Reading

There are many textbooks on Time Series analysis, mostly centred on the social sciences and the problems of forecasting. The foundational text, and still a definitive volume, is Box and Jenkins, first published in 1970, and now in its fifth edition as Box et al. (2015). Its rather technical though. A much simpler and very clear book is Chatfield (2003), and another good text is Brockwell and Davis (2016).

The field of stochastic processes tends to be dominated by rather abstruse and mathematical works. Two excellent books that are specifically aimed at Physicists are Lemons (2002) and Jacobs (2010). Some of the examples discussed and the derivations I have used are based on these two books, as described in the text. In this chapter, I have stopped short of fully explaining the Ito calculus or the equivalent Stratonovich method, which you need to solve Stochastic Differential Equations. Jacobs is particularly thorough on this issue, and also has a good description of applications in the financial world, including the notorious Black–Scholes equation. Some of the key original papers in the field of Brownian motion and related phenomena are Einstein (1905), Langevin (1908) and Uhlenbeck and Ornstein (1930). Note however, that Bachelier (1900) is generally credited as having first come up with the basic ideas of "Brownian Motion", as part of his Ph.D. thesis a "Theory of Speculation".

Throughout this chapter, I refer to "flicker noise" in a loose sense of any partially random time series. Some authors reserve this term specifically for the erratic fluctuations in electronic components, and some use it as a synonym for what is known as "1/f noise". A wide variety of erratic time series, when analysed in Fourier terms (see Sect. 12.2.4) show spectral density which seems to show a low-frequency divergent power law shape with frequency (ω or f), and specifically a power-law index of 1.0. For some time, this was seen as an intriguing mystery (e.g. Press 1978). However, just like with power-law probability distributions in general (see Chap. 6), a variety of methods can give such a shape, and actually it is likely that the shape is never truly $1/f$. A good example is the erratic fluctuations in the X-ray emission from the famous black hole candidate Cygnus X-1. Originally it seemed that these showed a power-law shape; but in fact when much more comprehensive data was assembled, it became clear that it shows a continuously curved spectral density—more like a random walk at high frequencies, roughly $1/f$ in the mid-range, and white noise at low frequencies (see e.g. Fig. 3 of Uttley et al. 2005).

Following the financial crash of 2008–9, there was discussion about whether the Black–Scholes equation was the cause—see for example the Guardian piece by Stewart (2012), or the Financial Times piece by James Weatherall (2013). Of course the model itself did not cause the crash, but its naive application did play a role. Stewart explains that effects such as the herd instinct were not taken into account; but essentially it boils down to assuming the behaviour can be modelled as a stationary process, when in fact it is not stationary. As I write, we are probably going through an analogous problem in trying to understand quasars. The optical light curves of quasars are often modelled as a "damped random walk" (e.g. MacLeod et al. 2010), which is another term for an AR(1) or OU process. This explains erratic variations of size around a few tens of percent; but every so often it seems some quasars suffer a dramatic "crash", dimming by an order of magnitude over a few years in a way that shouldn't happen according to the damped random walk model (MacLeod et al. 2016).

12.8 Exercises

12.1 Figure 12.2 illustrates the growth of dispersion versus lag for simulations of a typical flicker noise, contrasting the path-to-path dispersion of many simulated paths with the within-path dispersion of single simulated paths. In the examples shown, the within-path dispersion curves seem to be not just erratic, but also to systematically undershoot at low lags. Is this probably a fluke for the three examples shown, or a real effect? Based on what we learn later in the chapter, what lag do you need to get away from this effect?

12.2 The brightness of a quasar is measured on a monthly basis, and over the first fourteen months produces the following values: 31.6, 30.2, 28.7, 30.9, 31.4, 27.2, 33.4, 27.8, 29.3, 29.4, 30.6, 28.4, 30.6, 29.5. (The units don't matter). Calculate the value of the ACF at lag $k = 1$ month. At 5% significance, is there evidence that the variability of the quasar is anything other than random? Does this result agree with the visual judgement you might make by (a) plotting the light curve (i.e. the time series of brightness), or (b) by plotting successive values against each other?

12.3 A moving average (MA) process is given by $x_t = Z_t + \alpha(Z_{t-1} + Z_{t-2} + \cdots)$ with $\sigma_Z^2 = 1$. Is this a stationary process? Suppose we form a new process by taking the difference of successive x_t points. Is this stationary? Show that this process has an ACF with $\rho(1) = \alpha - 1/(2 + \alpha^2 - 2\alpha)$.

12.4 An AR(1) process has $\alpha = 0.8$. How accurate is the exponential approximation to the ACF for a lag of $k = 1, 3, 6$? In general, will the approximation be better or worse for smaller/larger α?

12.5 In the text, we discussed how a continuous pure noise process is pathological because the variance doesn't decrease as you make the time step smaller, so that the process is not differentiable, whereas for a flicker noise this is not in practice a problem because of correlations between neighbouring points. But is, say, an AR(1) process with a low value of α close to pathological in practice? Calculate the structure function for an AR(1) process with $\alpha = 0.1, 0.3, 0.6, 0.9$, and 0.99, and compare the $\sigma_{\Delta x}(k)$ values at $k = 1$ and $k = k_{\mathrm{ch}}$.

12.6 Show that the OU process can be approximated by an AR1 process with $\gamma = 1 - \alpha$.

12.7 A study has noted that in a specific location, if a day is rainy, there is a 25% chance that the next day will be rainy as well. On the other hand, if today does not have rain, there is an 80% chance that the next day will also be clear. The same study finds that the overall fraction of rainy days is 15%. Are these facts consistent with the idea that weather follows a Markov process? If not, why not?

References

Bachelier, M.L.: Théorie de la Spéculation. Annales Scientifiques de lcole Normale Suprieure **3**, 21–86 (1900)

Box, G.E.P., Jenkins, G.M., Reinsel, G.C., Ljung, G.M.: Time Series Analysis: Forecasting and Control, 5th edn. Chapman and Hall/CRC, Boca Raton (2015)

Brockwell, P.J., Davis, R.A.: Introduction to Time Series and Forecasting, 3rd edn. Springer, Berlin (2016)

Chatfield, C.: The Analysis of Time Series: An Introduction, 6th edn. Chapman and Hall/CRC, Boca Raton (2003)

Einstein, A.: Investigations on the theory of Brownian Motion. Annalen der Physik **17**, 549–560 (1905)

Jacobs, K.: Stochastic Processes for Physicists: Understanding Noisy Systems. Cambridge University Press, Cambridge (2010)

Langevin, P.: On the theory of Brownian Motion. Comptes Rendu **146**, 530–533 (1908)

Lemons, D.S.: An Introduction to Stochastic Processes in Physics. Johns Hopkins University Press, Baltimore (2002)

MacLeod, C.L., et al.: Modelling the time variability of SDSS stripe 82 quasars as a damped random walk. ApJ **721**, 1014–1033 (2010)

MacLeod, C.L., et al.: A systematic search for changing-look quasars in SDSS. MNRAS **457**, 389–404 (2016)

Press, W.H.: Flicker noises in astronomy and elsewhere. Comments Astrophys. **7**, 103–119 (1978)

Uttley, P., McHardy, I.M., Vaughan, S.: Non-linear X-ray variability in X-ray binaries and active galaxies. MNRAS **359**, 345–362 (2005)

Uhlenbeck, G.E., Ornstein, L.S.: On the theory of Brownian Motion. Phys. Rev. **36**, 823–841 (1930)

Websites (all accessed March 2019):

Guardian piece: Stewart (2012) https://www.theguardian.com/science/2012/feb/12/black-scholes-equation-credit-crunch

Financial Times piece: Weatherall (2013): https://www.ft.com/content/30fc4ece-760b-11e2-9891-00144feabdc0

Chapter 13
Probability in Quantum Physics

13.1 Outline of Content

- Unpredictability in the classical and quantum worlds
- Quantum probability amplitudes
- Measurement and state mixtures
- Probability consistency tests of hidden variable theories
- Whither Quantum Mechanics?

Quantum mechanics is the most successful theory we have. It is of enormous practical value, and makes very precise predictions about the behaviour of the physical world. Without it, there would be no electronics industry, no computers, and no internet. It explains why the Sun burns, and the shapes of crystals. We have had this amazingly powerful theory for almost a hundred years now; and yet still we find ourselves uncomfortable with its philosophical message. It seems to tell us that the world is irreducibly unpredictable at its heart, whereas our instinct as scientists is that the world is causal and deterministic, even though in practice we often need to take a probabilistic approach. In this short chapter we won't solve these deep problems, but we will look briefly at some key issues that involve the concepts of probability—intrinsic unpredictability; the appearance of microscopic unpredictability in the macroscopic world; the strange behaviour of probability amplitudes; the problem of measurement; and the question of whether "hidden variables" exist. We will assume that the reader has had a first course in quantum physics, but don't require a detailed knowledge of quantum mechanics. Let's start by contrasting randomness in the classical and quantum worlds.

13.2 Unpredictability in the Classical and Quantum Worlds

In Chap. 1 we discussed how our notion of randomness is really about the predictability of events. We listed various ways that a deterministic world, governed by strict laws of cause and effect, can nonetheless produce events that are, in prac-

A. Lawrence, *Probability in Physics*, Undergraduate Lecture Notes in Physics,
https://doi.org/10.1007/978-3-030-04544-9_13

tice, unpredictable—because of incomplete knowledge, large numbers, sensitivity to initial conditions, and influence by external factors. In fact, all these situations are variants on the problem of incomplete knowledge—for example, external factors only produce unpredictability if we don't know enough about those factors. The very fact that we call them "external" is just another way of saying that we don't know what they are. In the quantum world however, the unpredictability is not about the incomplete knowledge of a particular observer—it is intrinsic and inevitable.

13.2.1 Incomplete Knowledge and Observer Dependence

In Chap. 11 we discussed the relations between the concepts of probability, uncertainty, and information. This is another way of approaching the issue of incomplete knowledge; $h = \log_2(1/p)$ quantifies the amount of information we would need to supply in order to remove our uncertainty about the situation. One way of looking at quantum randomness is that, unlike in the classical view of the world, it is impossible to supply that missing information.

The problem of incomplete knowledge is also closely connected with the problem of subjectivity. As we stressed in Chap. 1, "subjective" does not mean "vague and woolly"—it just means "observer dependent". In many everyday circumstances, the observer dependence is quite extreme. Imagine witnessing a colleague tossing a coin and being asked to predict the outcome. Observer A may say "the probability of heads is 0.5". Observer B however happens to know that their colleague has a double-headed coin, and so says "the probability of heads is 1.0". The Bayesian approach handles this observer dependence well—according to Bayesians, the point of statistics is precisely to give us a way to reason objectively and quantitatively in the presence of incomplete knowledge. It is rational for Observer A to argue in a probabilistic fashion, even though we would like to be Observer B, know everything, and avoid all talk of probability.

At first glance, it seems as if this observer dependence is all about people—wet brains and consciousness—but actually it is not. You could conceive of building a machine that processes incoming data and then takes action, but it could be that different machines have different starting points. Likewise, various natural systems, acting differently in the presence of incomplete information, could have evolved in a world with no human beings at all.

In many of the situations we deal with in physics, the observer dependence is there in principle, but not in practice. Imagine a box full of (classical) particles, and a device that measures the velocity of a single particle. In principle, if we knew the starting conditions and all the relevant factors, we could precisely predict the result of the experiment. In practice this is not achievable. All *plausible* observers are in a state of equal ignorance. The theory of quantum mechanics however asserts that all *possible* observers are in a state of equal ignorance—events really are intrinsically unpredictable. Over the course of the last century, many people have been uncomfortable with this conclusion, and have felt that this must mean that quantum mechanics is not the ultimate theory. If only our Physics was better, they say, we would be like Observer B, and would be able to precisely predict microscopic events. Is that gut instinct, that quantum mechanics may be incomplete, correct? It seems not, as we shall explore in Sect. 13.7.

13.3 Macroscopic Effects of Microscopic Unpredictability

At the microscopic level, it makes no sense to ask e.g. "where is the electron?". Instead all we can do is specify a probability density distribution for where the electron might turn out to be once a measurement is made. The *correspondence principle* asserts that the behaviour of systems described by quantum mechanics reproduces classical physics in the limit of large quantum numbers. At a more technical level, *Ehrenfest's theorem* asserts that the expectation values of our probability distributions obey classical laws. You might think that this means that at a macroscopic level, observations are entirely predictable, that all the randomness is hidden away at the unseen microscopic level. But this is not the case. We can easily see the effects of microscopic randomness in the macroscopic world.

13.3.1 Unpredictability in the Two Slit Experiment

The most famous example of macroscopic unpredictability is the two slit experiment. In this experiment, we have a device that emits single electrons, a screen with two slits, and a detector the far side of the slits that records the position where each electron lands. In modern versions of this experiment, we can see the arrival of each electron at the detector as a separate event. Over a time, the pattern which emerges at the detector—the well known two-slit interference pattern—is completely repeatable and predictable. However, the arrival position of any one single electron is completely unpredictable. Nobody has built an experiment that enables us to predict where an individual electron will land. We will return to analyse the two slit experiment a little more in Sect. 13.4.

13.3.2 Unpredictability in Radioactivity

The phenomenon of *radioactivity* comes about when an atom spontaneously changes from one quantum state to another, and emits a particle—α, β or γ—i.e. Helium nucleus, electron, or photon—in order to conserve energy, angular momentum or other quantum numbers. The lifetimes of excited states of atoms and ions, i.e. the time before they spontaneously decay, varies enormously—from nanoseconds to billions of years, and is a well measured quantity for a specific isotope. However, the actual time of decay of a specific atom is completely unpredictable. A better way to characterise the situation mathematically is to say that there is a random probability per unit time, λ, of the decay occurring. This is the situation we considered in Chap. 6, where we showed that this leads to a probability distribution for the waiting time to the next event equation (6.4):

$$f(t) = \lambda e^{-\lambda t} = \frac{1}{\tau} e^{-t/\tau}.$$

where $\tau = 1/\lambda$ is the lifetime. For the case of radioactivity, there is only one event per atom, so $f(t)$ describes the probability distribution of the time we have to wait before any specific atom decays. If we start with N_0 atoms in the excited state, then the number remaining after time t is $N = N_0 e^{-\lambda t}$, so that after time $\tau = 1/\lambda$, the number remaining has declined by a factor $1/e$. For radioactive substances it is more normal to quote the "half-life" $t_{1/2} = \tau \ln 2 = 0.693\tau$ which is the time needed for half the atoms to decay.

Radioactive substances are very common. The human body itself is radioactive. A typical 70 kg human body contains about 160 g of Potassium, roughly 0.012% of which—about 0.019g—is the radioactive isotope ^{40}K. For ^{40}K the half life to beta-decay is very long—1.25 billion years. This means that for any one atom, the chance that it will decay within the space of a given second of time is very very small indeed. However, there are huge numbers of atoms. Even that tiny amount of ^{40}K corresponds to 3×10^{20} atoms, which will produce something like 4900 decays per second. Most of the β particles are absorbed inside the body, but 11% of the decays produce γ rays, around half of which escape the body—producing around 250 events per second detectable by a Geiger counter. Of course that rate is in all directions, and Geiger counters are not 100% efficient in detecting particles; in practice a Geiger counter held near a human body will produce perhaps ~20–30 counts per minute, i.e. one every two seconds or so. The rate of events is completely repeatable on average, but the gap between events is unpredictable. No experiment has been created which can predict when the next event will happen. Figure 13.1 shows an example of a sequence of events, and also shows that the waiting times follow the expected exponential waiting time probability distribution.

13.3.3 Random Number Applications

A source of random numbers is needed in a variety of practical applications—for example statistical sampling, simulations of physical systems, gambling, exploration of parameter space in model fitting (see Chap. 10), and generating keys for encryption systems. For many applications it is sufficient to use a *pseudorandom number generator*—an algorithm which is fed a seed value which determines its output. Such algorithms are very sensitive to the seed value, so repeated applications produce a sequence which is effectively random for most purposes. However, this method is not secure enough for cryptographic purposes.

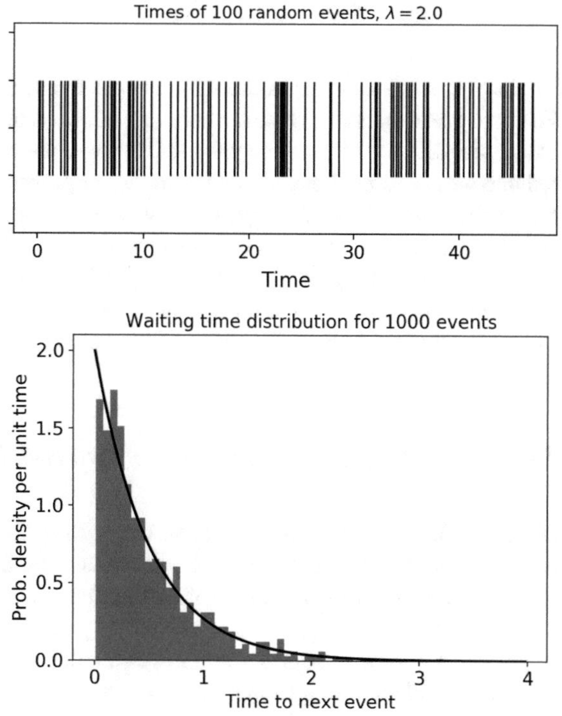

Fig. 13.1 Simulated sequence of radiation decay events, and distribution of waiting times

An alternative is to use a *hardware random number generator*. This could include rolling a die or spinning a roulette wheel. More usefully, it could mean a hardware device embodying a radio-active source, or thermal noise, or various other quantum effects. Sometimes atmospheric noise is used, which contains elements of both intrinsic quantum noise, and classical effective unpredictability due to unknown external factors. An analog to digital converter can then convert the random signal to a 1 or 0, and repeated application can produce a truly random binary string as long as we like.

However, it is difficult for quantum-based hardware methods to produce very long random strings, such as we need in security applications, at a fast enough rate. Rather than producing the sequence itself, quantum devices such as radioactive sources are normally used to produce a *random seed* as the input to an algorithmic pseudorandom number generator. This combination of random seed + algorithmic sequence enables the sequence to be produced at the desired rate, and allows one to adjust the probability distribution, often "whitening" the output so that the probability distribution is close to being flat.

Another application, of growing importance, is the generation of *shared keys* for secure encrypted applications, using methods such as quantum entanglement, which we will discuss later—see Sect. 13.6 and Exercise 13.6. This concept is known as *quantum key distribution*, but is often referred to as "quantum cryptography".

13.4 Probability Amplitudes

We have seen that events in the microscopic world are intrinsically unpredictable. For any given potentially observable quantity, quantum mechanics predicts only the probability of the possible values. However, the probabilities behave in a rather strange way. The mathematics of the theory concerns, not probabilities themselves, but *probability amplitudes*, via the wavefunction or equivalent mathematical construct. To understand what this means, let us start by looking at the famous two-slit experiment, assuming that electrons are particles—quantum bullets as it were. We will then contrast this with the way waves behave. In this section, we will follow the logic set out in the famous lectures by Feynman.

13.4.1 The Bullet Version of the Two-Slit Experiment

Imagine a device firing electrons at a screen with two slits, and recording the position at which the electrons arrive on a detector behind the slits. (See Fig. 13.2.) We know that the electrons come in lumps, as distinct bullets as it were, and furthermore we know that we have to treat their arrival at any given position on the detector as an unpredictable, probabilistic issue. We can model this situation classically as like a gun firing bullets. The gun may be a shotgun spraying the bullets erratically, and each bullet may scatter in a complicated way off the slit walls as it passes through the slit. If we knew all the mechanical factors precisely, we could in principle predict the trajectory of each individual bullet, but in practice those factors are unknown, and we have to treat the behaviour of the bullets probabilistically.

Imagine the horizontal position on our detector divided into a series of boxes. What is the probability that a specific electron will land in box number k? First, we assume that the electron must go *either* through slit A, *or* through slit B, or miss both and get absorbed in the screen. Suppose the probability of going through slit A is $P(A)$, and that of going through B is $P(B)$. The values of $P(A)$ and $P(B)$ will depend on how our electron gun works, how well we have lined up the geometry of the situation and so on. It doesn't matter as long as we don't touch the equipment—they will stay the same.

Next, we could consider only the electrons which have passed through slit A, by closing slit B. We could count the electrons landing in the various boxes, and get the probability of $P(k|A)$. It doesn't matter that we don't know how to calculate this distribution—we just measure it. Likewise we can close slit A and get $P(k|B)$. Finally, when we re-open both slits, and we count how many hits we get in the various boxes, we should find

$$P(k) = P(A) \cdot P(k|A) + P(B) \cdot P(k|B). \tag{13.1}$$

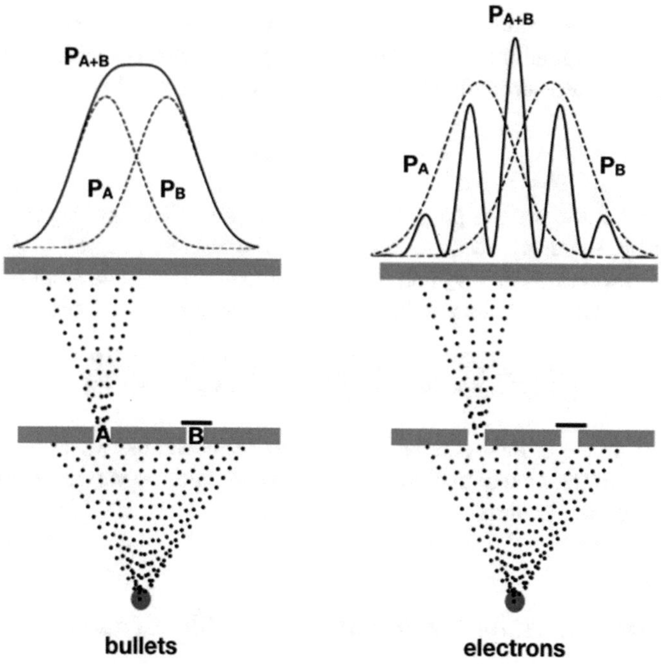

Fig. 13.2 Illustration of the two-slit experiment, contrasting the results with bullets and electrons. In each case P_A shows the probability distribution for where the bullets/electrons hit the screen, if we close slit B, and conversely for P_B. For bullets, when we open both screens, the probabilities add, but for electrons we get a wave-like interference pattern

The idea is illustrated in the left hand side of Fig. 13.2. If we do this experiment with actual macroscopic bullets, this is indeed exactly the result we will find.

13.4.2 The Wave Version of the Two-Slit Experiment

If instead of firing bullets at our slits, we send light waves, or water waves, then we see a result like that illustrated in the right hand side of Fig. 13.2, which is an *interference pattern*. Note that what produces the pattern is the way that the instantaneous *displacements* of the waves add; however, what we measure on the detector is the *intensity* of the light, corresponding to the square of the wave amplitude, i.e. the time averaged displacement. The point of course is that waves have both amplitude and phase, so that waves with differing phases can interfere either constructively or destructively. It is customary to think of a wave as the real part of a complex wave function

$$\psi(x,t) = Re\left[ce^{i\theta}\right] \text{ where } \theta(t) = (kx - \omega t).$$

Here k and ω are the spatial and temporal frequencies, and c is the *complex amplitude*. The real wave has amplitude a, and phase angle ϕ, referenced to some arbitrary time, but we think of these as combined into a single complex quantity

$$c = a(\sin\phi + i\cos\phi).$$

The Intensity of the wave is proportional to a^2. If we define the *complex conjugate* $\psi_0^* = a(\sin\phi - i\cos\phi)$, then we can recover the amplitude as $a^2 = \psi_0\psi_0^*$. Now consider two waves ψ_1 and ψ_2 that have the same real amplitude a but differing by phase δ. We add the complex amplitudes and find the intensity of the summed wave:

$$I \propto a_{\text{sum}}^2 = |\psi_1 + \psi_2|^2 = 2a^2(1 + \cos\delta). \tag{13.2}$$

For the two slit experiment, the phase difference between the waves arriving from the two slits is a function of position x, giving a continuously varying $I(x)$, the well known interference pattern.

13.4.3 The Quantum Version of the Two-Slit Experiment

When we do the experiment with electrons, we get features of both bullet and wave behaviour. Electrons behave like probabilistic bullets, arriving one at a time unpredictably, but the accumulated pattern over time looks like the light-wave interference pattern. Quantum mechanics proposes the following recipe as a solution.

- Any one electron is associated with a complex "wave function" ψ which varies with location in space
- We cannot measure the wave function itself, but if we make a measurement at some location, the probability of finding the electron there is given by $|\psi|^2$, which is a real number.
- The wavelength of the wave function is given by $\lambda = h/p$ where p is the momentum of the electron and h is Planck's constant. This allows us to calculate the *phase difference* between the electrons coming from the two slits.
- For two wavefunctions, we do not add the probabilities themselves, but the complex amplitudes: $P_{\text{sum}} = |\psi_A + \psi_B|^2$.

If we now consider closing slits A and B in turn, what we have is a little more complicated than equation (13.1):

$$P(k|A) \propto |\psi(k|A)|^2$$
$$P(k|B) \propto |\psi(k|B)|^2$$

$$P(k) \propto |\psi(k)|^2 = |\psi(k|A) + \psi(k|B)|^2. \tag{13.3}$$

So this is the second strange thing about probability in quantum physics—its not just that quantum situations are intrinsically, as opposed to effectively, unpredictable; it is also that the probabilities seem not to follow the usual laws of probability. It is the complex amplitudes that add, rather than the probabilities.

13.4.4 Quantum States and Complex Probability Amplitudes

Electromagnetic (EM) waves are not just a mathematical fiction; with radio antennas for example, we can detect the electric field displacements directly. On the other hand, writing EM waves as complex quantities is a calculational convenience. We would get the same answer, with more algebra, if we ignored complex numbers and added the real waves, attending to all the trig function algebra correctly. The essential point is that waves are characterised by *two numbers*, which we could see alternately as amplitude and phase, or a vector (a, b), or a complex number $a + ib$, as long as the appropriate mathematics is followed.

The observation of an interference pattern doesn't necessarily mean that electrons are actually waves, with something material waving. In the Dirac formalism, we see quantum states as vectors in some abstract space. We can express such an abstract state vector as $|\psi\rangle$, and attach to it a complex amplitude c, expressing the state overall as $c\,|\psi\rangle$. The probability associated with the state is then $|c|^2$. If we combine two states then the overall state can be expressed as

$$|\psi\rangle = c_1\,|\psi_1\rangle + c_2\,|\psi_2\rangle\,.$$

13.5 The Strange Behaviour of Measurement

As part of our "bullet" analysis above, a key assumption was that any given electron must go through *either one slit or the other*. However, quantum mechanics asserts that we can't make that assumption. What if we try to detect each electron as it passes through the slit? This is in principle possible, for example by having light scatter off the electron as it passes. Then we will *know* which slit the electron has gone through before we see it landing on our detector. When such an experiment is performed, the final probability distribution $P(k)$ is different from the distribution we get when we don't detect which slit the electron has gone through; and in fact, $P(k)$ obeys equation (13.1) perfectly. If we treat the electrons like bullets, they behave like bullets.

13.5.1 Collapse of the Wavefunction

It seems then that the basic laws of probability are not violated after all; the peculiar interference pattern comes about because the behaviour of electrons is not determined until we make a measurement. At first glance, this might seem no different from the way we have to treat some classical situations—we don't know the true state of a system, and so must treat it probabilistically. However, for a quantum system, it is not that we *don't know* the state of the system; it is that it is *not determined* until we make a measurement. This change from indeterminacy to a definite situation at the point of measurement is known as the *collapse of the wave function*.

We can illustrate this by returning to the coin analogy we used in an earlier section. A colleague tosses a coin, and after seeing whether it lands head or tails, puts it in a box, and hands it to you. Looking at the closed box, we don't know what the state of the coin is. A frequentist might say "if I imagine this same procedure being carried out many times, it is going to be heads 50% of the time". A Bayesian might say "if I try to attach a credibility to the hypothesis that it is heads, I might as well say 0.5, because I have nothing else to go on". For a quantum system, it is as if our colleague carefully placed the coin in the box *still spinning*. It will keep spinning until we open the box at which point it stops on either heads or tails. So it is not that we *don't know* whether the coin is heads or tails; it is *not fixed* until we make the measurement.

13.5.2 The Uncertainty Principle

The idea of the collapse of the wave function is closely connected to the famous Heisenberg Uncertainty Principle. This asserts that the position x and momentum p of a particle cannot be measured with arbitrary accuracy at the same time, and that the more accurately you measure one, the less accurately you will know the other. Suppose that the probability distribution for finding the particle at position x is $p(x)$. Then we characterise the uncertainty by the standard deviation σ_x. The Uncertainty Principle then states that

$$\sigma_x \sigma_p \geq \hbar/2, \qquad \text{where } \hbar = h/2\pi.$$

There are various other uncertainty pairs, such as energy and time, or the x and y components of electron spin. When we make a measurement of one of the two quantities, we partially collapse the joint wave function; then σ_1 becomes zero, and σ_2 becomes infinite—i.e. we have no information at all on its value. Before collapse, the two quantities can be seen as having a joint probability distribution, but measuring one quantity does not correspond simply to fixing its value and taking a slice across the joint distribution. It is the wave function that collapses rather than the probabilities.

13.5.3 Measurement Collapse and the Stern–Gerlach Experiment

Collapse at the point of measurement is seen particularly starkly in the Stern–Gerlach experiment, which involves electron spins. Considering the electron spins also prepares us nicely for the probability-based tests of hidden variable theories, which we will look at in Sect. 13.7.

Electrons have a magnetic moment μ (spin), which means they are deflected when passing through a magnetic field. In a classical world, you might imagine that the spins have random orientations, and so get deflected by various amounts. In fact, they get deflected by a fixed amount, either up or down with respect to the line defined by the magnetic field. We can understand this as follows. The x and y components of spin are an uncertainty pair, so we cannot know the value of μ_x and μ_y arbitrarily well at the same time. Their values are not just unknown; they are undetermined. When we measure the deflection parallel to the magnetic field we are collapsing our uncertainty on μ_x, and so at the same time can know *nothing* about μ_y. Seeing a spread of deflection sizes would however, statistically, tell us the spread of μ_y sizes, which would be inconsistent with the assertion that these are undetermined before the measurement.

Now suppose we have a second magnet, at angle θ to the first. We can think of the electrons as being *prepared* as all having spin along direction $\boldsymbol{n}$ parallel to the first magnetic field $\boldsymbol{B}_1$—but either "up" or "down" along $\boldsymbol{n}$—and then *measured* along direction $\boldsymbol{m}$ parallel to the second magnetic field $\boldsymbol{B}_2$. You might guess that the "up" electrons have a component along $\boldsymbol{B}_2$ of $\mu\boldsymbol{n}.\boldsymbol{m} = \mu\cos\theta$, and so give an upwards deflection of a smaller size than the first experiment. However, the deflections caused by the second magnetic field are of the same quantised size as the first experiment; and furthermore it is still unpredictable whether an individual electron goes up or down with respect to $\boldsymbol{m}$. What is determined by the angle θ is not the size of the deflection, but the *probability* of being deflected up or down.

$$P_{\text{up}} = \frac{1+\cos\theta}{2} = \cos^2\theta/2, \qquad P_{\text{down}} = \frac{1-\cos\theta}{2} = \sin^2\theta/2. \tag{13.4}$$

Note that the up and down probabilities sum to 1.0.

13.5.4 Indeterminacy and State Mixtures

Another way to look at the indeterminacy/collapse issue is to think of particles as existing in a *mixture of states*. When we have "prepared" an electron along $\boldsymbol{n}$, but before we have measured it along $\boldsymbol{m}$, the electron might be in either an up-state or a down-state. We could label these states $|\psi_{\text{up}}\rangle$ or $|\psi_{\text{down}}\rangle$. If there are two possible states associated with some system, and which have complex amplitudes c_1 and c_2, then the overall quantum state can be written

$$|\psi\rangle = c_1\,|\psi_1\rangle + c_2\,|\psi_2\rangle\,.$$

This should be seen as being a *mixture* of states. When we make the measurement, the probability of getting state-1 is $p_1 = |c_1|^2$ and the probability of state-2 is $p_2 = |c_2|^2$. The probability of getting either 1 or 2 is $p_{12} = |c_1 + c_2|^2$. If this is a complete list of possible states, we should find $p_{12} = 1.0$ of course. In the Stern Gerlach experiment, $c_1 = \cos(\theta/2)$ and $c_2 = i\sin(\theta/2)$, and you can easily confirm that in this case $p_{12} = 1.0$.

13.6 Quantum Entanglement and the Failure of Determinism

The theory of quantum mechanics is in some ways even stranger than the behaviour it attempts to explain—with the intrinsic unpredictability, the behaviour of complex probability amplitudes, and the indeterminacy until the point of measurement. Some scientists—notably Einstein—have argued that this strangeness must result because quantum mechanics is incomplete, because there are *hidden variables*. One could imagine that inside each particle there is a secret machinery which completely determines the behaviour of particles; its just that our current knowledge of physics doesn't have a theory for this secret machinery, and so in practice we have to treat the spins as random—just like in the rest of classical probability. If on the other hand, as quantum mechanics argues, the situation really is not deterministic, there are disturbing consequences for cause and effect. Such problems are clearest in experiments involving *quantum entanglement*, so lets start by explaining that idea.

13.6.1 Normal Entanglement and Quantum Entanglement

Sometimes physical states are connected to each other. We speak of them being "entangled" when there is also a probabilistic aspect to the situation, such that the states, and their probabilities, remain connected as our knowledge updates. It is quite easy to have a kind of entanglement in a classical world. Imagine experimenter Charlie placing an apple in one box and a banana in another, and giving a box each to the famous experimenters Alice and Bob. Before opening the box, Alice does not know whether she has an apple or a banana, and can only assign probability of 0.5 to either. However, as soon as she opens her box and sees a banana, she knows for sure that Bob has an apple in his box. The probabilities change together because the states are entangled.

Quantum entanglement however is stranger than this. It is as if Charlie has placed a still-spinning coin in each box. Whether the coin is heads or tails is not just unknown; it is not fixed until the box is opened and the coin stops spinning. When Alice opens her box and the coin stops on heads, she knows that when Bob opens his box, his spinning coin will stop on tails. In other words, it is not the entanglement per se that is strange; it is entanglement coupled with indeterminacy until the point of measurement.

13.6.2 *Entangled Spins*

Some atomic physics events (for example, some nuclear decays) lead to a pair of particles created with equal and opposite spins, so that angular momentum is conserved. They also move out from the event centre in opposite directions, conserving linear momentum. As they travel outwards, their orientation is undetermined; however they are linked together, in that whatever orientation one particle is eventually found to have when a measurement is made, the other particle will have precisely the opposite orientation. Their quantum states are *entangled*. They are like the spinning coins in the previous section.

Consider placing two magnets, with the same orientation, at either end of the particle travel, and measuring deflections. As in the regular Stern–Gerlach experiment, setting the magnetic field direction forces the undetermined orientation to collapse; all that is left undetermined is whether we find up or down. Suppose Alice is making the deflection measurements at one end. For any one particle, whether she gets up or down is both unpredictable and undetermined. When the deflection has occurred, and she finally knows whether the particle is up or down, the wave function has collapsed completely. However, at this point, Alice knows that the partner particle must have opposite spin. At the far end Bob is also measuring deflections, which as far as he is concerned are unpredictable. However, Alice can predict them. If she found a sequence $uududd$ she could write on a piece of paper "Bob gets $dduduu$". Note that she can predict Bob's events no matter how far apart Alice and Bob are.

13.6.3 *The EPR Paradox and Hidden Variables*

Is Alice's predictive power mysterious? If the particles were entangled in a classical sense, it would not be at all mysterious. Suppose Bob has an apple-banana sequence of boxes. It is already fixed; it's just that he doesn't know what the sequence is. Meanwhile Alice opens her boxes, and finds a sequence $bbabaa$. At this point everybody knows that Bob must have sequence $aababb$. There is no mystery. However, quantum mechanics asserts that Bob's sequence is not just unknown, but undetermined until he makes the measurement. So somehow it seems that Alice's measurements have *caused* Bob's measurements to come out the way they do. Now the distance between Alice and Bob becomes important; if they are far enough apart, and they make their measurements close in time, it seems that this causal influence can travel faster than the speed of light.

In the 1930s, Einstein, Podolsky and Rosen wrote a famous paper describing a thought experiment of this kind, and took it to show a paradox in the logic of quantum mechanics—we surely cannot have both indeterminacy and a distant causal link at the same time. Since then, a number of real "EPR experiments" have been performed which confirm both the expected entanglement and the apparent unpredictability, so that we are faced squarely with this apparently paradoxical situation. Let us summarise the problem arising from EPR style experiments.

- Quantum mechanics asserts that microscopic quantities are indeterminate—not fixed until measurement.
- At the point of measurement, the properties of particle pairs are strongly correlated, even at large distances.
- The previous two things can only both be true if causality is non-local.

In popular accounts, it is often suggested that Einstein was uncomfortable with quantum mechanics because of the first bullet point—that he objected philosophically to indeterminacy. In fact, his philosophical worry was with the third bullet point—action at a distance. He therefore *concluded* that indeterminacy must be wrong. His preferred route out was to suggest that quantum mechanics must be incomplete, because there are *hidden variables*.

13.7 Statistical Tests of Hidden Variable Theories

How do we test the general idea of hidden variables? That is, the possibility of a yet undiscovered deterministic theory that involves some variables we don't even know about yet? Normally in physics we test specific well posed theories, that make concrete predictions for measurements in particular circumstances. However in the 1960s John Bell showed that there are generic statistical consistency tests that *any* hidden variable theory must pass. Classical systems always pass these tests; certain quantum experiments fail them; but quantum theory correctly explains the observations.

Suppose we make a series of measurements on one or more variables $x, y, \ldots$. Each measurement gives a different result, but we see a consistent probability distribution $p(x, y, \ldots)$. In a deterministic theory, there may be other variables $\alpha, \beta, \gamma, \ldots$ that we can't track or even perhaps know about, and the apparent randomness in $x, y, \ldots$ is actually a consequence of our incomplete knowledge of $\alpha, \beta, \gamma, \ldots$. Do these assumptions produce constraints on the observed data $p(x, y, \ldots)$ that we can check? It turns out that the answer is yes, for specific kinds of situation involving three yes/no questions, as we will explain in the next section. Quantum entanglement experiments can be designed in just this way, to give answers to three yes/no questions. Before we look at the quantum experiments however, lets look at the statistical tests themselves, which have wider application.

13.7.1 Consistency Tests for Trivariate Yes/No Distributions

Consider a collection of objects which can be classified according to three properties, A, B, C, each of which is an either/or or a yes/no kind of thing. An example might be a survey of people who are asked three questions—did you go to university? do you own a car? do you believe in climate change? There are eight possible combinations of the answers to the questions A, B, C. We can express these neatly if we write A to mean the number (or normalised fraction) of people answering yes to question A, $\bar{A}$ to mean the number of people answering no to A, $AB\bar{C}$ to mean the number answering yes to A and B and no to C, and so on:

$$\begin{array}{cc} ABC & AB\bar{C} \\ A\bar{B}C & A\bar{B}\bar{C} \\ \bar{A}BC & \bar{A}B\bar{C} \\ \bar{A}\bar{B}C & \bar{A}\bar{B}\bar{C} \end{array}$$

In essence what we have is a $2 \times 2 \times 2$ trivariate distribution, with A and $\bar{A}$ along one axis and so on. Visualising a 3D block of numbers is hard, so quite often what we might have available is a set of three 2D summary tables, as illustrated in Fig. 13.3. In the survey example above, we might tabulate the number who did/did not go to university versus the number who do/do not own a car, regardless of whether they believe in climate change. What we are doing is marginalising over C to get a distribution over A and B.

Suppose we suspect there has been an error in compiling the three tables from the original raw data? Or perhaps the three 2×2 tables come from three different surveys, where for example we asked a number of people questions A and B without even asking them question C? How do we check that our three tables have in fact been drawn from some underlying $2 \times 2 \times 2$ dataset? Note that there are twelve cells in the various marginalised tables. Each of them is some different combination of the grid of eight raw cells. In Fig. 13.3, we label some of these combinations with symbols, to aid the discussion below.[1] We assume that we are dealing with the numbers in each cell, but the same logic applies if we are dealing with frequencies normalised by the total number in the sample.

Test-1. For *each* of the three 2×2 tables, the data in the cells should add up to same value, or be statistically consistent with the same values if the tables come from three separate surveys.

[1] If some of the discussion below seems a little abstract, trying working on the related problem in the exercises as you read.

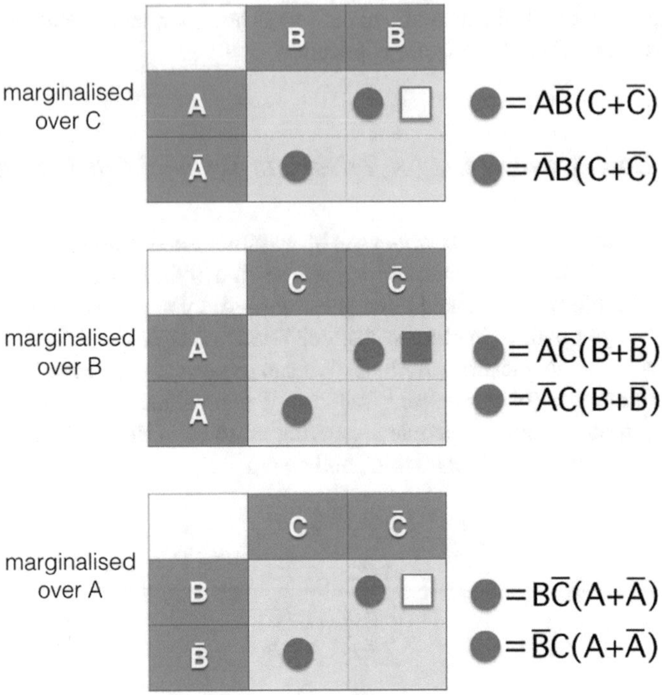

Fig. 13.3 Illustrating how consistency tests can be place on trivariate joint probability distributions. See text for explanation of symbols

Test-2. The overall number with property A can be obtained from adding the values in the first row of Table 1. It can also be obtained from adding the values in the first row of Table 2. So these two should agree with each other. They should agree within statistical errors if the tables represent distinct surveys; but if drawn from a single survey they should be identical. Likewise the value for B can be obtained from the first column of Table 1, or the first row of Table 3, and so on.

Test-3. This is the test originally considered by Bell. Consider the cells marked by squares. The sum of the values in the two cells with open squares must be at least as big as the value in the cell with the filled square.
Open squares: $A\bar{B}C + A\bar{B}\bar{C} + B\bar{C}A + B\bar{C}\bar{A}$.
Filled square: $A\bar{C}B + A\bar{C}\bar{B}$.

The open-squares list contains the same two terms as the filled square list, plus two other terms, so the former is clearly larger than the latter, unless those extra terms happen to be zero, in which case the open-square and filled-square values are equal. We can write

$$A\bar{B} + B\bar{C} \geq A\bar{C},$$

where we understand "$A\bar{B}$" to mean "the number with A and not with B, regardless of the value of C". This inequality must be obeyed for any set of 2×2 tables from a single trivariate distribution.

Test-4. This is another inequality, developed by Clauser et al. (1969), which is more complicated but is better suited to application to the physics experiments we describe in the next section. It can be stated as

$$(A\bar{B} + \bar{A}B) + (A\bar{C} + \bar{A}C) + (B\bar{C} + \bar{B}C) \leq 2. \tag{13.5}$$

where each of the two-symbol terms should be seen as marginalised over the third (missing) symbol. Proving this is simple but tedious. Each term corresponds to one of the cells with the filled circles. The values of these are indicated to the right of the relevant table. The next step is to multiply these out, so we get $A\bar{C}C + A\bar{B}\bar{C} + \bar{A}BC + \cdots$. Then we collect the terms. We find that all the 8 raw terms occur twice, except for two terms which don't occur:

$$\text{missing terms}: \quad ABC \quad \text{and} \quad \bar{A}\bar{B}\bar{C}$$

If all the 8 raw terms were there twice, our sum would have been exactly 2.0. If the two missing terms happen to be equal to 0.0, then our sum can still be 2.0; but more generally it must be less than 2.0. So the complete collection of terms must be less than or equal to 2.0, as required.

13.7.2 Application to Entangled Spin Experiments

So what do these consistency tests for joint probability distributions have to do with quantum mechanics? Imagine an experiment which is producing a stream of thousands of particles with some unknown spin. (For now we are back to thinking about a single particle, rather than entangled pairs.) We then have a Stern–Gerlach style setup to measure the spin-component of the particles in a chosen direction. For each particle, we can only get the answer "up" or "down". Suppose now we can rotate our magnet to various angles. Lets label one particular position A, and two other positions B and C, at, say, angles of 45 and 90° with respect to A. For a given particle, we could in principle ask three questions—"is the spin up or down with respect to position A?", "is the spin up or down with respect to position B?", and likewise for C. This then is an example of a collection of three either/or questions, as discussed in the previous section.

Next, let us suppose that the indeterminacy principle is wrong, and that there are indeed hidden variables such that the spin of each particle is fixed *at the point of production*, before the particles travel out on the journey towards the detector. However, we don't know these hidden variables. All we can do is imagine our stream of thousands of particles emerging with a probability distribution of spins. What we would like to do is to measure A or $\bar{A}$, B or $\bar{B}$, and C or $\bar{C}$ for each particle in turn, and then put the result into one of our 8 raw cells, i.e. $A\bar{B}C$ etc. As we accumulate counts in these boxes, we would end up with a well defined trivariate probability distribution.

However, there is a snag in carrying out such an experiment. For any one particle, we can only ask *one* of the three questions A,B,C at a time. If we could at least make *two* measurements at the same time, we could use one of the tests involving two questions at a time, marginalised over the third, as in (13.5). Note that we would have to run the experiment three separate times, first asking questions A and B, then asking questions A and C, and then asking questions B and C. However, this is ok—as long as the experimental set up is stable, the probability distribution should be repeatable.

But how do we ask two questions at the same time? This is where the entangled pairs come in. Instead of a stream of thousands of single particles, we produce a stream of thousands of entangled pairs moving in opposite directions. We have two magnets at the opposite ends, and can place these at differing angles. A counter-magnet at rotation $-45°$ is exactly equivalent to a magnet at $45°$, because the spins of the two particles are forced to be exactly $180°$ apart. So we can then run our experiment three times, with magnet pairs at (0, −45), (0, −90), and (45, −90), corresponding to the three question pairs above—AB, AC, BC.

The generic hidden variable hypothesis does not make a specific prediction for the trivariate probability distribution—different theories could make different predictions. However, it does predict a *limit* for the combination of marginal distributions specified in equation (13.5). Quantum mechanics meanwhile *does* make a specific prediction for the probabilities in equation (13.4). Recall that what we combine is the wave functions, rather than the probabilities themselves; the probability of finding a given measurement is given by the square of the modulus of the complex amplitude. As we described in Sect. 13.5.4, a given particle can be seen as existing in a mixture of possible states. The situation we have described here, with magnet and counter-magnet at two different angles, is equivalent to the process we described earlier "preparing" the particle in one orientation, and then measuring at an angle θ from that orientation. In this case the probabilities of getting "up" or "down" respectively are given by equation (13.4) as $1/2(1 \pm \cos\theta)$. Assuming angles 0, 45, and 90°, and wading through the combinations AB, AC, BC, accounting for the appropriate angle differences, the result is that the quantum mechanical prediction is that

$$(A\bar{B} + \bar{A}B) + (A\bar{C} + \bar{A}C) + (B\bar{C} + \bar{B}C) = 2\sqrt{2}. \tag{13.6}$$

which can be compared to the inequality of equation (13.5), which insists that this combined quantity must be less than 2.0. Which prediction corresponds to reality? The real experiments of course have all sorts of subtleties and complexities we have brushed over here, but the bottom line is that experiments agree with the quantum mechanics prediction. It seems that Nature is telling us that we cannot have both determinism and local causality. It remains technically possible that we could construct a hidden variable theory that is not local—where, at the moment of measurement, some unknown physics fixes the spin of the particle, and communicates it immediately to the other particle—but most people would find this just as uncomfortable as throwing away determinism.

13.8 Whither Quantum Mechanics?

The key mysteries in quantum mechanics are all closely related to probability—firstly that our calculations concern probabilities of events, rather than the behaviour of objects; secondly that probabilities combine in a very strange manner, via complex amplitudes; and thirdly, that outcomes are not just unknown, but undetermined until the moment of measurement (i.e. "wave function collapse"). The mathematical rules that physicists have assembled have been astonishingly successful, producing not just theoretical explanations, but extremely practical consequences for the everyday world. Nonetheless, debates still continue about the fundamental interpretation of quantum mechanics.

My summary in the paragraph above is a statement of the orthodox view, or "Copenhagen interpretation". Many scientists would feel that there is really nothing much else to say or do; all that is meaningful is to do the calculations. However, there have been a number of attempts to make the philosophical foundations of quantum mechanics more satisfying. We will briefly describe just a few, which have key probability related aspects.

13.8.1 The Many Worlds Interpretation

The many worlds interpretation denies the reality of wave-function collapse, but in a way that is philosophically controversial. All the possible outcomes are seen as happening in one of many "parallel universes". We are living through one particular branching track. It is a matter of debate whether probabilities should be attached to the various branches.

13.8.2 Bohmian Mechanics

Bohmian mechanics re-asserts the concrete reality of particles, which have definite positions at all times, and, following De Broglie's original suggestion, takes the wave function as an accompanying "pilot wave" that controls their behaviour. Along with Schrödinger's equation which determines the evolution of the wave function, there is a second *guiding equation* which determines particle tracks. The tracks are in principle deterministic, but in practice probabilistic, because they are very sensitive to initial conditions, which are unknown. (The related "stochastic mechanics" actually involves a kind of spacetime foam, which produces very erratic tracks.) Bohmian mechanics is a kind of hidden-variable theory; however because of the quantum entanglement results, we know that such a theory must be non-local, i.e. involve action at a distance.

13.8.3 Decoherence Theory

In decoherence theory, the key point made is that although we make measurements of specific local systems, in reality such systems are not isolated, but interact with their environments. Phase-coupling with the environment tends to make the system less coherent. The quantum nature of the system can be seen as "leaking" into the environment. Although one can argue that a universal wavefunction still exists, as the local system decoheres, it leads to a specific realisation, i.e. an effective collapse of the local wavefunction. There is a close analogy here with the behaviour of thermodynamic systems and the inevitable increase of entropy, which we will consider in Chap. 14. The phases of a large number of particles will become increasingly randomised and undergo an irreversible change from coherence to un-coherence. The decoherence process can be seen as sharing *information* between system and environment.

13.8.4 Quantum Bayesianism

When we say that "probability amplitudes" determine the probability of a given experimental outcome, what kind of probability do we mean? The frequency of possible outcomes in many imagined similar experiments? Or our degree of belief in/expectation of the various possible outcomes? In Quantum Bayesianism, often known as QBism, the central idea is that all probabilities are personal, rather than objective—i.e. they encode an experimenter's expectations. In this view quantum states, just like probabilities, are not objectively real, but are to do with an observer's knowledge of the situation. The infamous wave function collapse is then just the observer updating their beliefs after making a measurement. In a way, this is an extreme version of the Copenhagen interpretation. QBists suggest that even a probability of 1.0 should be considered personal, and argue that this way we can avoid non-locality in quantum physics.

These are deep waters, and the attitude of many physicists is that all of the above interpretations are metaphysics rather than physics—quantum mechanics just works. However, it remains possible that clarifying these foundational issues will produce new predictive power, in which case people will sit up and take notice. Decoherence theory is particularly important in this regard, as it involves issues that definitely have to be taken into account, regardless of whether or not doing so clears up our philosophical worries.

13.9 Key Concepts

Some of the key concepts from this chapter are:

- The realisation that microscopic randomness is clearly manifested at the macroscopic level, in phenomena like radioactive decay
- How calculating the probabilities of combined events involves adding complex probability amplitudes, rather than the probabilities themselves
- The importance of indeterminacy until the point of measurement, otherwise known as collapse of the wave function
- How quantum entanglement, as opposed to simple classical entanglement, seems to imply action at a distance
- The idea of consistency tests for trivariate probability distributions
- The application of consistency tests to quantum-entangled particle pairs
- How real quantum entanglement experiments rule out hidden variable theories that are both deterministic and causally local

The key formulae in this chapter include: the formula describing how probabilities should behave in a two slit experiment with bullets (13.1), with waves (13.2) and with quantum mechanical wave functions (13.3); the formulae for the "up" and "down" probabilities in a Stern–Gerlach experiment (13.4); the formula for the hidden variable consistency test used by Clauser et al. (13.5); and the quantum mechanical prediction for the same test (13.6).

13.10 Further Reading

There are of course many excellent textbooks on quantum mechanics. Rae and Napolitano (2015), which is a new edition of the book by Rae which has been popular for many years, is a short and clear mid-level undergraduate text, and in particular has a whole chapter on problems with the conceptual foundations of quantum mechanics. Binney and Skinner (2013) is the next step up in rigour, and has no messing about with waves—straight in with the Dirac formalism. Far from being too maths-centred, it is very good at making the physical point clear. At the other end of the scale, Susskind and Friedman (2015) is a very impressive semi-popular account of quantum mechanics—an attempt to give the intelligent public the "real thing" while not being too intimidating.

However, possibly still the clearest account of the fundamentals of quantum mechanics is Vol. III of the famous Feynman Lectures. The latest edition is Feynman et al. (2011). The text is also available at the Feynman Lectures website. Feynman's description of the two-slit experiment was actually a thought experiment—the historical experiment by Davisson and Germer which demonstrated the wave-particle duality in electrons involved scattering from the planes of crystals rather than slits. Feynman's experiment was finally carried out faithfully by Bach et al. (2013). A semi-popular description of that experiment, and the history leading up to is in the online article by Johnston (2013); this also includes a nice video illustrating how the interference pattern is built up one event at a time.

The topics of quantum entanglement and tests of hidden variable theories are covered in advanced textbooks, but the key original papers are all short, clearly written and well worth reading. It all started with Einstein et al. (1935); the idea that hidden variable theories had to pass a consistency test appeared with Bell (1964); the key paper on how to turn this into a real world experiment was Clauser et al. (1969); and the definitive experimental proof that quantum mechanics worked and local hidden variable theories didn't is in Aspect et al. (1982). An influential semi-popular account is D'Espagnat (1979), but probably the best informal account is by Bell himself, in a written version of a talk given in Paris in 1980 (Bell 1981). This has the wonderful title of "Bertlmann's Socks and the Nature of Reality".

When it comes to the conceptual foundations of quantum mechanics, probably the majority of physicists would agree with David Mermin, who once said his attitude was summed up as "shut up and calculate" (Mermin 1989; though in Mermin 2004 he muses on whether it was really Feynman who said it!). More recently Mermin has become a convert to Quantum Bayesianism (Mermin 2014). The ideas of Quantum Bayesianism were swirling around in the writings of Jaynes, Finetti and Ramsey, but have crystallised in the work of Fuchs, Caves, and Schack. A very lively summary with both accessible parts and very technical parts is given by Fuchs (2010). As for the other rival interpretations we discuss, the founding works are by Everett (1957) for many-worlds; Bohm (1952a, b) for Bohmian-mechanics; and Zeh (1970) for Quantum decoherence. To read further, the Wikipedia page on interpretations of quantum mechanics is a very good start. A livelier introduction can be found in a video recording of a panel debate held at the World Science Festival in New York in 2014.

13.11 Exercises

13.1 Twenty three percent of the human body is Carbon. Mostly this is ^{12}C, but about one atom in 10^{12} is ^{14}C, which is radioactive, with a half-life of 5370 years. Roughly how does the rate of Carbon events from the human body compare with that from Potassium?

The ^{14}C/^{12}C ratio in living beings is set by cosmic rate bombardment of the atmosphere. Why does this mean we can use radioactive events to date some archaeological remains?

13.2 The waiting time before a radioactive event should be given by (6.4), where λ is the mean rate of events. If we have just had a long gap between two events, does this change the probability that the next gap will also be long? Does the same logic apply to non-quantum sporadic events, such as the waiting time to the next No. 41 bus?

13.3 In a Stern–Gerlach experiment with an angle θ between the two magnets, what angle makes spin-up with respect to the second magnet twice as likely as spin-down?

13.4 A photon-entanglement experiment takes a millisecond to make a measurement of each pair. How far apart would you need to put the stations to make sure there isn't some subtle causal influence between Alice's measurement and Bob's?

13.5 A polling company carries out a survey on behalf of the Labour Party, asking a number of people three questions: (A) Are you a home-owner? (B) At the last election, did you vote Labour? and (C) Do you think the railways should be nationalised? Rather than sending the raw data, the polling company provide the results as three summary tables, as shown in Fig. 13.4. Party officials are concerned, as the results are somewhat different from what they expected; they privately worry that the numbers have been faked by the polling company. Test whether the numbers given have been drawn from a valid trivariate probability distribution.

13.6 Suppose two parties are exchanging cryptographic keys using an entangled-pair method, as suggested in Sect. 13.3.3. How could the Bell or Clauser inequalities be used to test for the presence of eavesdroppers?

Fig. 13.4 Summary results from the survey in question Exercise 13.5

Table 1	Labour	not Labour
home owner	345	133
not home owner	35	282

Table 2	nationalise	not nationalise
home owner	243	235
not home owner	191	126

Table 3	nationalise	not nationalise
Labour	284	96
not Labour	150	193

References

Aspect, A., Dalibard, J., Roger, G.: Experimental test of Bell's inequalities using time-varying analyzers. Phys. Rev. Lett. **49**, 1804–1807 (1982)

Bach, R., Pope, D., Liou, S-H., and Batelaan, H.: Controlled double-slit electron diffraction. New J. Phys. **15**, 1–7 (2013)

Bell, J.S.: On the Einstein Podolsky Rosen paradox. Physics **1**, 195–200 (1964)

Bell, J.S.: Bertlmann's socks and the nature of reality. J. Phys. Colloq. **42**, C2–41 (1981)

Binney, J., Skinner, D.: The Physics of Quantum Mechanics. Oxford University Press, Oxford (2013)

Bohm, D.: A suggested interpretation of the quantum theory in terms of "Hidden Variables" I. Phys. Rev. **85**, 166–179 (1952a)

Bohm, D.: A suggested interpretation of the quantum theory in terms of "Hidden Variables" II. Phys. Rev. **85**, 180–193 (1952b)

Clauser, J.F., Holt, R.A., Shimony, A., Horne, M.A.: Proposed experiment to test local hidden-variable theories. Phys. Rev. Lett. **23**, 880–884 (1969)

D'Espagnat, B.: The quantum theory and reality. Sci. Am. **241**(11), 128–166 (1979)

Einstein, A., Podolsky, B., Rosen, N.: Can quantum-mechanical description of physical reality be considered complete? Phys. Rev. **41**, 777–780 (1935)

Everett, H.: Relative state formulation of quantum mechanics. Rev. Mod. Phys. **29**, 454–462 (1957)

Feynman, R., Leighton, R., Sands, M.: The Feynman Lectures on Physics, Vol. III: Quantum Mechanics (New Millenium Edition). Basic Books, New York (2011)

Fuchs, C.A.: QBism, the Perimeter of Quantum Bayesianism (2010). arXiv:1003.65209

Mermin, N.D.: What's wrong with this pillow? Phys. Today **42**(4), 9 (1989)

Mermin, N.D.: Could Feynman have said this? Phys. Today **57**(5), 10 (2004)

Mermin, N.D.: Physics: QBism puts the scientist back into science. Nature **507**, 421–423 (2014)

Rae, A.I.M., Napolitano, J.: Quantum Mechanics, 6th edn. Routledge (2015)

Susskind, L., Friedman, A.: Quantum Mechanics: The Theoretical Minimum. Penguin (2015)

Zeh, H.D.: On the Interpretation of Measurement in Quantum Theory. Found. Phys. **1**, 69–76 (1970)

Websites (all accessed March 2019):

World Science Festival Debate on Quantum Physics and Reality: https://www.worldsciencefestival.com/programs/measure_for_measure/

Feynman Lectures website: http://www.feynmanlectures.caltech.edu

Johnston, H.: Feynmans double-slit experiment gets a makeover. Physics World online article (2013). https://physicsworld.com/a/feynmans-double-slit-experiment-gets-a-makeover/

Wikipedia page on interpretations of Quantum Mechanics: https://en.wikipedia.org/wiki/Interpretations_of_quantum_mechanics

Chapter 14
Entropy, Complexity, and the Arrow of Time

14.1 Outline of Content

- Thermodynamic Entropy
- Statistical Entropy
- Order, Entropy and Complexity
- Entropy and gravitating systems
- Probability and Time

In this final chapter we look at an issue that has caused much confusion amongst physicists, let alone the general public—the idea of entropy, and its relation to ideas of disorder, decay, and the arrow of time. Probability has been central to understanding the concept of entropy ever since Boltzmann proposed that it should be identified with the multiplicity of microscopic configurations of a gas. Some decades later, as we discussed in Chap. 11, Shannon developed a concept that he called "entropy" in the context of Information Theory—is this the same thing? And what do people mean by the entropy of a black hole? Do black holes really swallow information? Meanwhile, pinning down what we mean by "disorder" proves rather slippery. We instinctively feel it is the opposite of "structure", or "complexity", but how can we rigorously quantify these ideas? We often illustrate the second law of thermodynamics by noting that you can easily make an omelette from an egg, but not an egg from an omelette. However, Nature very certainly does make eggs. How do we square these observations?

Many of the issues outlined above remain controversial and are the subject of ongoing research. Our aim in this chapter is not to solve all the controversies, but to introduce the reader to the key concepts as clearly as we can, to make further reading in these choppy waters a little easier. As we step through the topics, there are two key things to keep in mind.

The first thing is distinguish carefully between equilibrium and non-equilibrium situations. Classical entropic, and indeed probabilistic, concepts apply only to equilibrium cases, whereas many of the most interesting phenomena in Nature involve constant evolution and/or a flow-through of energy. The second thing to pay attention

A. Lawrence, *Probability in Physics*, Undergraduate Lecture Notes in Physics,
https://doi.org/10.1007/978-3-030-04544-9_14

to is the role of subjectivity. Eugene Wigner once said that "entropy is an anthropomorphic concept", which sounds like an argument that it is just in our minds rather than really in Nature. However, as we have repeatedly seen in discussing probabilistic concepts, "subjective" doesn't mean "vague and woolly", but rather, "observer dependent".

14.2 Thermodynamic Entropy

Before we examine the role of probability, we should remind ourselves how the idea of entropy arises in classical Thermodynamics.

14.2.1 Entropy as a Thermodynamic State Variable

The equilibrium states of thermodynamic systems can be described by a number of *state variables*, or *thermodynamic co-ordinates* such as temperature T, pressure P, volume V, and internal energy U. Usually only two are needed; for a specific type of system, there will be equations linking the other variables. For example, if our box contains an ideal gas, then pressure, volume, and temperature are linked by the equation of state $PV = nRT$ where R is the gas constant and n is the number of moles of gas; and you can show that that $U = 3PV/2$. By analysing the behaviour of heat engines, nineteenth century physicists and engineers showed that around any closed reversible path, i.e. one we traverse quasi-statically so it can be re-traced in the opposite direction,

$$\oint \frac{\mathrm{d}Q}{T} = 0,$$

where $\mathrm{d}Q$ is the amount of heat taken in by a system, and T is the temperature at which the process happens. A quantity which doesn't depend on the path you take is by definition a state variable. So there must be some new state variable S—the entropy of the system—defined by

$$\mathrm{d}S = \frac{\mathrm{d}Q}{T} \quad \text{or} \quad S_a - S_b = \int_1^2 \frac{\mathrm{d}Q}{T}, \tag{14.1}$$

if we change the system between state 1 and state 2. Because S is a state variable, it doesn't matter which path we take between state 1 and state 2. This analysis only defines the difference in entropy between states; but it is possible to define a zero point for a given type of system, for example by requiring entropy to be zero at absolute zero temperature.

Entropy is a slightly peculiar state variable. Rather than being defined at a single point in time, it seems to be defined in terms of the change during a procedure—the transferring of heat. The way to make sense of this is invert the definition and think of $dQ = TdS$. Suppose we have two states of a system, which we can define by any conventional thermodynamic co-ordinates we wish—temperature, volume, etc. Now suppose we ask, how much heat would I need to put in to move the system from state a to state b? The answer is

$$Q = \int_a^b TdS.$$

So entropy seems to be a characteristic of a system which tells you how hard it is to change. Note that the units of entropy are Joules per degree Kelvin, which again is a slightly puzzling kind of unit for a state variable, until we realise that this is a historical accident due to the way temperature was first defined. Today we realise that temperature is really all about the mean energy of the particles constituting a system. If we had *defined* temperature as an energy scale, rather than as "degrees" between two fixed points, then the concept of entropy would still have emerged, but its units would have been Joules per Joule. In other words, in essence, entropy is a *dimensionless quantity* characterising a system which tells us how difficult it is to move it from one state to another. This makes much more sense when we consider statistical entropy.

14.2.2 Example Entropy Calculation

If we have an equation of state, and specify a particular change, then we can actually calculate the entropy, or at least its change. We will look at a simple example that we can revisit when looking at statistical entropy. Suppose a container holding a mole of an ideal gas expands isothermally from volume V_a to volume V_b. If we do this carefully by allowing a piston to move out slowly, then for each small change in volume dV the gas does work PdV. Because there is no change in internal energy, $dQ = PdV$. An expanding gas tends to cool down, so the missing heat needs to be supplied by the surroundings, which we can arrange by attaching the system to a heat reservoir at fixed temperature T. Finally, for one mole of gas we have $PV = RT$ and so we get

$$S_a - S_b = \int_a^b \frac{PdV}{T} = R\int_a^b \frac{dV}{V} = R\ln(V_a/V_b).$$

Alternatively, we could make our change in two stages; first let the system expand adiabatically; then slowly heat the system to return to the original temperature. As you can check in any thermodynamics textbook, this procedure gives the same answer as our isothermal expansion calculation.

14.2.3 Entropy and the Second Law of Thermodynamics

In studying the behaviour of heat engines, it was noticed empirically that although it is possible to turn mechanical work into heat, any attempt to turn heat into mechanical work is always imperfect—there is always some "waste heat". This is the original sense of the second law of thermodynamics. Clausius analysed heat engine changes in terms of the entropy, and expressed the second law more formally, as the observation that *total entropy never decreases*. To make that bold statement, we have to consider the entropy change in both our system and its environment. We also have to understand that thermodynamic entropy, as a state variable, is only meaningful when our system is in equilibrium. We can sum up the observed behaviour of entropy as follows:

- For a **closed reversible path**, $\Delta S = 0$.
- For a **reversible path** between two states of our system, $\Delta S(\text{sys}) \neq 0$
  but $\Delta S(\text{env}) = -\Delta S(\text{sys})$
  and so $\Delta S(\text{world}) = 0$.
- For an **irreversible path**, $\Delta S(\text{sys}) \neq 0$
  but $\Delta S(\text{env}) \geq -\Delta S(\text{sys})$
  and so $\Delta S(\text{world}) \geq 0$.

For example, if we take the classic irreversible example of free adiabatic expansion, by removing the barrier between two halves of a container, you can easily show that $\Delta S(\text{sys}) = nR \ln 2$, but $\Delta S(\text{env}) = 0$ because there is no heat exchange; then we get $\Delta S(\text{world}) = nR \ln 2$. On the other hand, suppose we pass heat Q through a container between two reservoirs at temperatures T_1 and T_2. Then $\Delta S(\text{sys}) = 0$, but $\Delta S(\text{env}) = \Delta S(\text{world}) = Q/T_2 - Q/T_1$.

It is important to keep in mind that it is the total entropy of system plus environment that increases. It is always possible to decrease the entropy in a system, at the expense of its environment. When chickens make eggs, they tend to spread chaos around them.

14.2.4 The Idea of Free Energy

An isolated system will find equilibrium by maximising its entropy. When placed in an environment, things can get more complicated because energy and particles can be exchanged. As well as tending to maximise its entropy, a system will tend to evolve towards the state of lowest potential energy, by shedding energy to its environment. The first law of thermodynamics tells us that $\mathrm{d}U = \mathrm{d}Q - \mathrm{d}W$, where by convention, $\mathrm{d}W$ is positive for work done *by* the system, and dQ is the heat put *in* to the system. We can relate this to the definitions of entropy ($\mathrm{d}S = \mathrm{d}Q/T$) and mechanical work ($\mathrm{d}W = P\mathrm{d}V$) and so state

$$\mathrm{d}U = T\mathrm{d}S - P\mathrm{d}V, \tag{14.2}$$

which is sometimes known as *the fundamental relation of thermodynamics*. Turning this round, we could write

$$P\mathrm{d}V = -(\mathrm{d}U - T\mathrm{d}S) = -\mathrm{d}F \quad \text{where} \quad F = (U - TS).$$

If it weren't for entropy considerations, if we were to extract internal energy $\mathrm{d}U$ from the system (i.e. change its internal energy by amount $-\mathrm{d}U$), we would be able to turn it all into work; but we fail to achieve this by the amount $T\mathrm{d}S$, which is therefore sometimes known as the "unavailable energy". We can construct a new state variable $F = U - TS$ which is known as the "Helmholtz free energy", which tells us the amount of energy which is actually available for conversion to work. More generally, we can define the "Gibbs free energy" $G = U - TS + PV$, which is also known as the *thermodynamic potential*.

Many problems in thermodynamics boil down to finding the equilibrium state which jointly maximises entropy and minimises potential energy. To do this, the quantities G or H can be seen as functions which we need to maximise.

14.3 Statistical Entropy

Having done a lot of physics, we need to return to probability. During the nineteenth century, atomism and kinetic theory were developing in parallel with thermodynamics, which made possible a statistical understanding of how matter behaves.

14.3.1 Boltzmann Entropy

Boltzmann's key insight was that entropy must be connected with the number of microscopic configurations available to a system, and he proposed the formula

$$S = k_{\mathrm{B}} \ln W, \tag{14.3}$$

where W is the number of microstates corresponding to the observed macrostate—the multiplicity—and k_{B} is the constant now named after Boltzmann in his honour. If the multiplicity of macrostate j is W_j and the total number of possible microstates is $W_{\mathrm{tot}} = \sum W_j$, then the probability of being in macrostate j is $P_j = W_j / W_{\mathrm{tot}}$. Maximising entropy is therefore the same as maximising probability, just as we discussed for simple cases in Chap. 3. Why did Boltzmann choose $\ln W$ rather than just W? Partly, it is because this choice, together with the choice of the normalising constant k_{B}, i.e. the Boltzmann constant, is what gives us numerical agreement with the thermodynamic definition of entropy (see Sect. 14.3.4). However, the logarithmic form is in general the best choice to make S additive. Suppose we have two sub-systems with entropy S_1 and S_2 and multiplicity W_1 and W_2. To get the total multiplicity of a macrostate of the combined system, we would need to consider each microstate of system 1, and pair it with every microstate in system 2. So $W_{\text{sys}} = W_1 \times W_2$ and so $S_{\text{sys}} = S_1 + S_2$.

In various circumstances, exactly what we mean by a "microstate" or by a "macrostate", may be rather different. It doesn't matter too much; the essence of the idea is multiplicity, and the concept lurking behind that is *distinguishability*—the idea that some things are the same as each other as far as an observer is concerned, whereas other things are distinguishable by the observer. The other key concept hiding behind the simple assumption that $P_i = W_i/W_{\text{tot}}$ is that our system will somehow "explore" all the possible microstates, spending equal time in each. This is the *ergodic* assumption, which we touched on briefly in Sect. 3.6.2. The idea that we will maximise entropy is therefore really the same as arguing that more time is spent in some configurations than others. (The idea is illustrated in Fig. 14.1.) We will discuss the issues of distinguishability and ergodicity in Sect. 14.4.

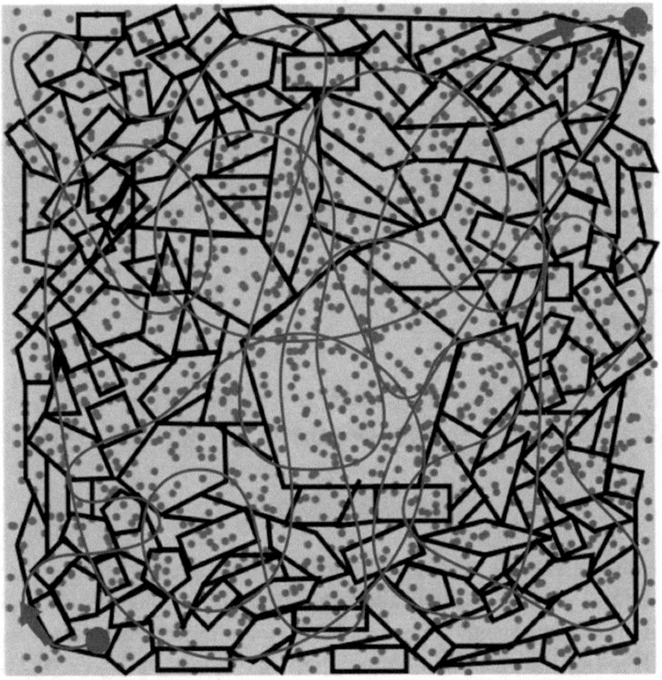

Fig. 14.1 Illustrating the concept of exploring the space of possible states. Each possible microstate is represented by a dot. Polygonal areas represent the possible macrostates, each of which corresponds to a number of microstates. The curved path shows the system exploring the space of possible states in a meandering fashion. The path explores the whole space in an unbiased way, but spends more time in macrostates that contain more microstates. In a real system, the difference between the multiplicity of the macrostates is much more dramatic

14.3.2 Gibbs Entropy and the Shannon Formula

The Boltzmann concept of entropy assumes that all the microstates are equally probable. In fact we don't need to make that assumption; if microstate i has probability p_i, we can define the *Gibbs entropy* as

$$S_G = -k_B \sum_i p_i \ln p_i, \tag{14.4}$$

which you can see is, apart from the base of the log, and the normalising factor, the same as the uncertainty or Shannon entropy from information theory (see Chap. 11). Note that here we are talking about the probability p_i of each microstate; be careful not to confuse this with the probability P_k of each macrostate, which we discussed above. The Gibbs formula reduces to the Boltzmann formula in the case where all the p_i are the same. The Boltzmann formula gives the right answer for the behaviour of ideal gases, i.e. with non-interacting particles. The Gibbs formula is more widely applicable to a range of different thermodynamic systems.

14.3.3 Counting Microstates

The idea of microstates, and the multiplicity/entropy associated with them, could be associated with a number of different properties of a system and its constituent particles. We could be considering the spatial positions of particles, which gives rise to the *locational entropy*, or to their velocities/energies, or to the six dimensional combination of position x, y, z and velocity v_x, v_y, v_z. In principle, we could completely specify a microstate by specifying the x, y, z, v_x, v_y, v_z co-ordinates of every particle. If there are N particles, you can think of the complete list as specifying a position in a space of $6 \times N$ dimensions, which is known as *phase space*. Alternatively, rather than positions and velocities, we might for example be considering the possible spin states of a collection of particles, and likewise calculate a corresponding *spin entropy*, and, listing all the particles, a "spin phase space" for the system. We will return in Sect. 14.4 to the issue of freedom of choice of which properties we are considering.

Sometimes the possible states constitute a discrete set, in which case the multiplicity W has a clear absolute meaning. Sometimes however, there is actually a continuous range of the relevant properties, for example spatial position. One approach is to divide imagine dividing space into cells. Then if we decrease the cell size to any size we wish, the absolute value of W will change, but our conclusions about entropy differences (multiplicity ratios) will not change. A more rigorous approach is to consider the *density of states* in a differential *phase space volume*. Again, this makes no essential difference to the probabilistic nature of what is going on. Finally of course, in a quantum view of the behaviour of a collection of particles, location and energy states are effectively discretised. This leads to definite and testable predictions for the measured entropy of quantum systems, but again, does not alter the probabilistic logic.

14.3.4 Equivalence of Statistical and Thermodynamic Entropy

How does the statistical entropy compare to the thermodynamic entropy we discussed earlier? It is possible to show that they are the same thing for any thermodynamic system. Here we will simply look at one specific case, the example of isothermal expansion considered in Sect. 14.2.2. Because the expansion is isothermal, we will assume that the distribution of the particles in energy does not change, and so will look at just the change in locational entropy. Consider then N molecules distributed into n cells. We can make the cells as small as we like, so imagine making the cells so small that each cell has either one molecule or none. Then the number of arrangements of N occupied cells and $n - N$ empty cells is

$$W = \frac{n!}{N!(n-N)!}.$$

Using Stirling's formula we have

$$\begin{aligned} \ln W &= n \ln n - N \ln N - (n-N) \ln (n-N) \\ &= -n \ln \frac{n-N}{n} + N \ln \frac{n-N}{N}. \end{aligned}$$

However, $n >> N$ so this becomes

$$\ln W = N + N \ln \frac{n}{N}.$$

Now consider the expansion from volume V_a to Volume V_b. If we keep our cells the same size, then the number of cells increases by the same ratio. So

$$\begin{aligned} \ln W_b - \ln W_b &= N \ln \frac{n_b}{N} - N \ln \frac{n_a}{N} \\ &= N \ln \frac{n_b}{n_a} \\ &= N \ln \frac{V_b}{V_a}. \end{aligned}$$

Now if we multiply by k_B, and N is the number of molecules in one mole, then

$$\begin{aligned} S_b - S_a &= Nk_B \ln V_b/V_a \\ &= R \ln V_b/V_a, \end{aligned}$$

which is exactly the answer we got in Sect. 14.2.2. Thermodynamic entropy and statistical entropy are therefore numerically the same.

14.3.5 Equilibrium and Non-equilibrium States

Thermodynamic entropy concerns only equilibrium states. It is possible to calculate the entropy change between the starting and ending points of an irreversible process, but as far as thermodynamics is concerned, the non-equilibrium states in between have no defined values of standard thermodynamic quantities at all. Statistical entropy is more general; we can calculate the multiplicity of any state, not just equilibrium states. Suppose we have n distinguishable particles, which can be distributed into k different energy levels. In this case, the set of population values $\{n_1, n_2, \ldots, n_k\}$ constitute the macrostate, and the specified energy levels of all n particles constitutes the microstate. We can calculate the multiplicity W of any set of $n_1, n_2, \ldots, n_k$ values, and think of this as the statistical entropy. However, for a fixed total energy E, the distribution which maximises W is the Boltzmann distribution, as shown in Chap. 3, Sect. 3.8.1, with $n_i \propto e^{-A_i}$ where the quantity A is essentially the temperature of the system.

There are two ways in which statistical entropy plays its role. The first is that over time, as the system meanders through the space of possible microstates (i.e. the phase space), it will spend nearly all of its time in the most probable regions of that phase space. If the system is initially not in such a high probability region, its randomly meandering path will nearly always take it closer to the maximum of probability/maximum entropy. Once reaching such a state, it is unlikely to leave; the system is therefore in equilibrium. Only at this point is the statistical entropy the same thing as the thermodynamic entropy.

Now suppose the system has arrived in equilibrium, but by our external action we cause our system to change to a new equilibrium state. We might do this reversibly—for example by slowly moving a piston—or we might do it irreversibly, for example by removing a dividing barrier and allowing the system to expand. In either case, we can calculate the multiplicity/entropy of the beginning and ending states and compare them. This is the situation to which the second law of thermodynamics applies. Note that when we carry out an irreversible change, we are almost always artificially creating a low entropy (improbable) configuration, and then letting the system settle into a new equilibrium. A "reversible" change is of course a theoretical abstraction. In any slow quasi-static change, the track is really made up of a sequence of miniature irreversible changes, each time settling into a new equilibrium.

In Sect. 14.2.4 we discussed how thermodynamic systems find a compromise between potential energy and entropy, maximising the free energy G or H. We can see how this works statistically. Suppose a box contains non-interacting gas particles, which can in principle occupy any part of the box. However, the box is in a gravitational field, which tends to make the particles settle towards the bottom of the box. If the gas is at very high temperature, gravity will be irrelevant, and the particles will tend to be distributed uniformly throughout the volume, therefore having a high value of locational entropy. However, if the gas can dissipate energy to its surroundings and cool down, the particles will tend to be found at the bottom of the box, a configuration which has a low locational entropy.

We can analyse such systems by finding the maximum of free energy. But we could also see them as examples of maximising multiplicity subject to one or more constraints, analogous to the manner in which we derived the Boltzmann distribution in Chap. 3. Another interesting example is strongly interacting particles, which can "stick", as random motions bring them together, to form molecules, or crystalline structures. Such regular structure may seem to have very low entropy; but in fact such a structure will represent the maximum of entropy consistent with the energetically favourable configuration.

14.4 Issues in the Interpretation of Entropy

14.4.1 The Ergodic Assumption and Relaxation Times

The statistical interpretation of entropy relies on the assumption that a system somehow "explores" all the possible microstates, so that the probability of being found in or near a particular macrostate is proportional to the volume of phase space in the vicinity of that state. Boltzmann initially assumed that probability was therefore equivalent to the fraction of time a system spends in a particular region, as it meanders around phase space. Gibbs suggested that instead we can imagine many possible similar systems, and consider the *ensemble average* of the properties of such systems. The question then becomes "if I pick a specific system at random, how likely is it I will find it in such-and-such macrostate?". The argument that the time average is the same as the ensemble average is the *ergodic hypothesis*.

Much time has been spent debating whether the ergodic hypothesis is formally correct, but the practical question is really how long a system will take to reach equilibrium if it starts out of equilibrium. This is often known as the *relaxation time*. Of course any statistical system will fluctuate about its mean state, so it will never be exactly in equilibrium, but we can get a rough idea of how long real systems will take to get somewhere near equilibrium.

For example, consider a gas similar to air at room temperature and density. Suppose the molecules are not in thermal equilibrium, i.e. not showing a Maxwell-Boltzmann distribution of velocities. Collisions between fast and slow moving molecules will tend to share out the momentum, and produce a Gaussian distribution in velocity space, as described in Chap. 5, after a few collisions per molecule. The mean free path for air molecules is around 70 nm, and they are typically moving at around 500 m s^{-1}. The typical timescale between collisions is therefore $\sim 10^{-10}$s. The gas will *thermalise* within a few nanoseconds. What about spatial arrangement? Suppose we consider the classic experiment of dividing a container into two compartments, with all the gas in one half, and then removing the divider. If the container is say 1m across, then molecules travelling to the far side can do so in a few msec, and the gas should become uniform within a fraction of a second. More generally a non-uniform structure in a gas may get randomised on something like the *sound-crossing time*, i.e. how long it takes pressure waves to travel across the region.

Suppose however we consider a slightly different experiment—the mixing of two gases. Instead of the second half of our container being empty, it contains a similar number of molecules of a different type. When removing the container, we start with a very low entropy/unlikely configuration, with all the type A molecules at one end, and all the type B molecules at the other end. Now, for a Type A molecule to travel to the far end, it cannot travel unimpeded, but must diffuse across in the manner we considered in Chap. 6. From equation (6.3), the time this will take will be $t = D^2/2v\lambda$ where D is the size of the container, v the molecular velocity, and λ is the mean free path. For air molecules of two similar but different types mixing across a 1m box, this gives a timescale of $\sim$4 h. The time to reach equilibrium can therefore be quite significant in some circumstances.

In astronomical situations, relaxation times can be much longer. Interstellar gas clouds are of very low density—a few atoms per cubic metre—can be very cold, and also quite large. Thermalisation timescales can be a fraction of a second, and spatial re-arrangement timescales can be extremely long—hundreds of thousands of years. To some extent, we can consider clusters of stars, or even clusters of galaxies, as gas-like statistical systems. They may suffer physical collisions very rarely, but they interact with each other gravitationally, allowing us to calculate relaxation timescales. For star and galaxy clusters, it is a contentious point whether or not we expect such systems to thermalise or become uniform within the current age of the Universe.

Let us look at a much simpler earth-bound example. Consider a row of coins in a box that are all heads-up. That is a very low-entropy configuration. But nothing is going to change. The system will not explore parameter space all by itself. On the other hand if you shake the box, you can soon randomise the head-tail distribution. In other situations, shaking a container can encourage non-random effects. Consider a box of muesli containing oat flakes, raisins, and nuts. Shaking the box tends to bring the nuts to the top, producing what is apparently a low-entropy structure—but of course it may be that this structure is energetically favourable, and the environment may have increased its entropy.

All in all, it is clear that although many systems reach equilibrium extremely quickly, others do not, so we must be very careful applying the concepts of equilibrium thermodynamics. However, we can often employ a statistical approach to the behaviour of a system, including calculating its statistical entropy, even if it is not in equilibrium. The subject of *non-equilibrium thermodynamics* is of growing importance. The key point is really about the competition of timescales. Random processes such as collisions, interactions, turbulent motions and so on will tend to make systems gradually more uniform or at least make them occupy more high-entropy configurations. On the other hand, external influences, whether man-made or natural, can easily be driving systems towards more low-probability configurations—they are not random processes. So systems are often experiencing a competition between structuring and randomisation, and it is a question of which process is faster. From this point of view, living organisms are particularly interesting. By consuming food, they are using a constant stream of energy available for restructuring.

14.4.2 Subjectivity of Entropy Values

Along with the exploration of the space of microstates, the second key assumption lurking in Boltzmann's original idea is that states can be grouped together in a unique way. The various microstates may in principle be distinguishable, but in practice all we can distinguish is the various macrostates. Some microstates correspond to one macrostate, and some to another. However, a thought experiment shows that this micro-macro correspondence is subjective, i.e. observer dependent.

Consider Fig. 14.2, which shows molecules distributed across two boxes. The macrostate is the pair of numbers n_A, n_B where n_A is the number of molecules in box A and n_B the number in box B, and the total is $n = n_A + n_B$. The microstate is the list of which molecules are in which box—$M_1 = A, M_2 = B, M_3 = B, \ldots$. In the top-left figure, $n_A = n, n_B = 0$. There is only one microstate which gives this macrostate. The top-right figure shows the uniform macrostate $n_A = n/2, n_B = n/2$. There are many microstates which correspond to this same macrostate. If we start in the left-hand situation and then remove the divider, it is clear that the system will soon evolve to the state with larger multiplicity/higher entropy.

Now consider the situation in the lower part of Fig. 14.2. Here, half the molecules are dark-grey ones, and half of them are light grey ones. At bottom left, all the dark grey ones are in box A, and all the light grey ones in box B. At bottom right, the different shades are uniformly distributed. The situation is the same as in the top row; there is only one microstate corresponding to the macrostate at bottom-left, but many corresponding to the macro-state at bottom right.

However, this conclusion assumes that the observer can distinguish the different shades of molecule. Perhaps the molecules are red and green, and the observer is red-green colour-blind. If the types of molecule are not distinguishable, then the situations at bottom-left and bottom-right are the *same*. Different observers will assign different entropy values to these situations. However, they will still agree on the laws of Physics. The non-colour-blind observer says "oh look, a very unlikely state, but it is evolving towards higher entropy, just like I expected". The colour-blind observer says "The system is already at maximal entropy. Nothing much is changing, just like I expected".

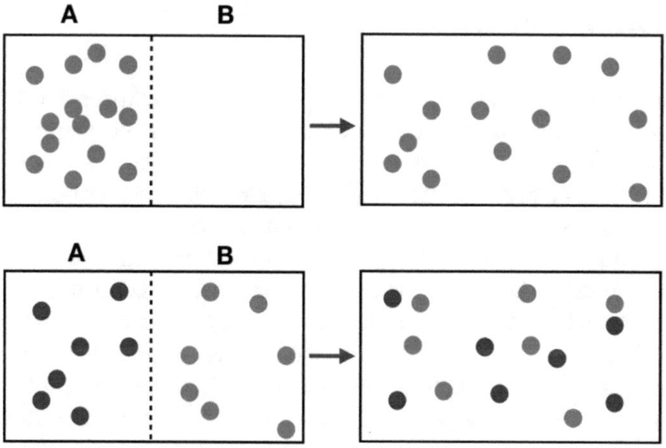

Fig. 14.2 Illustrating subjectivity of entropy. In the upper row, particles spread when the divider is removed. In the lower row, two types of particle mix when the remover is divided. An observer who cannot, or chooses not to, distinguish the light grey and dark grey particles, will give a different value for entropy, but will still agree that the second law is obeyed

The dependence on the state of knowledge of the observer is exactly the same as we have discussed for probability in general. Probability is about reasoning in the presence of incomplete knowledge. The statistical approach to entropy is really just an example of probabilistic reasoning in action. In many situations, our knowledge is highly observer dependent, and so correspondingly are the probabilities we assign. In other situations, although our knowledge is in principle observer dependent, in practice every plausible observer is in the same state of ignorance. Although we don't know the microscopic state of a system, every observer will attribute the same value of temperature, pressure etc to a thermodynamic system.

We have so far described the subjectivity issue as being to do with whether the observer is *able* to distinguish certain qualities, such as the colour of the particles. However it is also to do with whether we *choose* to distinguish those properties. If we are trying to study the re-distribution of particles in the remove-divider experiment, and if the particles are identical apart from their colour, and the colour of a particle has no effect on its dynamics, then we may decide to ignore the colour of the particles. It has no bearing on our problem. On the other hand, if we are actually studying the mixing of gases, we may very much want to know which particles are which. Suppose we are studying some particles, and one day, by looking closer, we discover that some of them have a small black dot and some don't. Will this invalidate our earlier experiments? It will not. But it will matter very much if later we wish to study the physics of small black dots.

Entropy is determined by the experiment we choose to do, or rather, the way we choose to analyse the results. Each set of properties can have its own characteristic entropy, and the second law will be obeyed in each case, because of the inexorable logic of probability. Entropy is both completely subjective and very concrete.

14.4.3 Order, Disorder, and Structure

In popular writing, and very often in textbooks too, you will see entropy equated with “disorder”. Generally speaking, authors do not define or quantify what they mean by order or disorder, but tellingly, the word is often used in a combined phrase such as “disorder or randomness” as if it was clear that they are the same thing. Another popular term is “spreading”—the second law of thermodynamics is seen as a tendency for systems to spread out, becoming more uniform. However, a regular crystalline grid could be seen as being extremely uniform, but in fact has a low entropy. In normal English, “order” is seen as being to do with either showing a pattern, or following some kind of rule. We have an instinctive sense that some patterns are more richly structured than others, but this sense is hard to pin down. Although everybody might agree that a marching squad of soldiers is more ordered than pedestrians milling about on a busy shopping street, people might find it difficult to rank different groupings of people—for example a single troop of soldiers versus a sequence of colourful bands in a Mardi Gras parade—in “degree of orderliness”, or “degree of structure”, and indeed they might give different answers for those two questions.

There may be a variety of ways to capture what we mean in natural language by order or structure, and each of them may be useful for answering various different questions, but the only concept that clearly matches entropy is the one that we have already discussed in Chap. 11 and in this Chapter—the average uncertainty, or Shannon entropy, associated with a probability distribution. Imagine 1D space divided into many small cells, with particles occupying the cells. Our structure can be defined by the average occupation per cell $n(x_i)$, with the probability of a specific particle being in a specific cell given by $P_i = n(x_i)/N$ where N is the total number of particles. The apparently uniform regular grid can now be seen as a kind of honeycomb structure. A set of cells at regular spacing have high probability, but there are also many completely empty cells in between. We can contrast this with a truly uniform distribution where all cells have the same probability. In the language we used in Chap. 3, a specific set of occupation numbers n_i can be seen as a “partition” or as a “microstate” of the collection of particles.

Consider a specific labelled particle, and take a specific cell at x_i. If you suggest “my particle is in cell i”, how often would you be wrong? Then repeat the test many times, each time for a different cell. How often would you be wrong on average? Alternatively you could ask how many yes-no questions you would need to ask on average in order to pin down where your particle is. This is given by equation (11.3), which specifies the average uncertainty as $H = \sum_i P_i \log_2 P_i$. As we saw in Chap. 11, Sect. 11.3.2, if there is no other constraint, H is maximised when all the P_i are the same; but if for example, we fix the mean and variance, then the maximum H is given by a Gaussian shape for P_i. Alternatively, one could ask how any different ways—W—we could place our collection of N particles in the cells that correspond to the same set of n_i values. In Sect. 11.5.1, we have already shown that these two methods—maximising H and maximising $\log W$, are the same thing.

All in all, the concepts of order, structure, or spreading are useful only if we define and quantify them in just the way we have already done for average uncertainty in information theory.

14.4.4 Complexity

Random processes seem to erase structure and statistically homogenise systems. On the other hand, structure and patterns seem to emerge spontaneously in Nature—for example crystallisation, or the forming of snowflakes, or the growth of leaf patterns. It seems as if the possibility for such structures is hidden inside matter, which just needs some encouragement. This idea is known as *emergence* or *self organisation*. Some have argued that self-organisation normally involves a continuing flow of matter through a system (this idea is particularly connected with Buckminster Fuller), and others that fluctuations/noise can prod a system into finding structured configurations (this idea is particularly connected with Ilya Prigogine).

Mathematically, the emergence of structure can be connected with the existence of an *attractor* within the phase space of a system—quasi-stable solutions to its evolution. The simplest attractor could be a fixed point, but much more rich and complicated attractors can exist even for very simple systems. The key point for our discussion in this chapter is that such an attractor occupies a very small fraction of the total possible phase space for a system, so the system does not over time explore all the possible microstates.

Over recent decades, there have many different attempts to quantify the complexity of the spatial and temporal structures that emerge in natural processes. Two approaches seem particularly relevant to the issues in this chapter. One approach is to see a structure as a kind of probability distribution, and calculate the Shannon entropy, as we discussed above in Sect. 14.4.3. The other approach is to ask "what sequence of instructions could reproduce this structure?". This idea leads to what is known as the *algorithmic complexity* or sometimes the *Kolmogorov complexity*—the length in bits of the computer programme that could produce the structure as its output. The string "aaabbbaaabbbaaabbbaaabbb..." could be output by the instruction "keep repeating a and b three times each", whereas "x7bf62nfjhd7fkt;6k8f'ksgsy8..." can only be output by specifying each character. In other words, a long random string has much higher Kolmogorov complexity than a patterned one. This is closely connected to the Shannon sense of information. A random string carries more information, in the sense that it is more surprising, or that you would need to provide more information to specify it.

Once again, this is a technical sense of complexity that may not agree with other natural language senses. An interesting example is shown in Fig. 14.3. This shows a beautiful fractal pattern, which to the human eye seems astonishingly rich and complicated. Specifying the image pixel by pixel would take an enormous amount of information. But in fact the image is uniquely generated by a relatively short piece of computer code.

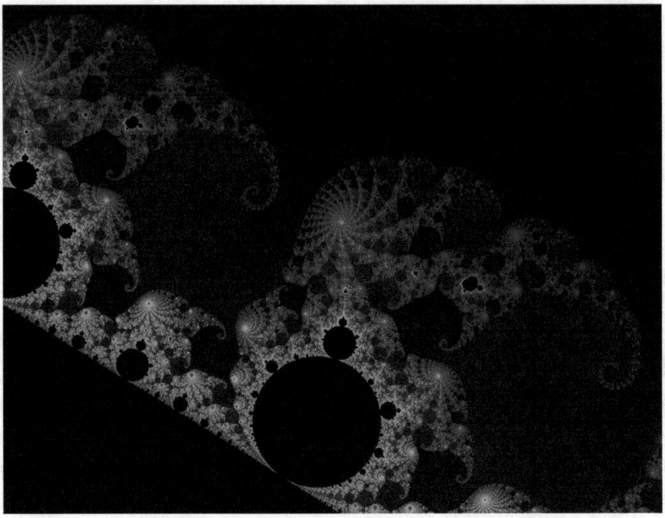

Fig. 14.3 Illustration of the Mandelbrot set, taken from the Wikipedia page on Kolmogorov complexity. Although seeming rich and complex to the human eye, the structure is defined by a relatively simple algorithm, requiring only a relatively short piece of computer code to create the image. (From a figure originally made by Wikimedia user Reguilee)

14.4.5 Entropy in Gravitating Systems

For particles of ideal gas in a box, we can consider the locational entropy, the velocity entropy, or both combined, but location and velocity structure are decoupled. The maximum entropy solution has a uniform spatial structure, but a Gaussian velocity structure. Once we introduce potential energy into the picture, and the possibility of transferring energy to the environment, things can get more interesting, for example crystallisation into a grid-like structure. Bringing gravity into the picture is even more interesting because of its long-range Nature.

Let us first consider a system with an external gravitational field—for example, a container of gas sitting on the surface of the Earth. If particles are dropped into the box, they will descend and lose potential energy (V); collisions between particles will thermalise the gained kinetic energy (T), and eventually there will be a balance between V and T, satisfying the virial theorem $2T + V = 0$. Once T and V are in balance, the total T for all the particles will be constant, and as the system explores the available kinetic energy microstates, the distribution of the whole collection of particles $N_i(T_i)$ will be that which maximises W subject to the constraint $\sum N_i T_i$ = const, leading to the exponential Boltzmann distribution, as shown in Chap. 3, Sect. 3.8.1. (Alternatively we could argue that in each of v_x, v_y, v_z, maximising the Shannon entropy subject to fixed mean and variance gives a 3D Gaussian, which re-assuringly gives an exponential distribution in energy.[1]) The total V of the collection of particles will also be constant, but particles at different heights above the bottom of the container have different individual values of V. As the particles explore the possible locational microstates, then maximising W with the constraint of fixed total V leads to another Boltzmann distribution, but this time in height. The system therefore settles into an exponential density distribution.

Note by the way that we have implicitly assumed that the system is isothermal—i.e. the gas temperature is the same at all points in the box. This might be reasonable if for example the system is embedded within an external reservoir that maintains a constant temperature. Alternatively it could be that the temperature is vertically stratified. This is the case for example for the structure of the Earth's atmosphere, sitting in the Earth's gravitational field. We can find a solution which both maximises entropy and satisfies that balance between T and V, but we won't pursue that any further.

Now let us consider a *self-gravitating* system, such as a star. There are two key issues. The first is that there is no longer a natural "bottom" to the potential well, so that in principle there is nothing to stop a self-gravitating system settling into a radial structure with a central singularity of infinite density. As the gas thermalises, pressure holds the proto-star up against gravity. However, the second key issue is that a hot proto-star will *radiate* energy away, shedding energy to infinity, and encouraging a continuing collapse. Counter-intuitively, the more energy a proto-star sheds, the hotter it gets, because it collapses some more. At each stage in the slow collapse, the proto-star will have the temperature and density structure which maximises entropy for that particular total energy; and the total entropy of star+environment will increase. An initially uniform gas therefore has a tendency to clump. Total entropy is increasing as it does so, but note that this evolution is also producing what most people would recognise as "structure".

Star clusters, including galaxies if we ignore the dissipative gas components, are more subtle yet again. Stars in a cluster may seem like the molecules in a gas, but in fact stars essentially never collide. They can have one-on-one gravitational interactions, but even those interactions are rare enough that star clusters do not thermalise locally. Instead, the dynamics of any star is dominated by the potential produced by the whole cluster, i.e. all the other stars combined. However, as the stars move under the influence of that potential, the potential itself is evolving on the same timescale as the stars are moving, and the system has a kind of feedback loop known as "violent relaxation". Studying the phase-space structure, energy, and entropy of such systems is a very active research subject, which we don't have time to pursue here; suffice it to say that it is not obvious that the system explores all of the micro-state space.

[1] It is interesting to note that this equivalence only works if space has three dimensions.

14.4.6 Black Hole Thermodynamics

Returning to stars, does anything stop the inexorable collapse, as the star radiates its energy away to infinity? For a long time, the collapse is halted by nuclear burning.[2] Eventually when the nuclear fuel runs out, smaller stars can end up as white dwarfs or neutron stars, but it is expected that for larger objects nothing halts the collapse and the object becomes a *black hole*.

In recent decades, there has been talk of *black hole thermodynamics* and the entropy of black holes. What does this mean? It stems from the idea that you can extract energy from a rotating black hole by the *Penrose process* which (crudely speaking) involves throwing in counter-rotating rocks in a carefully designed way. The mass of the black hole increases, and its surface area increases, but its angular momentum decreases, and it loses energy, which is donated to the person throwing the rock in. You can show (see for example Caroll, Sect. 6.7) that the mass, area, and angular momentum are related by

$$\Delta M = \frac{\kappa}{8\pi G}\Delta A + \Omega_H \Delta J,$$

where M, A and J are the mass, surface area, and angular momentum of the black hole, κ is the surface gravity at the event horizon, and Ω_H is the angular velocity at the event horizon. (Warning: that equation is in "geometrical units" with $h = c = 1$.) Hawking and Bekenstein pointed out the similarity with equation (14.2)

$$\mathrm{d}U = T\mathrm{d}S - P\mathrm{d}V,$$

and argued that we can then treat black hole mass, area, and surface gravity (M, A, κ) as equivalent to energy, entropy, and temperature (U, S, T). In particular, just as entropy always increases, black hole area always increases. This may seem to be just an analogy, and it is far from clear what the appropriate micro-states are, in order to relate this analogous "entropy" to a standard statistical entropy. However, by separate reasoning, Hawking showed that quantum effects allow black holes to radiate at a temperature $T = \kappa/2\pi$, producing "Hawking radiation", suggesting that we take the analogy seriously. Bekenstein proposed a generalized second law:

$$\Delta\left(S + \frac{A}{4G}\right) \geq 0, \tag{14.5}$$

so that the combined entropy of matter and black holes always increases. Of course, even if we stubbornly insist that thermodynamic/statistical entropy and black hole "entropy" are conceptually distinct things, the above statement about the Universe may still be true.

[2]When the star gets hot enough the switch on nuclear burning, then we have a "star" as opposed to a "proto-star".

14.5 Probability and Time

Outside of thermodynamics, the laws of physics are time reversible. However, our human experience is that there is an inexorable flow of time in one direction. The second law of thermodynamics says that entropy increases with time—so can it in fact *explain* our experience of an "arrow of time"?

14.5.1 Reversibility of Physics

The basic equations of physics are time reversible. If you change t to $-t$ the equations look the same. You can start with the initial conditions for a system, turn the handle on the equations that describe it, and predict the state of the system at any future time. However, you can also start with the current state, and using the same equations, calculate what the past state of the system was, at any time in the past. A popular way of getting this across is to imagine making a film of a system evolving, and then running the film backwards. According to Physics, an observer should not be able to tell you whether the film is running forwards or backwards. Indeed for many simple systems this is the case. We can comfortably watch a recording of the planets circling around the Sun in either direction.

However, for most moderately complex real world systems it is very obvious indeed if the film is being run backwards. If we see a broken egg leap up from the frying pan and re-assemble itself, we know that this is something that does not spontaneously happen. If gas particles were to spontaneously gather at one end of a container, we would likewise be very suspicious.

The word "spontaneously" is the key one here. We know of course that actually Nature **does** make eggs, but that they are laboriously assembled; complex structures can arise when they are energetically favoured, or when there is energy exchange with the environment; and we can always decrease entropy locally at the expense of the environment. Despite all this, it is clear that the gas molecules never gather at one end all by themselves.

If you could zoom in on a just few molecules and watch them move around and collide with each other, you would see that behaviour follows the basic laws of physics, and is in fact completely time reversible; if you made a film of that handful of molecules and then showed it backwards, it would look perfectly fine. So where does the irreversibility come from?

14.5.2 Monotonicity of Time

Since the development of General Relativity (GR), physicists' time is not quite as simple as it was in the Newtonian era. We see "space-time" as a four-dimensional block. Observers will disagree about how to divide that block into space plus time, and can even disagree about the order of events. However, any one observer can define a unique local time, which we refer to as "proper time", and place events in a clear sequence. In other words, time is *monotonic*; but there is no sense of the *direction* of that monotonic sequence—nothing that says what is plus or minus time, what is forwards or backwards. GR, like the rest of fundamental physics, is time reversible.

14.5.3 Reversibility of Probability

The statistical view of entropy tells us that the gas molecules never gather at one end because this is extremely improbable. So is the origin of the second law simply that probability itself is time-asymmetric? The language in which we normally frame probability problems encourages that thought. We imagine ourselves about to roll a die, and ask ourselves "what is the probability I will roll a six?". If you take a frequentist approach, you will see the probability as telling us what fraction of future experiments will in fact result in a six being rolled; if you take a Bayesian approach you see the probability as quantifying our degree of belief in the proposal that the experiment we are about to perform will result in a six being rolled.

But we can just as easily make these arguments about the past. If our colleague tells us that they did a die-rolling experiment last week, then the probability quantifies our degree of belief that the result must have been a six. Or alternatively, it tells us what fraction of all past experiments will have given us a six. Probability is symmetric in time; so this is not the source of the second law.

14.5.4 The Importance of Initial Conditions

The second law is in fact entirely about initial conditions. Once an isolated thermodynamic system is in equilibrium, there is no directionality of entropy. The system will fluctuate around the equilibrium, with small changes in entropy, but there is no particular tendency for entropy to increase rather than decrease with time. If the system is not isolated, and can exchange energy and/or particles with its surroundings, then entropy can change substantially—in either direction.

All the familiar examples of entropy increasing involve some kind of relatively low probability starting point. It could be that some external event has knocked a system out of equilibrium—for example a shock wave from a passing jet plane injects energy into the particles of the atmosphere, which are then not in thermal equilibrium. The gas particles will rapidly find a new equilibrium. Alternatively, natural processes have assembled some highly structured object, such as an egg. When you drop this on the floor, you are starting from a very well ordered low probability state, so it has only one way to go. Another common situation is that we remove a constraint. Imagine a container of gas with a divider half-way across, where all the gas is initially in one half of the container. The gas is all in one half because *that is where we put it for the experiment*. Given its constraints, the system is already at maximum entropy. But now we remove the divider. Suddenly there are more microstates available to the system, but it finds itself starting in what is now a very low probability state from the point of view of the enlarged system.

When you observe a system evolving from a more ordered state to a less ordered state, the question you should ask is—how did it get ordered in the first place? Many natural processes produce order, but they normally involve a flow of energy, long timescales, human intervention, or all of those things, and we don't tend to see them as "spontaneous". Given that many systems are put into low probability configurations by such processes, when they are allowed to evolve *randomly*, we will see them spontaneously relax back into more probable states.

If we assume an arrow of time as an external given, then the production of low-probability states by non-random processes, and the production of high-probability states by random processes, is perfectly sensible. But these considerations do not of themselves create an arrow of time.

14.5.5 Cosmological Arrow of Time

Space has no intrinsic directionality. If you film some molecular collisions happening in outer space, and then rotate or flip the film, you can't tell the difference. However, on the surface of the Earth, it is very obvious which way is up and which way is down. The symmetry is broken by an externally imposed field, and the result is an "arrow of space". Is there some similar imposed effect which produces a global gradient of entropy, and hence an arrow of time?

As far as we know, the laws of thermodynamics are the same everywhere, so any such gradient needs to be universal. A number of scientists have proposed that the evolution of the Universe as a whole produces exactly such a gradient—a cosmological arrow of time. In Sect. 14.5.2 we emphasised that each observer has a distinct local "proper time". However, in cosmology, on the largest scales, where we can assume spatial homogeneity, we can go further and identify a unique global cosmic time; furthermore, because of the expansion of the Universe, cosmic time does have a directionality. The additional idea is that the *initial condition* of the Universe happened to be a very low entropy state, and that the Universe as a whole has been systematically increasing in entropy ever since, producing a systematic gradient.

Is it reasonable that the Universe started in a low entropy state? Standard cosmology assumes that the Universe started as a very hot and almost completely uniform "particle soup". Imposed on this uniformity were unavoidable quantum fluctuations, which, as the Universe cooled and formed atoms, acted as the seeds for clumping of matter, which then gradually formed into stars and galaxies. For most laboratory physical systems, a hot uniform state is one of high entropy; but as we saw in Sect. 14.4.5, self-gravitating systems inexorably clump.

This idea, that the evolution of the Universe causes the arrow of time is a very appealing possibility, but much work needs to be done to show that it works. It is not clear that the basic assumptions behind statistical entropy can be used when considering the whole Universe. (i) Does the Universe as a whole explore all the possible microstates? Note that we cannot appeal to an "ensemble" of Universes. (ii) Do local systems "know" about the cosmic gradient, or are they dominated by local conditions and forces? (iii) We might expect that eventually all structure will be erased, and the Universe will suffer a "heat death". However, our observations so far seem to show the richness of structure in the Universe increasing. But of course other effects will be producing this structure; perhaps the Universe will go through a maximum of complexity, and then gradually become a warm sludge.

14.6 Key Concepts

Some of the key concepts from this chapter are:

- The idea of entropy as a thermodynamic state variable
- The observation that the total thermodynamic entropy of system plus environment never decreases (i.e. the Second Law of Thermodynamics)
- The concept of free energy, and how finding the equilibrium of a system involves a compromise between potential energy and entropy
- The key idea that a system spends more time in regions of phase space that have a higher density of related states
- The demonstration that, for equilibrium systems, thermodynamic and statistical entropy are the same thing
- The importance of the competition between structuring and randomisation
- The idea that absolute entropy values are subjective, but that the second law is obeyed regardless
- The idea that one could define order, structure, and complexity in a variety of ways, but that disorder is most sensibly defined in terms of Shannon entropy
- The idea of quantifying complexity by the length of code needed to produce the structure in question (Kolmogorov complexity)
- How self-gravitating systems trying to reach equilibrium will inexorably become more clumped rather than more uniform
- The idea that the surface area of a black hole is analogous to entropy, in that it always increases
- The fact that probability is reversible in time, so that probability itself is not the origin of the arrow of time
- The possibility that the arrow of time is caused by the Universe starting in a low entropy state, leading to a cosmological gradient of entropy

Key formulae in this chapter include: the definition of thermodynamic entropy in terms of heat input corresponding to a state change (14.1); the relation of internal energy change to work done and change of entropy (14.2); the Boltzmann (14.3) and Gibbs (14.4) definitions of statistical entropy; and Bekenstein's formula for the generalised second law including black hole entropy (14.5).

14.7 Further Reading

Entropy in thermodynamics and statistical physics is of course covered in many standard textbooks. I would recommend the same three textbooks listed in Chap. 3—Baierlein (2010), Ford (2013), and Brown (1968). It is worth getting hold of Brown's book as it is written so clearly. Going back further in time, many researchers still value the original textbook by Gibbs (1902), and the research papers by Jaynes (1957, 1965) were extremely influential. In particular, Jaynes (1965) quotes E. P. Wigner as stating that "entropy is an anthropomorphic concept", an idea which I hope I have explained in this Chapter. Also very good on the anthropomorphic Nature of entropy is Coles (2010), which is overall a very lively semi-technical explanation of many of the key issues in this book. Also very lively is Ben-Naim (2012), who makes a very strong sales pitch for thermodynamic entropy being a subset of what he refers to as the "Shannon Measure of Information", i.e. what we have called "average uncertainty" or "Shannon entropy". This is one of a series of books by Ben-Naim at a variety of levels. They are full of colourful detail, partly because of some difficult battles he has clearly had with other authors. I have been strongly influenced in my own views of entropy by all of Jaynes, Coles, and Ben-Naim.

Readers interested in the key issues of this chapter may want to read more about non-equilibrium thermodynamics and complexity theory. Non-equilibrium thermodynamics is covered in a number of typically rather advanced books. The key early book, and still useful, is Prigogine (1962). Prigogine was a prolific character who wrote many works—research papers, textbooks, and popular books. One popular book—Prigogine and Stengers (1993)—was influential for both public and scientists alike. Complexity, self organisation and so forth are tricky areas to pursue, as writings on these topics contain a mixture of the very technical, the somewhat woolly and the almost mystical. Holland (2014) is a nice short introduction; Ball et al. (2013) is a very good collection of articles; and if you want to sample the wacky end, you should try Fuller and Applewhite (1976). Yes, thats the same Buckminster Fuller who popularised the geodesic dome.

To read more about the statistical behaviour of self-gravitating systems in astrophysics, the "bible" is Binney and Tremaine (2008). The famous research paper which came up with the idea of "violent relaxation" was Lynden-Bell (1967). To read about black hole entropy, a good starting point is the textbook on general relativity by Carroll (2013).

Finally, the infamous question of the arrow of time. People have fretted about the mystery of time ever since St Augustine, and before, and Boltzmann was certainly very vexed by it, but the phrase "arrow of time" was actually coined by Eddington (1928). Eddington's book remains one of the best ever public expositions of Entropy, Relativity, and Quantum Mechanics, and contains many deep insights and quotable passages. In more recent times, the books by Price (1997) and Zeh (2007) are excellent discussions of possibilities for explaining the arrow of time. As well as entropy and cosmology, which we have discussed in this chapter, the other areas of physics where an arrow of time seems to appear are in radiation (waves emerge from sources; they never converge on sinks), and in quantum mechanics (the wavefunction collapses; it never un-collapses). Zeh's book is more technical, and Price's book more philosophical, but they are both very stimulating. Zeh (who sadly died in 2018) had an interesting website with other links which does seem to be still available; also worth looking up on the web is the website of the wonderfully named "Centre For Time" at the University of Sydney.

A popular exposition of the key theories of time is given in Carroll (2011). More recently, Smolin (2014) has argued that all attempts to "explain" time as an illusion in a fundamentally timeless world are fruitless; we should recast physics on the assumption that time is real. My own instinct is that this is indeed a promising approach.

14.8 Exercises

14.1 Nitrogen molecules at standard atmospheric pressure and temperature (i.e. 1 bar and $T = 273.15$K) have a mean free path of 5.9×10^{-8}m. (a) If a box full of Nitrogen molecules are disturbed so that they temporarily have a non-Maxwellian velocity distribution, roughly how long will they take to return to thermal equilibrium? (b) Suppose a box 2 m long has a central divider separating two different isotopes of Nitrogen. If the divider is removed, roughly how long will the two isotopes take to mix?

14.2 What happens to the entropy of the system in the two cases above?

14.3 Suppose the Nitrogen in the problem above is pumped down to a high vacuum, so that the pressure is $10^{-}5$ mbar. According to standard kinetic theory, the mean free path should be inversely proportional to pressure. What happens to the thermalisation timescale and the mixing timescale?

14.4 Consider a box divided into two compartments initially containing N particles each. Once every millisecond a particle is chosen at random from one compartment or the other and transferred to the other compartment. How long will it be before there is an even chance that the population of one compartment differs from its starting point by at least 1%, if $N = 1000$, $N = 10^6$, or $N = 10^9$, or $N = 10^{12}$?

14.5 The random walk sample paths shown in Fig. 6.1 are clearly diverging forward in time. Does this show that probability is intrinsically time asymmetric, and hence a possible origin of the arrow of time?

14.6 As described in the text, although space has no intrinsic directionality, a local "arrow of space" is imposed by the Earth's gravity. Suppose the Earth's gravity were to change—by accretion of matter perhaps. How would surrounding matter know a change had happened? Is there an equivalent process for the proposed cosmological arrow of time?

References

Baierlein, R.: Thermal Physics. Cambridge University Press, Cambridge (2010)
Ball, R., Kolokoltsov, V., MacKay, R. (eds.): Complexity Science: The Warwick Masters Course. Cambridge University Press, Cambridge (2013)
Ben-Naim, A.: Entropy and the Second Law: Interpretation and Misss-interpretationsss. WSPC, Singapore (2012)
Binney, J., Tremaine, S.: Galactic Dynamics, 2nd edn. Princeton University Press, Princeton (2008)
Brown, A.: Statistical Physics. Edinburgh University Press, Edinburgh (1968)
Carroll, S.M.: Spacetime and Geometry: An Introduction to General Relativity. Pearson, London (2013)
Carroll, S.M.: From Eternity to Here: The Quest for the Ultimate Theory of Time, 2nd edn. Oneworld Publications, London (2011)
Coles, P.: From Cosmos to Chaos: The Science of Unpredictability. Oxford University Press, Oxford (2010)
Eddington, A.: The Nature of the Physical World. Cambridge University Press, Cambridge (1928)
Ford, I.: Statistical Physics: An Entropic Approach. Wiley-Blackwell, Hoboken (2013)
Fuller, B., Applewhite, E.J.: Synergetics: Explorations in the Geometry of Thinking. MacMillan Publishing Company, London (1976)
Gibbs, J.W.: Elementary Principles of Statistical Mechanics, developed with special reference to the rational foundation of thermodynamics, Charles Scribner's Sons (1902)
Holland, J.H.: Complexity: A Very Short Introduction. Oxford University Press, Oxford (2014)
Jaynes, E.T.: Information theory and statistical mechanics. Phys. Rev. **106**, 398–620 (1957)
Jaynes, E.T.: Gibbs vs Boltzmann entropies. Am. J. Phys. **33**, 391–398 (1965)
Lynden-Bell, D.: Statistical mechanics of violent relaxation in stellar systems. MNRAS **136**, 101–121 (1967)
Price, H.: Time's Arrow and Archimedes' Point: New Directions for the Physics of Time. Oxford University Press, Oxford (1997)
Prigogine, I.: Non-equilibrium Statistical Mechanics. Wiley, Hoboken (1962)
Prigogine, I., Stengers, I.: Order Out of Chaos: Man's New Dialogue with Nature, re-issue, Flamingo (1993)
Smolin, L.: Time Reborn: From the Crisis in Physics to the Future of the Universe. Mariner Books, Boston (2014)
Zeh, H.D.: The Physical Basis of the Direction of Time, 5th edn. Springer, Berlin (2007)

Websites (all accessed March 2019):

University of Sydney Centre for Time: https://sydney.edu.au/arts/our-research/centres-institutes-and-groups/centre-for-time.html
H. Dieter Zeh's web page: http://www.rzuser.uni-heidelberg.de/~as3

Solutions

Some of the solutions provided are very brief. Some are more explanatory, as the exercise makes a point that expands on the text.

Exercises from Chap. 1: Randomness and Probability

1.1 Drawing Aces. (a) When the card is replaced, the probability is $1/13 \times 1/13 = 1/169$. (b) When not replaced, the probability is $1/13 \times 3/51 = 1/221$.

1.2 Rolling dice. For one die, 3 and 6 are divisible by 3, so the answer is $2/6 = 1/3$. For two dice, we could have $T = 3, 6, 9, 12$ and they have frequencies 2/36, 5/36, 4/36, 1/36. So overall the probability is $12/36 = 1/3$.

1.3 The farmer's will. The fractions add up to 17/18. Although this is a silly example, in more complicated problems it can be easy to miss the fact that your frequencies don't add up to 1.0.

1.4 Three non-exclusive events. We represent the events as C = Cleaning, F = Filling, and E = Extraction. First we add $P(C) + P(E) + P(F)$, then we subtract each of the intersection regions, $P(C \text{ and} F)$ etc. (See Fig. A.1). However we have then subtracted the central region one time too many, and need to put it back. The result is

$$\begin{aligned}&P(C \text{ or } F \text{ or } E) = \\ &P(C) + P(F) + P(E) \\ &- P(C \text{ and } F) \quad - P(C \text{ and } E) \quad - P(F \text{ and } E) \\ &+ P(C \text{ and } F \text{ and } E)\end{aligned}$$

which gives $0.44 + 0.24 + 0.21 - 0.08 - 0.11 - 0.07 + 0.03 = 0.66$.

1.5 Airplane parts. If R is "ready for shipment on time" and D is "ready for shipment **and** delivered on time", then we are told $P(R) = 0.80$ and $P(R \text{ and } D) = P(R, D) = 0.72$, and what we want to know is $P(D|R)$. So we have $P(D|R) =$

A. Lawrence, *Probability in Physics*, Undergraduate Lecture Notes in Physics,
https://doi.org/10.1007/978-3-030-04544-9

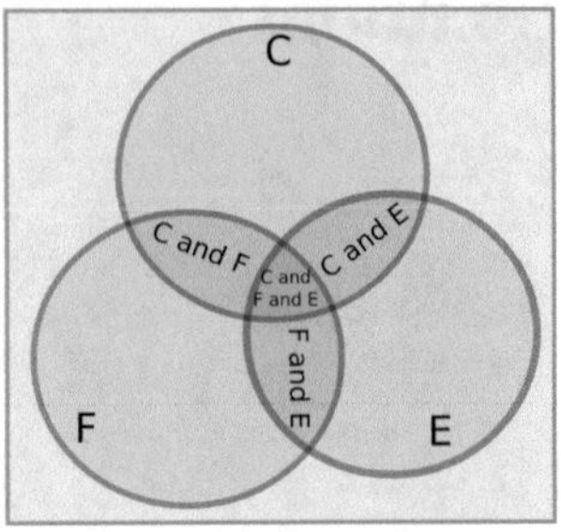

Fig. A.1 Dental Venn diagram

$\frac{P(R,D)}{P(R)} = \frac{0.72}{0.80} = 0.90$ (Note that if we were asked for $P(R|D)$, this would not be possible to calculate from the information we have. We would need to know $P(D)$.)

1.6 Cancer Test. Write C for the event of having cancer, N for not having cancer, and $+/-$ for getting a positive/negative test result. Then we have:

$p(+|C) = 0.9$ i.e. if you do have cancer, there is a 90% chance you will get a positive test result.

$p(-|N) = 0.9$ i.e. if you do not have cancer, there is a 90% chance you will get a negative test result. Note that $p(+|N) = 0.1$.

$p(C) = 0.01$ i.e. there is a 1% overall chance that you have you cancer, regardless of tests. Note that $p(N) = 0.99$.

What we want is $p(C|+)$, the probability that we really do have cancer, given that we have a positive test. Using Baye's theorem,

$$p(C|+) = \frac{p(+|C)p(C)}{p(+)}$$

and we can also see that the overall probability of getting + is

$$p(+) = p(+|C)p(C) + p(+|N)p(N)$$

i.e. we could get a positive test with large probability if we really do have cancer, or with a small probability if we don't have cancer. So we get

$$p(C|+) = \frac{0.9 \times 0.01}{0.9 \times 0.01 + 0.1 \times 0.99} = \frac{0.009}{0.009 + 0.099} = 1/12.$$

So even if you get a positive test, the odds are still reasonably in favour of you being healthy. You can also see this problem graphically, as shown in Fig. A.2. The conditional probabilities are traced as branches, where the number by the branch indicates the probability given the previous branch, and the probabilities of the end nodes are given multiplying the probabilities of the chain of branches. From this you can see that there are two ways of getting +, and adding these, $p(+) = 0.099 + 0.009 = 0.108$. Of the two ways, the route through C has $p = 0.009$, and so $p(C|+) = 0.009/0.108 = 1/12$.

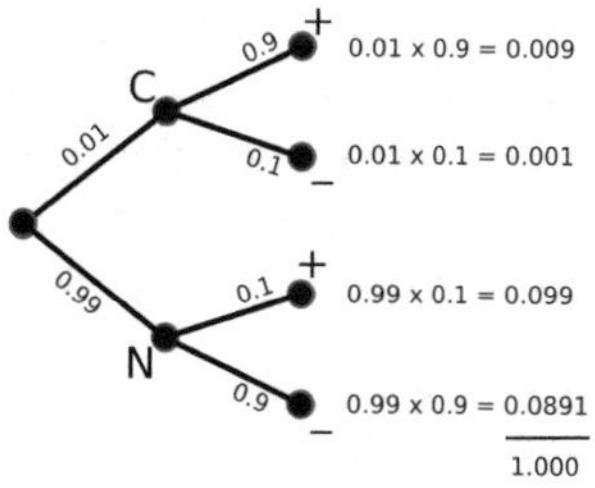

Fig. A.2 Tree diagram for solving the cancer-test problem

1.7 Selecting socks. This is just a question of enumerating the possibilities BB, BG, GB, GG. For example

$$BB \,:\, P = \frac{5}{8} \cdot \frac{4}{7} = \frac{5}{14}$$

BG and GB will give us two different colours. These both have $P = 15/56$, so the probability of two different colours is $15/28$.

1.8 The three doors: do you switch? This is often known as the "Monty Hall Problem", after a real American game show host. You should definitely switch! There are various ways of looking at this problem, but I think the simplest is as follows. If we write C or car and T for toy, then the only possibilities are (a) CTT, (b) TCT, and (c) TTC, and they are equally probable. For (a), you lose by switching; for both (b) and (c) you win by switching. Therefore the chance of winning by switching is 2/3.

1.9 Russian Roulette. Suppose the probability that Player-1 loses is q. This is also the probability that the Player-2 wins. But we can look at the probability that Player-2 wins another way. Either Player-1 loses on the first shot ($p = 1/6$), or Player-1 survives ($p = 5/6$) and Player-2 wins the rest of the sequence. At this point, Player-2 is in the same situation as Player-1 at the start of the game, so the probability of losing is q and of winning is $p = 1 - q$. So finally we have

$$q = \frac{1}{6} + \frac{5}{6}(1 - q)$$

Solving this gives $q = 6/11$. If the gun has n chambers, the solution is $q = n/(2n - 1)$.

Exercises from Chap. 2: Distributions, Moments, and Errors

2.1 Estimating Head size. The marginalised distributions are the very bottom row and very rightmost column in the Table. As we have only a binned version of the dataset, to estimate the mean we take the the number in each bin as representative of the value at the centre of the bin, and e.g. for headlength use

$$(1 \times 16.25 \;+ 2 \times 16.75 \;+ 14 \times 17.25 \;+ \;\cdots\; 3 \times 21.25)/3000. = 19.22$$

To estimate the median, add the numbers in successive bins up to the sixth bin, which gives 1033. To get halfway, we need 1500; so we have another 467 to go. The next bin has a value 1026, so we need to go 46% of the way through a bin of width 0.5; so finally our estimate of the median is 19.23, pretty close to the mean. The mode is also quite close; the highest bin is 19.0–19.5, and the bins either side are fairly similar, so the peak is about halfway through. The same exercise with head breadth gives mean, median and mode of 15.10, 15.09, and about 15.17.

2.2 Integrating PDF. First we find the value of k from the requirement that the integral of $f(x)$ is 1.0, which gives $k = 3$. Then the probability between 0.5 and 1.0 is

$$\int_{0.5}^{1.0} 3e^{-3x}\mathrm{d}x = \left[-e^{-3x}\right]_{0.5}^{1.0} = -e^{-3} + e^{-1.5} = 0.173$$

2.3 Marginal PDFs. To find the marginal density for x, we integrate over all values of y. So

$$g(x) = \int_{-\infty}^{\infty} f(x, y)\mathrm{d}y = \int_0^1 \frac{2}{3}(x + 2y)\mathrm{d}y = \frac{2}{3}(x + 1)$$

and likewise

$$h(y) = \int_{-\infty}^{\infty} f(x, y)\mathrm{d}x = \int_0^1 \frac{2}{3}(x + 2y)\mathrm{d}x = \frac{1}{3}(1 + 4y)$$

2.4 Dice function expectation. We have

$$E[g(x)] = \sum_{x=1}^{6}(2x^2 + 1)p(x)$$

but the probability $p(x) = 1/6$ for all values of x. The expectation value is therefore

$$\frac{1}{6}\left[(2.1^2 + 1) + \cdots (2.6^2 + 1)\right] = \frac{94}{3}$$

2.5 Handy variance formula. The variance is defined as the expected value of the square of the deviation from the mean. The mean is just $E(x)$, so $\sigma^2 = E[(x - E(x))^2]$. So then

$$\begin{aligned}
\sigma^2 &= E[(x - E(x))^2] \\
&= E[x^2 - 2xE(x) + E(x)^2] \\
&= E(x^2) - E(2xE(x)) + E(x)^2] \\
&= E(x^2) - 2E(x)E(x) + E(x)^2] \\
&= E(x^2) - 2E(x)^2 + E(x)^2] \\
&= E(x^2) - E(x)^2
\end{aligned}$$

which is the result required.

2.6 Moment about arbitrary point. The n_{th} moment with respect to $x = x_a$ is

$$\mu_n^a = \frac{1}{N}\sum_{i=1}^{N}(x_i - x_a)^n$$

For $n = 0$, each sum in the term is just 1, so $\mu_0^a = 1$. For $n = 1$ we get

$$\mu_1^a = \frac{1}{N}\sum x_i - \frac{1}{N}\sum x_a = \bar{x} - \frac{1}{N}\cdot Nx_a = \bar{x} - x_a$$

For $n = 2$ we have

$$\begin{aligned}\mu_2^a &= \frac{1}{N}\sum (x_i = x_a)^2 \\ &= \frac{1}{N}\sum x_i^2 - \frac{1}{N}\sum 2x_i x_a + \frac{1}{N}x_a^2 \\ &= \frac{1}{N}\sum x_i^2 - 2x_a\bar{x} + x_a^2\end{aligned}$$

but $\sum x_i^2/N = s^2 + \bar{x}^2$ (see Problem 2.5) and so we get

$$\mu_2^a = s^2 + \bar{x}^2 - 2x_a\bar{x} + x_a^2 = s^2 + (\bar{x} - x_a)^2$$

2.7 Average tosses before heads. To get heads on exactly the nth toss, we need a sequence of $n - 1$ tails and 1 head, but each has $P = 1/2$, so the probability distribution for n is $P(n) + 1/2^n$. The expected value of n is therefore

$$E[n] = \sum_{i=1}^{\infty}\frac{n}{2^n} = 2.0$$

To get a six on exactly the nth roll, we have $p = 6\ n - 1$ times, and $p = 1/6$ once, and the expected value of n is

$$E[n] = \sum_{i=1}^{\infty} n\cdot\frac{5^{n-1}}{6^n} = 6.0$$

For dungeons and dragons fans, you can likewise show that the the average number of rolls before getting the maximum value on a d-sided die is just d.

2.8 Errors on measurement comparisons. The difference is $\Delta m = m_B - m_A =$ 234 eV. For $ax + by$ we have $\sigma_f^2 = a^2\sigma_x^2 + b^2\sigma_y^2$ with $a = 1$ and $b = -1$, which gives $\sigma = 195.3$. The difference is therefore similar to the error, and our instinct would be that the evidence for the offset is not strong. To make this more rigorous, we need the techniques of Part III of the book, on hypothesis testing. For the ratio

$f = x/y$ we have

$$\left(\sigma_f/f\right)^2 = (\sigma_x/x)^2 + \left(\sigma_y/y\right)^2$$

So $f = 1560/1326 = 1.176$ and $\sigma_f = 1.176 \times \left((150/1326)^2 + (125/1560)^2\right)^{1/2} = 0.133$. So the ratio differs from 1.0 by an amount similiar to its error; this is also not strong evidence.

2.9 Magnitude errors. The relevant formula is

$$f = a \ln(\pm bx) \qquad \Longrightarrow \qquad \sigma_f = a(\sigma_x/x)$$

Note that the formula applies to log to the base e, whereas the magnitude formula uses log to the base 10. So we can rewrite the magnitude formula as

$$m = k \times \frac{\ln(F/F_0)}{\ln 10}$$

so we can use the standard formula with $x = F$ and $a = k/\ln 10$. We therefore get

$$\sigma_m = \frac{k}{\ln 10} \cdot \frac{\sigma_F}{F}$$

To get σ_m equal to σ_F/F we therefore need $k = \ln 10 = 2.303$. (Or of course –2.303). With the standard $k = -2.5$ we get within 10% of this handy feature.

Exercises from Chap. 3: Counting the Ways: Arrangements and Subsets

3.1 Dinner party arrangements. In the text we saw that n distinguishable objects have $n!$ arrangements. However, if we arrange distinguishable objects in a cycle, there is an extra subtlety. You can rotate the objects around the cycle and it looks the same. In other words, abc is different from acb but the same as cba. One way to look at this is that you take away the first choice. You could put the first person anywhere round the circular table. After that, you have $n - 1$ choices for the neighbouring seat, $n - 2$ for the next, and so on. The number of distinct ways is therefore $N = (n - 1)!$ which in this case is $N = 4! = 24$.

3.2 Anagrams of book. First let us imagine the two o's to be distinct—labelling them o_1 and o_2. The number of permutations is $n! = 4!$. Now if we remove the subscripts, for each one of those permutations, the o_1 and the o_2 can be swapped—ko_1bo_2 is the same as ko_2bo_1. So the number of distinguishable arrangements is 4!/2! = 12. (Note that this includes the option "book". One might argue that this shouldn't count, in which case there are 11 anagrams.) More generally for n items of which m are identical, the answer is $N = n!/m!$. If the designer had been using "books', then $n = 5$ and $m = 2$, and so the number of distinguishable anagrams is $5!/2! = 60$.

3.3 Departmental committee. Consider picking the men and the women separately. The men can be picked $\binom{11}{3}$ ways = 11!/8!3! = 11.10.9/3.2 = 165. The women can be picked $\binom{5}{3}$ ways = 5!/2!3! = 5.4/2 = 10. For each male-pick we can match any of the female-picks. So the total number is 165.10 = 1650.

3.4 Picking teams. This question is equivalent to partioning into three groups—two groups of 5 and one group of 2. The answer depends on whether we care which is Team A and which is Team B. Lets assume that we do. So we have three buckets which we can label A, B, and L for the leftovers. The number of ways to partition n things into k buckets is

$$N = \frac{n!}{n_1!n_2!...n_k!} = \binom{n}{n_1, n_2, ...n_k}$$

So in this case $N = 12!/5!5!2! = 16,632$. Now suppose we don't care which team is which. Note that we will still be able to distinguish the left-over group. So we can swap which team is A and which team is B and this reduces the number of distinguishable partitions by $2! = 2$, so $N = 8,316$.

3.5 Approximating partitions. The exact answer is of course $N = 16!/11! \times 5!$ The exact calculation gives $N = 4368$. If we use the better version of Stirling's approximation, $n! \sim \sqrt{2\pi n}(n/e)^n$ we get $N = 4451.8$, which is off by 1.9%. If we use the more approximate version of Stirling's approximation, which is the one physicists most often use, we have $\ln n! \sim n \ln n - n$ or $n! = \exp(n \ln n - n)$. This gives $N = 20,690$—way off! We should only use the second version when n is very large.

3.6 Dungeons and Dragons. This is much the same as the the molecules-all-at-one-end problem. On each roll, the probability of all $n - 1 = 5$ players getting the maximum is $p = (1/20)^5$. Of course this *might* occur anytime, but typically we would expect it to take 20^5 rolls times 10 min, which gives almost 61 years. Some games of Dungeons and Dragons certainly feel that long.

Exercises from Chap. 4: Counting Statistics: Binomial and Poisson Distributions

4.1 Recovery from cancer. The probability follows the binomial distribution

$$f_{np}(x) = \binom{n}{x} p^x (1-p)^{n-x} \text{ with } p = 0.8, n = 10 \text{ and } x = 7 \rightarrow f =\sim 0.20$$

For $x = 3$, $\binom{10}{3}$ is the same as $\binom{10}{7}$ and we get $f = 0.00079$. For $x \leq 7$ we want $1 - (f(8) + f(9) + f(10)$ which is

$$\binom{10}{8}0.8^8 0.2^2 + \binom{10}{9}0.8^9 0.2^1 + \binom{10}{10}0.8^{10} 0.2^0$$

which is $45 \cdot 0.0067 + 10 \cdot 0.027 + 1 \cdot 0.107 = 0.68$ and so $P(\leq 7) = 0.32$.

4.2 Die coefficient of variation. The coefficient of variation α is the standard deviation divided by the mean. For binomial $\mu = np$ and $\sigma^2 = np(1-p)$ so

$$\alpha^2 = \frac{np(1-p)}{(np)^2} = \frac{1-p}{np}$$

which gives us

$$n = \frac{1-p}{p} \cdot \frac{1}{\alpha^2}$$

In this case the probability of success per trial is $p = 1/6$ and we require $\alpha < 0.1$. Putting these numbers in, we require $n > 500$.

4.3 Passing driving test. This is an example of the geometric distribution, or binomial waiting time distribution for the case of success on the xth trial. From Sect. 4.4 the probability is $g(x) = p(1-p)^{x-1}$ where p is the probability of success for each trial and x is the number of the trial on which we get success. With $p = 0.75$ and $x = 4$ we get $g = 0.75(0.25)^3 = 0.0117$.

The geometric formula is actually fairly obvious from first principles. To get a hit on the fourth go, we need to fail 1, and fail 2, and fail 3, and get a hit on 4, and multiply each of those probabilities together. The implied assumption is that the probability of passing on each attempt is independent. This is unlikely to be true. A learner will have a better chance of passing the second time and so on (assuming that they learn from experience!)

4.4 Lorry inspection. This is an example of binomial sampling but without replacement, with the probability given by the hypergeometric distribution. Initially, the probability of picking a defective lorry is 4/24, but the probability on the next pick depends on whether the first pick found a duff lorry or not. In Sect. 4.4 we stated (without proof) that the probability of r successes out of n picks is

$$P(r) = \frac{\binom{W}{r}\binom{N-W}{n-r}}{\binom{N}{n}}$$

n is the number of picks; $n = 6$ in this case
r is the number of successes (finding a defective lorry in this case)
N is the total pool of things to pick from ($N = 24$ lorries in this case)
W is the starting pool of possible successful things
($W = 4$ defective lorries in this case).
So what we want is

$$P(0) = \frac{\binom{4}{0}\binom{20}{6}}{\binom{24}{6}} = 1 \times \frac{20!}{14!6!} \times \frac{18!6!}{24!} = \frac{18.17.16.15}{24.23.22.21} = 0.288$$

4.5 Fabric defects. The number of defects for a given length should follow a Poisson distribution. At any spot, the chance of there being a defect is very small, but are many possible such spots where a defect might occur. If the mean rate of defects is 0.2/m, then on a 30 m length, the expected number of defects is 6. So the probability of r defects is

$$p(r) = \frac{\mu^r}{r!} e^{-\mu}$$

with $\mu = 6$. To get the probability of four or more defects, we just do $P(> 4) = 1 - [p(0) + p(1) + p(2) + p(3)]$. This gives us

$$P(> 4) = 1 \quad e^{-6} \times \left[1 + 6 + 6^2/2! + 6^3/3!\right] = 0.8488$$

Rejecting a long length with four or more defects is therefore probably overly strict.

4.6 Proving Poisson mean. The population mean is the expectation value of x, $E(x) = \sum x f(x)$ so for Poisson we have

$$\begin{aligned} E(x) &= \sum_{x=0}^{\infty} x \cdot \frac{\mu^x}{x!} e^{-\mu} \\ &= \sum_{x=0}^{\infty} \frac{\mu^x}{(x-1)!} e^{-\mu} \\ &= \mu e^{-\mu} \sum_{x=0}^{\infty} \frac{\mu^{x-1}}{(x-1)!} \\ &= \mu e^{-\mu} \sum_{y=0}^{\infty} \frac{\mu^y}{y!} \end{aligned}$$

where in the last step we have changed from x to $y = x - 1$, and assume that $y = -1$ is undefined and can be ignored. However recall that

$$e^x = 1 + x + \frac{x^2}{2!} + \frac{x^3}{3} + \cdots$$

which is the same as our series in y. So we have

$$E(x) = \mu e^{-\mu} e^{\mu} = \mu$$

4.7 Tree covering factor. The key to this question is working out the mean number of trees in the line of sight, μ, and taking this to be Poisson distributed. Then for the bird to be *not* hidden, in its direction there has to be zero trees in the line of sight, which has probability $P(0) = e^{-\mu}$. The fraction of sight lines where the bird is hidden is therefore $1 - P(0)$.

The second key insight is understanding that the trees can overlap in the line of sight and that this is what gives us the mean number of trees in the line of sight. Consider a section of the clump with horizontal width W. The depth of the clump is D. The area of ground concerned is therefore $A = DW$. If the surface density of trees is σ, then the number of trees in our section is $N = A\sigma = DW\sigma$.

Each tree has width d. The summed width of trees is therefore $T = Nd = DW\sigma d$. If we compare this to the actual width W, then the mean number of trees in the line of sight is $\mu = T/W = DW\sigma d/W = D\sigma d$.

We are given $D = 57$ m, $\sigma = 0.033$ m^{-2}, and $d = 0.7$ m, so $\mu = 1.32$. Then the covered fraction is

$$C = 1 - e^{-\mu} = 0.73$$

There is a 73% chance of the bird being hidden. One might ask whether the assumption of the Poisson distribution is justified. The answer is reasonably so. If you any individual tree in the whole clump, the probability of it being in our sight line is small. However there is a large number of trees that could potentially be in our sight line. The mean number in the line of sight is a middling number.

Exercises from Chap. 5: Combining Many Factors: The Gaussian Distribution

5.1 Equating Gaussians. This is simply a question of finding the value of x where the two functions are equal. Both functions have $\sigma = 1$ so we just equate the exponents, i.e. set $(x - \mu_A)^2 = (x - \mu_B)^2$. We have $\mu_a = 0$ and $\mu_B = 3$ and so we want the solution of $x^2 = (x - 3)^2$. This has three solutions: $x = +\infty$, $x = -\infty$, and $x = 3/2$. So the answer is that p_B exceeds p_A when $x > 3/2$.

5.2 Width of Gaussian. It is simplest to work in terms of the standard form of the Gaussian, $f(z) = \frac{1}{\sqrt{2\pi}} e^{-z^2/2}$. The maximum, at $z = 0$, is $f_{\max} = 1/\sqrt{2\pi}$. In terms of the height scaled to the maximum, $r = f/f_{\max}$ we then have

$$r(z) = e^{-z^2/2} \quad \text{and so} \quad z^2 = -2\ln(r)$$

So for 50% of maximum, $r = 1/2$ and so $z = \sqrt{-2\ln(1/2)} = 1.1774$. The full width is at $\pm z$, so we get FWHM $= 2.355$. For 20% of maximum, $W_{20} = 2 \times \sqrt{-2\ln(1/5)} = 3.588$.

5.3 Gaussian gradient. Using the standard form $f(z)$, the slope is $\mathrm{d}f/\mathrm{d}z$, and we want to find z where $\mathrm{d}^2 f/\mathrm{d}z^2 = 0$.

$$\begin{aligned}
f &= \frac{1}{\sqrt{2\pi}} e^{-z^2/2} \\
\frac{\mathrm{d}f}{\mathrm{d}z} &= \frac{1}{\sqrt{2\pi}} \cdot -z e^{-z^2/2} \\
\frac{\mathrm{d}^2 f}{\mathrm{d}z^2} &= \frac{1}{\sqrt{2\pi}} \left(z e^{-z^2/2} . \frac{-2z}{2} + 1.e^{-z^2/2} \right) \\
&= \frac{1}{\sqrt{2\pi}} .e^{-z^2/2}(1 - z^2) \qquad\qquad = 0 \implies z = 1
\end{aligned}$$

So $z = 1$, i.e. at 1σ, is where the slope is steepest. To find the intercept, note that at $z = 1, \mathrm{d}f/\mathrm{d}z = -(1/\sqrt{2\pi}).e^{-1/2}$, and the height of the curve is the same except for the minus sign $f = (1/\sqrt{2\pi}).e^{-1/2}$. If we write the equation of the tangent line as $f = az + b$ where $a = \mathrm{d}f/\mathrm{d}z$ is the slope, then at $z = 1$ we have $f = -a$ so that the equation becomes $-a = a + b$ which gives us $b = -2a$. Then if we ask where is $f = 0$ we get $0 = az - 2a$ from which we find $z = 2$. In other words, the tangent at the 1σ point intercepts the x-axis at the 2σ point.

This can be quite handy, as a rough approximation for the central part of the Gaussian is a triangle which passes through the centre and the $\pm 2\sigma$ points.

5.4 Approximating Binomial with Gaussian. The exact answer comes from the binomial distribution for $n = 16$, $p = 1/2$, and $x = 6$. So we get

$$f_{n,p}(x) = \binom{16}{6} \cdot \left(\frac{1}{2}\right)^6 \left(1 - \frac{1}{2}\right)^{16-6} = \frac{8{,}008}{65{,}536} \sim 0.122$$

where right at the end we have approximated the exact fraction to three decimal places. The mean of the binomial is $\mu = np = 8$ and the standard deviation is $\sigma = (np(1-p))^{1/2} = 2$. To look at the Gaussian approximation, we use the same values of μ and σ in the Gaussian formula. If we are looking to approximate $x = 6$ we can take any values greater than 5.5 and less than 6.5. From the normal distribution spreadsheet, or by any other tabulation, $P(<5.5) = 0.10565$ and $P(<6.5) = 0.22663$ so $P \approx 6 = 0.121$ to three decimal places. This differs from the exact answer by only 0.001—so for $n = 16$ the binomial is already very close to a Gaussian.

5.5 Integrated Binomial approximated with Gaussian. An exact answer sums the binomial probabilities for 70, 71, 72 and so on. It is easier to use the Gaussian approximation, and find the integrated probability above 69.5. The mean for the binomial is $\mu = np$ and here we have $p = 0.75$ and $n = 100$ and so $\mu = 75$. The variance is $\sigma^2 = np(1-p) \to \sigma = 4.33$. The value $x = 69.5$ is therefore at $z = (x - \mu)/\sigma = (69.5 - 75)/4.33 = -1.27$. Gaussian tables or routines show $P(z > 1.27) = 0.102$ so here we want $1 - P = 0.898$; there is roughly a 90% chance that at least 70 mosquitos will be killed.

5.6 2D Gaussian profile. Integrating the radial profile of (5.4) we have

$$\begin{aligned} P(<R) &= \int_{r=0}^{r=R} \frac{r}{\sigma^2} \exp\left[\frac{-r^2}{2\sigma^2}\right] \mathrm{d}r \\ &= \left[-e^{-r^2/2\sigma^2}\right]_{r=0}^{r=R} \\ &= 1 - e^{-R^2/2\sigma^2} \end{aligned}$$

and so

$$\frac{R}{\sigma} = \sqrt{-2\ln(1-P)}$$

For $P = 0.68$ the 1D Gaussian will give this in the range $\pm 1\sigma$.
For the 2D Gaussian, we get $P = 0.68$ for $R < 1.51\sigma$.
For $P = 0.90$ the 1D Gaussian will give this in the range $\pm 1.64\sigma$.
For the 2D Gaussian, we get $P = 0.90$ for $R < 2.15\sigma$.
For $P = 0.95$ the 1D Gaussian will give this in the range $\pm 1.96\sigma$.
For the 2D Gaussian, we get $P = 0.90$ for $R < 2.45\sigma$.

5.7 Toothpaste stock. To solve this question, we use the Gaussian approximation to the Poisson distribution. For a mean of $N = 150$, the standard deviation is $\sigma = \sqrt{N} = 12.25$. From Gaussian tables, to get $P(>z) = 0.05$ we need $z = (x - \mu)/\sigma = 1.64$. Note that we want the one-tailed integral—we are not interested in cases where the chemist sells less than average. So we want the value $x = \mu + z\sigma$, where we use $z = 1.64$, $\mu = N = 150$, and $\sigma = \sqrt{N}$. This gives $x = 170.1$. So the chemist should keep at least 170 tubes in stock each week.

5.8 Star counting accuracy. Again, this question is about using the Gaussian approximation to the Poisson distribution. First we ask, what z range contains 95% of the probability? In this case we want the two tailed integral, i.e. the integral within $\pm z$—we don't want our star-count to be either too low or too high. Gaussian tables show us this requires $|z| < 1.96$ Next, if the count in our sample is N, the standard deviation will be $\sigma = \sqrt{N}$, but in this case we want 1.96σ. On each side of N we want that 1.96σ range to be equivalent to $0.1N$. So we want $1.96\sqrt{N} = 0.1N$ and so $\sqrt{N} = 19.6$. So to get 10% accuracy with 95% confidence we want $N > 384$.

5.9 Gaussian M.A.D. Writing $y = x - \mu$, the mean of the absolute deviation $|y|$ is given by its expectation value when subject to a Gaussian PDF:

$$E[\,|y|\,] = \int_{-\infty}^{+\infty} |y|\; p(y) \mathrm{d}y \qquad \text{with} \qquad p(y) = \frac{1}{\sigma\sqrt{2\pi}} e^{-y^2/2\sigma^2}$$

However, $p(y)$ is an even function, so

$$\begin{aligned} E[\,|y|\,] &= 2\textstyle\int_0^{+\infty} y\; p(y) \mathrm{d}y \\ &= 2\tfrac{1}{\sigma\sqrt{2\pi}} \textstyle\int_0^{\infty} y e^{-y^2/2\sigma^2} \mathrm{d}y \\ &= \tfrac{1}{\sigma}\cdot\sqrt{\tfrac{2}{\pi}} \left[-\sigma^2 e^{-y^2/2\sigma^2}\right]_0^{\infty} \\ &= \sigma\sqrt{\tfrac{2}{\pi}}\; (0 - (-1)) \\ &= \sigma\sqrt{\tfrac{2}{\pi}} \sim 0.8\sigma \end{aligned}$$

5.10 Maxwell–Boltzmann peaks. (a) We take $p(v)$ from the equation in Sect. 5.9.1 and differentiate. This gives the maximum at position $v = \sqrt{kT/m}$. (b) The energy

E is $mv^2/2$, and so the most probable v corresponds to energy $E = kT/2$. (c) To find the most probable energy E we need to convert to the probability density per unit E, using $f(E)\mathrm{d}E = p(v)\mathrm{d}v$. We then find that $f(E) \propto E^{3/2} \exp(-E/kT)$. Differentiating this, we find the maximum at $E = 3kT/2$. The most probable energy does not correspond to the energy of the most probable velocity.

Exercises from Chap. 6: Distributions Arising from Processes: Random Walks, Shot Noise, Lorentzians, and Power-Laws

6.1 Perfume spreading time. The molecules follow a three-dimensional random walk. Using the peak of $p(R)$ as given in Chap. 6, after time t we have $l = \sqrt{2v\lambda.t}$, and so the time taken for distance l is $l^2/2v\lambda$. Here $l = 1$ m, the molecular speed is $v = 490$ ms^{-1}, and he mean free path is $\lambda = 1.3 \times 10^{-7}$ m. This gives $t = 7849$ s, or 2.2 h.

6.2 Perfume spreading-2. Substituting $\sigma^2 = na^2$ into (6.2) gives

$$p(R) = \frac{R^2}{\sigma^3}\sqrt{\frac{2}{\pi}}e^{-R^2/2\sigma^2}$$

which is the same as (5.5). If we now further substitute $R = z\sigma$ we get

$$p(z) = \frac{z^2\sigma^2}{\sigma^3}\sqrt{\frac{2}{\pi}}e^{-z^2\sigma^2/2\sigma^2} \propto z^2 e^{-z^2/2}$$

as requested. Differentiating, we get

$$\frac{\mathrm{d}p}{\mathrm{d}z} = 2ze^{-z^2/2} - z^3 e^{-z^2/2} = 0$$

One solution is at z = 0 but thats a minimum; the maximum is where $2z = z^3$ and so $z = \sqrt{2}$. The density per unit volume, as opposed to per unit radius, goes as $\rho(z) \propto e^{-z^2/2}$. Then the density per unit volume at $z = 5$ is down from $z = 0$ by factor $e^{-5^2/2} = 3.7 \times 10^{-6}$. Even a very small volume of a substance contains huge numbers of molecules, but the human nose can detect a handful of molecules, so it is quite likely that it can detect a density of molecules a million times lower than the peak density. We can therefore detect the spreading vapour a long time before the radial density peak reaches us.

6.3 Mean of radial density. The radial density can be written

$$p(R) = Ae^{-R^2/2\sigma^2} \text{ where } A = \frac{4\pi R^2}{(2\pi\sigma^2)^{3/2}}$$

which is then the same as $A \times I_3$ with $h = 1/2\sigma^2$ and $\sigma^2 = na^2$. Substitution and re-arrangement then gives the result, $R = a\sqrt{8n/\pi}$.

Note: The odd-numbered error integrals I_n can be recursively solved by differentiating inside the integral. So

$$I_1 = \int_0^\infty xe^{-hx^2}\mathrm{d}x = \frac{1}{2h}$$

by straightfoward integration, and then

$$I_3 = \int_0^\infty x^3e^{-hx^2}\mathrm{d}x = -\frac{\partial}{\partial h}\int_0^\infty xe^-hx^2\mathrm{d}x$$
$$= -\frac{\partial}{\partial h}\left(\frac{1}{2h}\right) = \frac{1}{2h^2}$$

Likewise, it is easy to find that $I_5 = 1/h^3$.

6.4 Double Events. For events occuring at rate λ, the probability of an event in time Δt is $p_1 = \lambda\Delta t^2$. The probability of two events is $p_2 = p_1^2$. So we want $p_1/p_2 = 1/p_1 \geq 100$ which for $\lambda = 10$ gives $\Delta t \leq 1$ ms.

6.5 Thorium decay. The distribution of waiting times should follow $f(t) = \lambda e^{-\lambda t}$ where λ is the rate of events. We can estimate this because the total number of events is 766 in 2496 s, so $\lambda = 0.3069\ \mathrm{s}^{-1}$. We can then calculate the expected number of events in each of the bins, watching out for the fact that they are not the same size, and taking the position of each bin as halfway. This gives the following, with predicted numbers rounded to nearest integer. The match is pretty good. To decide whether it's *good enough* requires the techniques of Chap. 7.

Time interval (s)	Observed frequency	Predicted frequency
0 - 1/2	101	109
1/2 - 1	98	93
1 - 2	159	149
2 - 3	114	110
3 - 4	74	80
4 - 5	48	59
5 - 7	75	76
7 - 10	59	54
10 - 15	32	28
15 - 20	4	6
20 - 30	2	2
Over 30	0	0
Total	766	766

6.6 Lorentzian versus Gaussian width. From Exercise 5.2, for the Gaussian $W_{20}/W_{50} = 3.588/2.355 = 1.52$. For the Lorentzian, we use the standard form with $\mu = 0$ and $\Gamma = 1$ which is the Cauchy distribution $f(x) = 1/(\pi(1+x^2))$. The maximum, at $x = 0$, is $f_{\max}=1/\pi$. At $f_{\max}/N$ we have $f = 1/N\pi$. Solving for x we have

$$x^2 = \frac{1}{\pi f} - 1 = \frac{1}{\pi} \cdot N\pi - 1 = N - 1$$

So for $N = 2$ we get $W_{50} = 2 \times \sqrt{2-1} = 2$, and for $N = 5$ we get $W_{20} = 2 \times \sqrt{5-1} = 4$. The ratio $W_{20}/W_{50} = 2$, considerably bigger than for the Gaussian. If we tried 10, 5% etc, the difference would get even larger.

6.7 Underestimating wings. First, assuming a Gaussian FWHM $= 0.86\,\text{GeV}$ implies $\sigma = 0.365\,\text{GeV}$. An event seen at $+1.16\,\text{GeV}$ is therefore at a deviation of $z = +3.18\sigma$. Looking up standard tables, $P(> z = 0.00074)$ i.e. a chance of 1 in 1351. If the distribution is Lorentzian, there is no σ, but in terms of the FWHM Γ we are at $x - \mu = a\Gamma$ with $a = 1.16/0.86 = 1.349$. To estimate the probability $P(>a)$ we can calculate the probability density at $x - \mu = a$ and then extrapolate as x^{-2}. In terms of Γ and a the Lorentzian pdf is

$$f(a) = \frac{1}{2\pi\Gamma} \cdot \frac{1}{a^2 + \frac{1}{4}}$$

With $\Gamma = 0.86$ and $a = 1.349$ we get $f = 0.089$. Approximating $f = ka^{-2}$ then putting in these values of f and a we get $k = 0.16$. The integrated probability is

$$P(>a) = \int_a^\infty ka^{-2} = \left[ka^{-1}\right]_a^\infty = \frac{k}{a}$$

and so finally we get $P(>a) \sim 0.12$. This is enormously different from the Gaussian estimate, so the scientist could be making a big mistake claiming a discovery!

6.8 Wealth distribution. The integrated number of people above some value w is

$$N(w) = \int_w^\infty n(w)\mathrm{d}w = \left[\frac{Ax^{1-\alpha}}{\alpha - 1}\right]_w^\infty = \frac{Aw^{1-\alpha}}{\alpha - 1}$$

The total number is $N_{\text{tot}} = N(w_{\text{min}})$, and above the dividing line $W_{1/2}$, we have $N_{1/2} = N_{\text{tot}}/2 = N(w_{1/2})$. Making the subsitutions and re-arranging gives the result $w_{1/2} = w_{\text{min}} 2^{1/(\alpha-1)}$. On the other hand, the total wealth above w is

$$W(w) = \int_w^\infty wn(w)\mathrm{d}w = \frac{Ax^{2-\alpha}}{\alpha - 2}$$

which of course only converges if $\alpha > 2$. Inserting $w = w_{\text{min}}$ and $w = w_{1/2}$ the fraction of the wealth above $w_{1/2}$ is found to be

$$F = 2^{-(\alpha-2)/(\alpha-1)}$$

For $\alpha = 2.1$ this gives $F = 0.94$. Note that the top-heaviness of the wealth distribution is very sensitive to the value of α near the critical value of 2.0.

Exercises from Chap. 7: Hypothesis Testing

7.1 Yotta mass and one-tailed versus two-tailed tests. This is a good example of why you need a clear alternative hypothesis, as well as a null hypothesis. In both cases, the null hypothesis is that all four points are drawn from the same Gaussian with $\mu = 1375$ and $\sigma = 65$. However, the alternatives are quite different.

(a) If we restricting our attention to a single point, and furthermore, expecting it to be high, then we can use the one-tailed Gaussian z-test. A point at 1512 compared to mean 1375 and error 65 is at a deviation of $z = 2.11$. From Gaussian tables, we can see that $P(>z) = 0.017$. So we would indeed reject the null hypothesis at 5% significance.

(b) If we are prepared to be interested in any discrepant point, there are two subtleties. The first is that although we spotted a high point, we would have been equally impressed if we had seen a low point. The probability of seeing *either* a high *or* a low point is twice as big, $P = 0.034$. The second subtlety is that we would have been impressed by seeing any of the four points being discrepant. So just like with the jelly bean question, we have to ask "whats the probability of seeing at least one discrepant point out of four?". With $p = 0.034$ and $n = 4$ we have $P(\geq 1) = 1 - (1 - p)^4 = 0.129$. So there is roughly a 13% chance of getting at least one dodgy-looking point, which would not cause us to reject the null hypothesis.

7.2 Which urn was fetched? We can use F = Fred's hypothesis (10 red 20 black), and J = Jane's hypothesis (15 red 15 black). We start with priors $\pi_F = \pi_J = 1/2$.

The data is $D = 3$ reds out of 30. For Fred's hypothesis the likelihood of D is $L_F = (10/30)^3 = 1/27$; on the other hand $L_J = (15/30)^3 = 1/8$. (Note we know that there 30 balls in total; also note that the balls are replaced.)

The marginalised likelhood is $E = L_F \cdot \pi_F + L_J \cdot \pi_J = \frac{35}{27\times 16}$. Then for the posteriors we get

$$P_F = \pi_F \cdot \frac{L_F}{E} = \frac{1}{2} \cdot \frac{1}{27} \cdot \frac{27 \times 16}{35} = \frac{8}{35}$$

$$P_J = \pi_J \cdot \frac{L_J}{E} = \frac{1}{2} \cdot \frac{1}{8} \cdot \frac{27 \times 16}{35} = \frac{27}{35}$$

So Jane has becomes more than three times as likely to be right. Note that the credibilities add up to 1.0. For the second experiment, we can use the posteriors after the first experiment as our new priors: $\pi_F = 8/35$, $\pi_J = 27/35$. The second experiment gets two blacks, so we get likelihoods $L_F = (2/3)^2 = 4/9$ and $L_J = (1/2)^2 = 1/4$. Working through the same calculation we end up with $P_F = 0.345$, $P_J = 0.655$. The odds have improved in Fred's favour, but Jane is still more likely to be right. You can imagine that as we add more and more data, this oscillation would stop, and it would gradually become more and more clear who was right.

7.3 Too many points above the line? The null hypothesis is that the fitted line correctly describes the data points. In the absence of more complete knowledge, we should consider that each point has an equal chance of being above or below the line.

So its like tossing a coin six times. Being above the line is "success" with $p = 1/2$. With $n = 6$ trials, the binomial distribution gives us

$$f(5) = \frac{6}{2^6} \quad \text{and} \quad f(6) = \frac{1}{2^6}$$

so the probability of getting at least five points above the line is $7/2^6 = 0.109$, which we would not reject at 95% confidence, or even at 90% confidence. If the student is being even more careful, they might have asked whether they would have been equally impressed with five points out of six below the line; then the chance is twice as large.

7.4 Is the mosquito spray faulty? The calculation is intrinsically a binomial one, but the normal approximation is easier to calculate. We have $n = 100$, $p = 0.75$, so in the exact binomial version $\mu = np = 75$ and the standard deviation is $\sigma = (np(1-p))^{1/2} = \sqrt{100 \cdot 0.75 \cdot 0.25} = 4.33$. So the deviation is $z = (65 - 75)/4.33 = -2.31$. From our usual spreadsheet or tables, doing a one-tailed test, we find that $P(z < -2.31 = 0.0104)$. Probably here a two-tailed test is appropriate. We would therefore reject the null hypothesis at 95% confidence but not at 99% confidence. If we had $x = 60$ on the other hand, we have $z = (60 - 75)/4.33 = -3.46$, which has $P(z < -3.46 = 0.0003)$, which is a highly significant deviation. This shows how sensitive the normal deviation test is around the 2–3σ range.

Now suppose instead of 65/100, we had found the same fraction in a larger sample, 650/1000. Now the mean is $\mu = 750$ and standard deviation is $\sigma = (np(1 - p))^{1/2} = \sqrt{1000 \cdot 0.75 \cdot 0.25} = 13.69$. Our result of 650 then represents a deviation $z = (650 - 750)/13.69 = -7.30$, which is an extremely unlikely deviation. So sample size matters a lot.

7.5 How long do we need to run the Thorium experiment? We can take the standard theory as the null hypothesis. It predicts rate λ_0 so after time t will accumulate $N_0 = \lambda_0 t$ counts. The Poisson error on this is $\sigma = N_0^{1/2}$, and we will assume that we get enough counts that we can treat this as a Gaussian error. (We will justify this when we see the answer). If the alternative hypothesis is rate λ_1 predicting counts $N_1 = \lambda_1 t$, then what we are really asking is "assuming mean N_0 and $\sigma = N_0^{1/2}$, what is the probability of getting at least N_1?". For that probability to be 5% we need 1.65σ. So we require

$$N_1 - N_0 = 1.65 N_0^{1/2} \quad \text{or} \quad \lambda_1 t - \lambda_0 t = 1.65(\lambda_0 t)^{1/2}$$

from which we find

$$t = \frac{2.72\lambda_0}{(\lambda_1 - \lambda_0)^2}$$

With $\lambda_0 = 0.35$ and $\lambda_1 = 0.37$ this gives $t = 2380$ s. (This gives $N_0 = 833$ counts, so the Gaussian approximation is fine.)

7.6 What does a small χ^2 mean? The most common way this can happen is if the experimentalists have *overestimated* their error bars for some reason. Remembering that $\chi^2 = \sum (x_i - \mu)^2/\sigma_i^2$ you can see how this happens.

7.7 Likelihood versus χ^2 For a single data point x the likelihood is

$$L_i(\mu, \sigma) = \frac{1}{\sigma\sqrt{2\pi}} \exp\left[-\frac{1}{2}\left(\frac{x_i - \mu}{\sigma}\right)^2\right]$$

The joint likelihood of the N points is given by multiplying the probabilities:

$$L(\mu) = \left(\frac{1}{\sqrt{2\pi}}\right)^N \prod_i \left(\frac{1}{\sigma_i}\right) \cdot e^{-\frac{1}{2}\sum\left(\frac{x_i-\mu}{\sigma_i}\right)^2}$$

Looking at the three terms, the first two terms do not involve μ and so can be treated as constant. For the third term, the expression is almost the same as χ^2:

$$\chi^2 = \sum_i (x_i - \mu)^2/\sigma_i^2$$

and so you can see that as required $\ln L(\mu) = \text{const} - \chi^2/2$. Quite often in the literature you find the "likelihood function" defined as the joint likelihood without all the constant terms, so that you may see simply $\ln L = -\chi^2/2$.

7.8 Testing ball-drop time. The sample mean is $\bar{x} = \sum x_i/N$. We don't have the individual x_i values, but we can take the number of objects in each bin n_j and use the centre of each bin $T_j = 0.595$ etc, and then we get

$$T_{\text{obs}} = \bar{T} = \frac{1}{N}\sum_{j=1}^{10} T_j n_j = (1/50) \times 2 \times 0.595 + 2 \times 0.605... = 0.6334$$

The sample standard deviation is given by

$$\sigma_{\text{obs}} = s = \frac{1}{N^{1/2}}\left[\sum n_j (T_j - \bar{T})^2\right]^{1/2} = 0.0197$$

From simple physics, the time to fall height h from rest with acceleration g is $T_f = (2h/g)^{1/2} = 0.6386$ s. Now we test the Gaussian curve against the data histogram. Given $\mu = T_f$ and $\sigma = \sigma_{\text{obs}}$ the probability density of a given T value is given by $f(T)$ where f is the usual Gaussian function. As T, T_f and σ are all in units of seconds, the probability density is per second. The width of each bin is 0.01 s, and the total number of objects is 50, so the predicted number in a given bin is $n_j = 50 \times 0.01 \times f(T)$. We can take the square root of this as a Poisson error on that bin, and will assume it is a Gaussian error. (This may be the weakest assumption). Finally, we can treat the value in each bin as a data point, and so get χ^2 by accumulating values of the squares of the deviations of each bin from its predicted, divided by its individual variance. This is built up as in the table below.

T_j	n_j	$n_{\rm pred}$	σ_j	$(n_p - n_j)^2/\sigma_j^2$
0.595	2	0.88	0.94	1.43
0.605	2	2.37	1.54	0.06
0.615	11	4.94	2.22	7.42
0.625	6	7.97	2.82	0.49
0.635	12	9.94	3.15	0.43
0.645	8	9.59	3.10	0.26
0.655	4	7.16	2.68	1.39
0.665	3	4.13	2.03	0.31
0.675	1	1.84	1.36	0.39
0.685	1	0.64	0.80	0.21
χ^2				**12.56**

Because we have estimated σ from the data, the degrees of freedom is $\nu = N - 1 = 9$. Looking up these values on the spreadsheet $P(\chi^2 > 12.56) = 0.184$ so in fact the expected Gaussian is a perfectly good fit. Note however that we have use the $\sqrt(N)$ approximation for some rather small numbers, so this result may not be very accurate.

Exercises from Chap. 8: Parameter Estimation

8.1 Mean brightness of star. The straight sample mean is $\bar{x} = \sum x_i/N = 243$. The weighted mean is $\sum(x_i/\sigma_i^2)/\sum(1/\sigma_i^2)$ which gives 224.03. The error on the mean is $1/\sum(1/\sigma_i^2) = 4.94$.

To get 95% confidence, with a two-tailed test, we want the value of z which gives $P(>z) = 0.025$, which is $z = 1.96$. The 95% limits are then $224.03 \pm 1.96 \times 4.94$ which gives 214.35–233.71. This range does not include the simple unweighted mean. The unweighted mean is poor because one of the points has a very large error; the weighted mean plays down that point.

The high point is many multiples of σ_μ from the weighted mean. However, in terms of its own error, it is only 1.7σ from the mean. It is probably not that this point is "dodgy"—it just happens to have a large error. So sigma-clipping is not justified. If we did remove the high point, the straight mean would be 221, quite consistent with the weighted mean. However this is not surprising, because the weighted mean is dominated by the other two points.

8.2 Biased *versus* unbiased standard deviation. The ratio of the two s.d. estimates is

$$r = \frac{\sigma_{un}}{s} = \sqrt{\frac{N}{N-1}}$$

Solving for N we have $N = \frac{r^2-1}{r^2}$. So for $r = 1.05$ we get $N = 10.75$ and therefore we need at least 11 data points.

8.3 Confidence regions for Yotta mass. The sample mean is $\bar{m} = \sum m_i/N = 86.34$. The sample variance is $s^2 = \sum(m_i - \bar{x})^2/N = 5.41$ and the error on the mean is $\sigma_\mu = s/\sqrt{N} = 2.42$.

For a z-test, we would compare the predicted mean m_s with the observed mean, and need to assume a value for σ. If as the question suggests, we just use the sample variance, we then have $z = (\bar{m} - m_s)/\sigma_\mu = (86.34 - 91.93)/2.42 = -2.31$. A two-tailed test is appropriate here. Looking up $P(< -z, > z)$ in a Gaussian table we get $P = 2.1\%$, a very significant result.

However, because we have estimated both mean and variance from the same data, we should use the t test. The value of t is the same as the z we just calculated, i.e. $t = 2.31$, but now we need to look up in a t-table, with degrees of freedom is $\nu = N - 1 = 4$. Looking up $P(< -t, > t)$ we get $P = 8.0\%$, a very different result. So for small samples, using t as opposed to z really matters.

To get 95% confidence, we want $P(>t) = 0.025$ for a two tailed test. Playing with t tables with $\nu = 4$ we find we need $t = 2.78$; so with $\sigma_\mu = 2.42$ we want $\bar{m} \pm 2.78 \times 2.42$ i.e. 79.61–93.07.

8.4 ***t versus*** **Cauchy**. Putting $\nu = 1$ into the equation from Sect. 8.3.4. we have

$$f_1(t) = \frac{\Gamma(1)}{\pi^{1/2}\Gamma(1/2)}\left[1 + t^2\right]^{-1}$$

However, $\Gamma(1) = 0! = 1$ and (not so obvious, but true) $\Gamma(1/2) = \pi^{1/2}$. So we get

$$f_1(t) = \frac{1}{\pi(1 + t^2)}$$

which is the Cauchy distribution, as required.

8.5 Maximum likelihood estimate of decay rate. The exponential distribution is $f_\lambda(t) = \lambda e^{-\lambda t}$. The joint likelihood of N points t_i is therefore

$$L = \prod_{i=1}^{N} \lambda e^{-\lambda t_i} = \lambda^N e^{-\sum \lambda t_i}$$

and so $\log L = N \log \lambda - \lambda \sum t_i$. To get the maximum,

$$\frac{\partial \log L}{\partial \lambda} = 0 = \frac{N}{\lambda} - \sum t_i$$

and so finally we get the maximum likelihood estimate of λ as

$$\lambda_{ML} = \frac{N}{\sum t_i}$$

8.6 Weighted *versus* unweighted mean. If all $\sigma_i = \sigma$, the numerator in the weighted mean formula gives $\sum x_i/\sigma_i^2 = \frac{1}{\sigma^2}\sum x_i$, and the denominator gives $\sum 1/\sigma_i^2 = N/\sigma^2$. We then get $\hat{\mu} = \sum x_i/N$ which is the standard $\bar{x}$ formula.

8.7 Estimating coin bias. (a) The natural estimate for p is just $r/n = 0.75$. If the value of r is drawn at random from a binomial distribution which does indeed have probability p per success, then the dispersion is $\sigma_r = \sqrt{np(1-p)} = \sqrt{4 \times 0.75 \times 0.25} = 0.866$. Because $p = r/n$, following the standard propagation of error formulae, we get $\sigma_p = \sigma_r/n = 0.217$.

(b) The Likelihood for seeing $r = 3$ given $n = 4$ and hypothesised probability p is just given by the binomial distribution. Because we are assuming a uniform prior, the posterior is proportional to the Likelihood, so

$$P(p) = \text{const} \times \frac{n!}{r!(n-r)!} p^r (1-p)^{n-r} = \text{const} \times 4 \times p^3(1-p)$$

The maximum posterior/likelihood is then where $\mathrm{d}p/\mathrm{d}p = 0$ which gives $3p^2 = 4p^3$ and so $p = 3/4$, exactly in agreement with the natural estimate of part (a).

The maximum posterior is where $p = 3/4$. Ignoring the constants we get $P_{\text{max}} = p^3 - p^4 = 0.75^3.0.25 = 0.1054$. The half-max points are then at the solution of $P_{\text{max}}/2 = p^3 - p^4 = 0.0527$. Trial and error on a calculator pretty soon gives $p = 0.460$ and $p = 0.936$ giving FWHM $= 0.476$. If we assume a Gaussian approximation, we would have FWHM $= 2.355\sigma$ so we get $\sigma = 0.202$. This is close to the part (a) estimate, but definitely different. This seems reasonable, because with $n = 4$ the approximation of the binomial to Gaussian won't be that good. Indeed, you can see that the two values of p we found for $P_{\text{max}}/2$ are not symmetrical about the maximum at $p = 0.75$.

(c) Because we are just looking for the maximum, we can once again ignore the constants, and have

$$P(p) \propto \pi(p).L(p) \propto (p - p^2)(p^3 - p^4) = p^6 - 2p^5 + p^4$$

The maximum is where $\mathrm{d}p/\mathrm{d}p = 0$ and so $6p^5 - 10p^4 + 4p^3 = 0$
but assuming $p^3 \neq 0$ we have $6p^2 - 10p + 4 = 0$
which factorises to $2(3p - 2)(p - 1) = 0$
and so $p = 2/3$ or $p = 1$. The latter is in fact a minimum of the function. The solution for maximum posterior is $p = 2/3$. This is clearly different from the uniform prior estimate, but well within the 1σ error for that estimate.

(d) For a uniform prior, the maximum posterior estimate will once again be the same as the natural estimate, so $p = 65/100$. The likelihood $L(r|p)$ is binomial with $\mu = np = 0.65$ and dispersion $\sigma_r = \sqrt{np(1-p)} = \sqrt{100 \times 0.65 \times 0.35} = 4.770$. We then have $\sigma_p = \sigma_r/n = 0.0477$. For $n = 100$ the distribution will be a very good approximation to a Gaussian, and for a uniform prior this will also be the shape of the posterior distribution. Assuming then a Gaussian distribution, 95% overall credibiliy will fall within $z = \pm 1.96\sigma$, which gives the range $0.557 < p < 0.743$. Note that this does not include $p = 0.5$—getting 65 heads out of a 100 is pretty strong evidence of bias.

8.8 Three point Gaussian solution. Following the notes, we use the formulae

$$Q_0 = Q_3 - \Delta Q \left[\frac{y_3 - y_2}{y_1 - 2y_2 + y3} + \frac{1}{2} \right] \qquad \sigma_Q = \Delta Q \sqrt{(2y_2 - y1 - y3)^{-1}}$$

with $Q_1 = 16.0$, $Q_2 = 17.0$, $Q3 = 18.0$ and $y = \log P$ with $P_1 = 0.0372$, $P_2 = 0.0489$, $P_3 = 0.0416$ and $\Delta Q = 1.0$. Plugging these numbers in we get $Q_0 = 17.128$, $\sigma_Q = 1.516$.
Another method to calculate the s.d. is using the curvature formula.

$$\sigma_Q^2 = -\left(\frac{\mathrm{d}^2 P}{\mathrm{d}Q^2} \right)^{-1}$$

Calculating the differences,

$$\Delta_{12} = y_2 - y_1 = 0.2735 \qquad \Delta_{12} = y_3 - y_2 = -0.1616 \qquad \Delta(\Delta) = -0.4351$$

which gives $\sigma_Q = 1.5159$, exactly as before.

8.9 Relation between posterior and χ^2. Prior, posterior and likelihood are linked by Bayes's formula:

$$P(\mu) = \pi(\mu) \cdot \frac{L(D|\mu)}{E}$$

where E is the marginalised likelihood. Each data point is drawn from the same Gaussian, so the joint likelihood of the dataset is

$$L(\mu) = \left(\frac{1}{\sigma\sqrt{2\pi}} \right)^N \exp\left[-\sum \frac{1}{2} \left(\frac{x_i - \mu}{\sigma} \right)^2 \right]$$

Taking (natural) logs we get

$$\log L = -N \log(\sigma\sqrt{2\pi}) - \frac{1}{2} \sum \left(\frac{x_i - \mu}{\sigma} \right)^2 = \text{const} - \frac{1}{2}\chi^2(\mu)$$

Taking logs of the Bayes formula, we have $\log P(\mu) = \log \pi(\mu) + \log L(\mu) - \log E$ and we note that E is a constant, i.e. not a function of μ. Combining these results, we get

$$\log P(\mu) = \text{const} + \log \pi(\mu) - \frac{1}{2}\chi^2(\mu)$$

as required. For the minimum of χ^2 to be in roughly the same place as the maximum of P, we need $\pi(\mu)$ to change more slowly than $\chi^2(\mu)$, i.e. so to a good approximation it can be seen as a uniform prior.

8.10 Gaussian approximation to χ^2. The χ^2 distribution has mean ν and variance 2ν, where $\nu = N - m$ is the number of degrees of freedom. Those formulae for mean and variance are always correct, but for large enough ν we can treat the variance as corresponding to a Gaussian variance. For a Gaussian, 95% confidence is at 1.65σ—note that a one-tailed test is appropriate. The corresponding approximate value of χ^2 is therefore

$$\chi^2 = \nu + 1.65 \times \sqrt{2} \times \nu^{1/2} = (N - m) + 2.33(N - m)^{1/2}$$

To see how accurate this is, we can use a χ^2 table to compare to the correct value which gives $P(> \chi^2) = 0.05$.

$N = 3$, $\nu = 2$ True $\chi^2 = 5.99$; Gaussian approximation $\chi^2 = 5.30$, which gives real $P = 7.1\%$

$N = 10$, $\nu = 9$ True $\chi^2 = 16.92$; Gaussian approximation $\chi^2 = 16.00$, which gives real $P = 6.7\%$

$N = 50$, $\nu = 49$ True $\chi^2 = 66.34$; Gaussian approximation $\chi^2 = 65.33$, which gives real $P = 6.0\%$.

Exercises from Chap. 9: Inference with Two Variables: Correlation Testing and Line Fitting

9.1 Perfect correlation. For perfect correlation $y_i = a + bx_i$ for all values of i. Then

$$\bar{y} = \frac{\sum y_i}{N} = \frac{\sum a + bx_i}{N} = a + \frac{\sum bx_i}{N} = a + b\bar{x}$$

which means that $y_i - \bar{y} = a + bx_i - a - b\bar{x} = b(x_i - \bar{x})$

Taking the definition of r from Sect. 9.4.1, the numerator then gives

$$\sum (x_i - \bar{x})(y_i - \bar{y}) = b \sum (x_i - \bar{x})^2$$

and the denominator gives

$$\sqrt{\sum (x_i - \bar{x})^2 . b^2 \sum (x_i - \bar{x})^2} = b \sum (x_i - \bar{x})^2$$

This is the same as the numerator, so $r = 1$.

9.2 Maths/Physics/Art correlations. Using the formula from Sect. 9, we want to calculate the deviation from the mean for each data point for each of Maths (M), Physics (P) and Art (A). First we get the means: $\bar{M} = 39.1$, $\bar{P} = 30.0$, $\bar{A} = 35.5$. Then writing dM for $M_i - \bar{M}$, to test Maths versus Physics we accumulate sums as follows:

Sum of d$M \times$ d$p = 327.0$
Sum of d$M^2 = 496.9$
Sum of d$p^2 = 466.0$
Putting these into the formula we get $r = 0.68$ and $t = 2.78$. For 9 degrees of freedom this gives $P < 1.1\%$ (one tailed test). So there is pretty good evidence of correlation.

For Maths versus Art we get
Sum of d$M \times$ d$A = -99.55$
Sum of d$M^2 = 496.9$
Sum of d$A^2 = 582.7$
This gives $r = -0.19$ and $t = -0.6$, which gives $P < 28\%$ (one tailed test). So there is no sign of any correlation.

9.3 Regressing two different ways. Given that we don't have errors on the individual marks, we just use the least-squares formulae, equations (9.8) and (9.9). If we take x as the Maths mark and y as the Physics mark, and accumulate the necessary sums, we get $b = 0.658$ and $a = 4.276$. However if we take x as the Physics mark and y as the Maths mark, we get $b = 0.702$ and $a = 18.039$. These two lines are plotted on the figure along with the data. Note that above we have re-defined x as the independent variable and y as the dependent variable in each case. So the solutions we have are

$$P = a + bM \quad \text{where} \quad a = 4.276\ b = 0.658$$

$$M = c + \mathrm{d}p \quad \text{where} \quad c = 18.039\ d = 0.702$$

If we want to compare the P versus M slope in the two cases then we have (Fig. A.3)

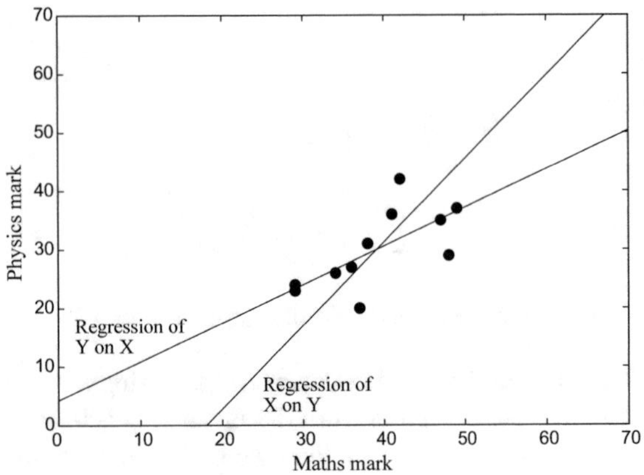

Fig. A.3 Plot of Maths mark versus Physics mark, and the two regression lines

$$P = a' + b'M \qquad \text{where} \qquad a' = -\frac{c}{d} = -26.08 \qquad \text{and} \qquad b' = \frac{1}{d} = 1.425$$

The slopes are therefore very definitely not the same. But is the difference significant?

9.4 Error on regression slopes. With all $\sigma_i = \sigma$ (9.11) becomes $\sigma_b^2 = N/(\Delta\sigma^2)$. We also have $\Delta = S_2S_5 - S_3^2$, and following the definitions and likewise putting in $\sigma_i = \sigma$ we end up with

$$\sigma_b = \sigma.\sqrt{\frac{N}{N\sum x_i^2 - (\sum x_i)^2}}$$

Here x = maths score, $N = 11$, $\sum x_i = 439.0$, and $\sum x_i^2 = 17306$. This gives $\sigma_b = 0.224$. The error on the other slope is similar. The difference between the Physics-on-Maths and Maths-on-Physics slopes is therefore significant at a few sigma.

9.5 Regression slopes and correlation coefficient. We could fit either $y = a + bx$ or $x = e + fy$ and get solutions

$$b = \frac{\sum \mathrm{d}x\mathrm{d}y}{\sum \mathrm{d}x^2} \qquad\qquad f = \frac{\sum \mathrm{d}x\mathrm{d}y}{\sum \mathrm{d}y^2}$$

where $\mathrm{d}x$ is shorthand for $x_i - \bar{x}$ etc. If we invert this latter solution to $y = a' + b'x$ we have $b' = 1/f$. Now we can see that

$$\frac{b}{b'} = bf = \frac{\left(\sum \mathrm{d}x\mathrm{d}y\right)^2}{\sum \mathrm{d}x^2 \sum \mathrm{d}y^2}$$

However, the correlation coefficient is

$$r = \frac{\sum \mathrm{d}x\mathrm{d}y}{\sqrt{\sum \mathrm{d}x^2 \sum \mathrm{d}y^2}}$$

so we see that $r = \sqrt{b/b'}$. This makes good qualitative sense. If we have a perfect correlation, with a thin strip of points, the two lines will be the same, so $b = b'$ and $r = 1$. On the other hand completely uncorrelated points would be a circular blob of points, and the two lines would be at right angles, so that $r = 0$.

9.6 Correlation *versus* catchment. The two correlations have $r_1 = 0.67$ and $r_2 = 0.43$. We transform to Fisher's $z = \frac{1}{2}\ln[(1+r)/(1-r)]$, giving $z_1 = 0.8107$, $z_2 = 0.4497$ and so a difference $\Delta z = 0.361$. If the correlations are the same, Δz should be consistent with zero. What is the variance of our Δz estimate? We have $\sigma_z^2 = 1/(N-3)$ with $N = 75$ and $N = 63$ respectively. and so $\sigma_{\Delta z} = \sqrt{\sigma_{z_1}^2 + \sigma_{z_2}^2} = 0.1749$.

So our $\Delta z = 0.361$ is 2.064 standard deviations away from the mean, which is just significant at 95% confidence.

9.7 Spearman rank correlation test. First, we need to convert the scores into ranks. Below, tied ranks have been allocated at random. Some textbooks suggest giving all tied ranks the same, possibly with half-rank values so that the sum stays the same. Either procedure is ok. We then collect values of $X_i - Y_i$ in another column.

Student	Maths mark	X_i =maths rank	Art mark	Y_i =art rank	$X_i - Y_i$
A	41	5	38	4	1
B	37	7	44	2	5
C	38	6	35	6	0
D	29	10	49	1	9
E	49	1	35	7	–6
F	47	3	29	10	–7
G	42	4	42	3	1
H	34	9	36	5	4
I	36	8	32	8	0
J	48	2	29	9	–7
K	29	11	22	11	0

Using (9.5) we get $r_s = -0.54$, i.e. there is if anything an anti-correlation. Then transforming with (9.6) we get $t_s = -1.92$ for $\nu = N - 2 = 9$. Looking up in t-tables, we should use a two-tailed test—we didn't know whether to expect a correlation or anti-correlation and could be interested in either. This gives $P(> t) = 0.087$, a very interesting but not definitive result.

Exercises from Chap. 10**: Model Fitting**

10.1 Errors from covariance matrix. The off-diagonal elements are zero, so the parameters are not correlated. The top-left element of the matrix is σ_p^2; so the conditional error is $\sigma_p = 3.11$. To get 90% we need to go 1.64σ; so $p = 19.7 \pm 5.1 = 14.6 - 24.8$.

For the joint error, with two parameters, we need to go to $k\sigma$ where $k = 1.52$. For 90% we need $k \times 1.64 \times \sigma_p$, so $p = 19.7 \pm 7.8 = 11.9 - 27.5$. Note σ_q is irrelevant because the parameters are not correlated.

10.2 Estimating joint error. The parameters are clearly correlated, so we need the bounding box for the contour. This is sketched in Fig. A.4. The annotation gives an estimate for best fit $a = 464$, range ± 192, but we are in 2D so we need $k = 1.52$, so we have ± 292. Finally we have $a = 464 \pm 292 = 172 - 756$.

10.3 Error on four parameter fit. It is easiest to answer this the other way round. For $m = 4$ parameters, $\Delta\chi^2$ will be distributed as χ_4^2, i.e. χ^2 with $\nu = 4$. Looking up in a χ^2 table, we find $\Delta\chi^2 = 4.72$. The multiple of σ is given by $k = \sqrt{\Delta\chi^2} = 2.17$.

Fig. A.4 One sigma contour with bounding box sketched on the figure

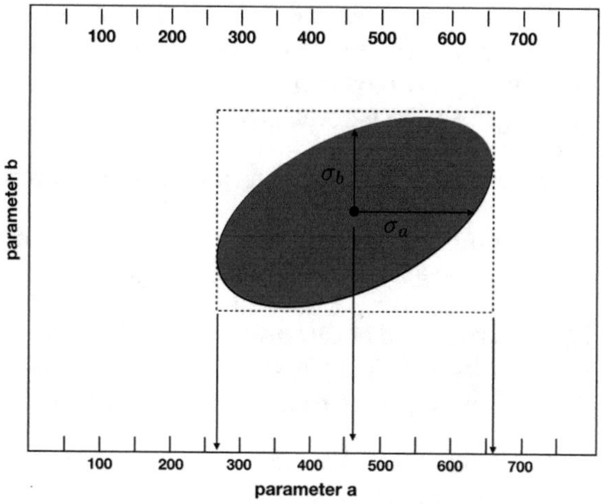

10.4 χ^2 for maths-physics correlation. If M is the maths mark, then we predict the physics mark as $P_{\text{pred}} = a + bM$ with $a = 4.276$ and $b = 0.658$. Then for each student we do the deviation $P - P_{\text{pred}}$, and do the sum of squares divided by σ^2. Using the table of numbers in Chap. 9, for $\sigma = 5$ this gives $\chi^2 = 10.03$. For $N = 11$ and two parameters fitted, this is $\nu = 9$ which gives $P(> \chi^2) = 0.35$ which is a perfectly acceptable fit.

However for $\sigma = 3.5$ $\chi^2 = 20.47$ which gives $P(> \chi^2) = 0.015$ which would be rejected at better than 95% confidence, and almost at 99% confidence.

The general lesson is that one must get the errors right—as well as testing whether we have the right shape, the χ^2 test is very sensitive to the assumed error size.

10.5 Extra parameter test. Model-1 with $\chi^2 = 11.6$ and $\nu = 9 - 2 = 7$ degrees of freedom, gives $P(>\chi^2) = 0.12$. This looks rather poor, but you wouldn't reject it. Model-2, with $\chi^2 = 6.2$ and $\nu = 9 - 3 = 6$ degrees of freedom, gives $P(>\chi^2) = 0.40$, a good fit. We can test the significance of the improvement can be assessed with the $\Delta\chi^2$ test.

We have $\Delta\chi^2 = 11.6 - 6.2 = 5.4$. This should be distributed like χ^2 with $7 - 6 = 1$ degrees of freedom. We then find $P(>\Delta\chi^2) = 0.02$, a very significant result, arguing that the second model is indeed better.

10.6 Model comparison with F-test. With $\chi_1^2 = 11.6$, $\nu_1 = 7$, $\chi_2^2 = 6.2$, $\nu_2 = 6$, we get $F = 1.604$. Looking up in a standard F-table, $P(>F = 1.60) = 0.29$. So it seems the second model is better, but the difference is not very significant.

There is no reason the two methods should give the same answer—they are mathematically different tests, asking different questions. In this case the $\Delta\chi^2$ test is much more sensitive. It can be tempting to try different tests until you get a significant result. but then you are falling into the multiple tests trap. Your result is not as significant as you think.

10.7 Many-parameter fit. This follows the method described in Sect. 10.5.6. For 15 parameters, $\Delta\chi^2$ will be distributed like χ^2 with $\nu = 15$. Looking up a χ^2 tables, to get $P(>\chi^2) = 0.05$ needs $\chi^2 = 25.0$. If $k = r/\sigma$ then $k = \sqrt{\Delta\chi^2} = 5.0$. So we have to go to 5σ to get 95% confidence.

Exercises from Chap. 11: Information, Uncertainty, and Surprise

11.1 Units of uncertainty. Uncertainty in Shannons or bits is $h = -\log_2(p)$. Here $p = 1/6$ so $h = 2.85$. In bans, $h = -\log_{10}(p) = 0.301$.

11.2 Playing card uncertainty. (a) $h(\text{spade}) = -\log_2(1/4) = 2.00$; (b) $h(\text{Ace}) = -\log_2(1/13) = 3.70$ (c) $h(\text{Ace of spades}) = -\log_2(1/52) = 5.70$. We can see that $h(\text{Ace of spades}) = h(\text{spade}) + h(\text{Ace}) = 2.00 + 3.70 = 5.70$.

11.3 Biased coin. For the two-box distribution, $H = p\log_2(p) + (1-p)\log_2(1-p)$; so with $p = 0.6$ we get $H = 0.97$, c.f. 1.0 for a balanced coin. Solving for the p needed for a given H can be done by numerical experiment, which quite easily shows that we need $p = 0.89$ to give $H = 0.5$.

11.4 Maximum H for binomial and Gaussian. The maximum Shannon entropy for the binomial distribution is given by $H_{\max} = \frac{1}{2}\log_2\left(\frac{n\pi e}{2}\right)$. We can re-express in terms of variance $\sigma^2 = np(1-p)$, but note that the maximum occurs when for $p = 1/2$, so we use $\sigma^2 = n/4$. This gives us

$$H_{\max} = \frac{1}{2}\log\left(\frac{4\sigma^2\pi e}{2}\right) \qquad = \log(2\pi\sigma^2 e)$$

which is exactly the same as the Gaussian result in the text.

11.5 Information in quantum numbers for new particle. First, we add the nine numbers in each grid (76, 59, 110 for A,B,C respectively), then we divide each cell by that total to give probabilities. Next we add $P_{ij}\log_2 P_{ij}$ for each of the 9 cells to give the joint uncertainty: $H_A = 3.0243$, $H_B = 2.7784$, $H_C = 3.0284$ bits. Then we add each column to give the marginal distribution in $X = Q_1$ and each row to give the marginal distribution in $Y = Q_2$. Next, in each case we add the three values of $P_i\log_2 P_i$ to give the marginal uncertainties $H_A(X) = 1.50$, $H_A(Y) = 1.5243$, $H_B(X) = 1.5372$, $H_B(Y) = 1.5372$, $H_C(X) = 1.5848$, $H_C(Y) = 1.5848$. Summing $HX + HY$ in each case and comparing to the total H, we find that the mutual information is $I_A = 0.0$, $I_B = 0.296$, and $I_C = 0.1412$. These calculations then show that model A produces a distribution with independent variables, whereas in models B and C the variables are dependent.

We can assign -1, 0 and $+1$ as numerical values, and then get the means. All the distributions give mean zero. We can then use this to calculate the covariance, and get $\sigma^2_{XY}(A) = 0.0$, $\sigma^2_{XY}(B) = 0.2712$, $\sigma^2_{XY}(C) = 0.0$. So distribution A is independent and uncorrelated; distribution B is dependent and correlated; and distribution C is dependent but not correlated.

11.6 Message coding efficiency. The information per symbol is $H_{\text{sym}} = \log_2 6 = 2.58$ bits. The channel capacity is 4 bits, so the efficiency is 2.58/4.0 = 62.5%. If we take two symbols at a time, there are 36 possibilities so $H_{\text{sym}} = 5.17$. So this can't be done with 4 bits. If we use 8 bit words, the efficiency is 64.6%, no better than before. If however, we take 3 symbols at a time, there 216 options, so $H_{\text{sym}} = 7.75$, and using 8 bits words is pretty good.

Exercises from Chap. 12: Erratic Time Series

12.1 Undershoot of variance within single path. The "undershoot" effect is real, because neighbouring points are correlated, so the variance is smaller. The sequence is made with $\alpha = 0.95$. Based on (12.8), $\tau = 1/(1-\alpha) = 20$, so one would need a lag at least several times this.

12.2 ACF of quasar light curve. Our estimate of the ACF at lag = 1 is just the correlation coefficient between successive points. The values given lead to $r = -0.632$. With $N = 13$ and this corresponds to $t = -2.06$ for $\nu = 11$, which has $P(> t) = 3.2\%$. So at 5% significance, there is a measurable autocorrelation. If you just plot the light curve, you'd be hard put to see this, but if you plot successive points against each other, the anti-correlation is fairly clear.

12.3 Stationarity of MA. This is MA with $c = 0$, $\beta_0 = 1$ and $\beta_1 = \beta_2 = \cdots = 1$. The variance is $\sigma_z^2 \times \sum \beta_i^2$ so with all the terms equal to 1, this is infinite and the series is not stationary—it just keeps growing. But if you do $y_t = x_t - x_{t-1}$ you get $y_t = Z_t + (\alpha - 1)X_{t-1}$ and all the other terms cancel. So this is a $q = 1$ series with $\beta_1 = \alpha - 1$. From the notes $\rho(1) = \beta/1 + \beta^2$ so we get $\rho(1) = \alpha - 1/2 + \alpha^2 - 2\alpha$ as required.

12.4 Exponential approximation for ACF of AR process. The correct version is $\rho(k) = \alpha^k$ and the approximation is $\rho = e^{-k/k_{\text{ch}}}$ with $k_{\text{ch}} = 1/(1-\alpha)$. Putting in $\alpha = 0.8$ for $k = 1, 3, 6$ the difference is 2, 7, 15%. To get the approximation we used $\log(1+x) = x - x^2/2 + \cdots$ with $x = \alpha - 1$, and used only the first term. The ratio of the second term to the first term is $-x/2$, so we want x as small as possible. This needs α close to 1.0. Small α gives a poor approximation, but so does $\alpha > 1$.

12.5 Growth of variance for AR(1) versus α. We use

$$\sigma_{\Delta x}^2(k) = 2\frac{\sigma_z^2}{1-\alpha^2} \cdot \left(1 - e^{-k/\tau}\right) \qquad \text{with} \qquad \tau = 1/(1-\alpha)$$

Then we get
$\alpha = 0.1$: $\sigma_1/\sigma_{k_{\text{ch}}} = 0.97$
$\alpha = 0.3$: $\sigma_1/\sigma_{k_{\text{ch}}} = 0.89$
$\alpha = 0.6$: $\sigma_1/\sigma_{k_{\text{ch}}} = 0.72$
$\alpha = 0.9$: $\sigma_1/\sigma_{k_{\text{ch}}} = 0.39$
$\alpha = 0.99$: $\sigma_1/\sigma_{k_{\text{ch}}} = 0.13$
You can see that as expected, low values of α are quite close to pure noise—the variance hardly decreases—whereas for $\alpha \sim 1$ the small-lag variance is a lot less.

12.6 OU versus AR. On a finite timescale Δt the OU difference equation becomes $\Delta x = -\gamma x \Delta t + Z$ where Z is a Gaussian process with σ_Z determined by the OU parameter σ. The AR1 process is defined so that $x_t = \alpha x_{t-1} + Z_t$ which can be re-arranged as $\Delta x = x_{t-1}(\alpha - 1) + Z_t$. This is equivalent to the OU equation with $\Delta t = 1$ and $\gamma = 1 - \alpha$.

12.7 Weather as a Markov process. If we have S_1 = clear and S_2 = rainy, then we have $A_{12} = 0.2$ and $A_{21} = 0.75$. Then the expected steady state probabilities are $P_1 = 0.79$ and $P_2 = 0.21$. The latter is inconsistent with the observed $P_2 = 0.15$. It could be that the input transition values are a bit off. For example $A_{12} = 0.15$ and $A_{21} = 0.85$ would get agreement. But it could also be that daily weather really isn't a simple Markov process—it does have some memory. However, modelling it as a Markov chain is not too bad.

Exercises from Chap. 13: Probability in Quantum Physics

13.1 Carbon decay rate. We need to look up the mass of a Carbon atom, which is 39.083 a.m.u. Then a 70 kg body will contain 8.07×10^{14} atoms of ^{14}C. With a half life of 5370 years, this gives 3302 events per second, broadly comparable to the rate from ^{40}K. While an organism is alive, and either absorbing carbon from the air or eating plants, it will have a constant rate of events per kg. Once it dies, the amount of ^{14}C will slowly decline, and so the event rate decreases, at a predictable rate.

13.2 Event waiting times. For radioactive events (or any truly random events) the waiting time to the next event does **not** depend on how long the previous wait was. For sporadic events in general, such as bus waiting times, there may be such a dependency, because they may not be truly random.

13.3 Stern–Gerlach state mixtures. From Sect. 13.5.3, $P_{\text{up}} = \frac{1}{2}(1 + \cos\theta)$ and $P_{\text{down}} = \frac{1}{2}(1 + \cos\theta)$. So if $P_{\text{up}} = 2P_{\text{down}}$ we get $\theta = 70.53°$.

13.4 Entanglement distance. The fastest a causal influence could travel is the speed of light c. So if the measurement takes time t, to avoid causal influence, we need the distance between measurements to be $d > ct$. For $t = 1$ ms this is 300 km. At the time of writing, the longest entanglement experiment covered 143 km, but longer ones are planned. Of course, you can also try to make the measurement time faster.

13.5 Testing survey results. Note that because the tables supposedly derive from a single survey in which everybody was asked the same questions, the consistency tests of section 13.7.1 should be obeyed exactly, rather than with in statistical errors. Test-1 is passed, as the sum of cells in each table adds to the same number, 795. Test-2 is also passed—the two different ways of summing the A-values both give 478; likewise the test for B and C give 380 and 434 respectively. However the data fails Test-3 (the original Bell test). The number answering yes on A and no on B is 133, and the number answering yes on B and no on C is 96. These values sum to 229, which according to Test-3 must be greater than the number that said yes to

A and no to C. However, the latter is 235. These tables therefore *cannot* have been drawn from a single $2 \times 2 \times 2$ dataset.

Of course the polling company might have made a simple error, but passing Test-2 does tend to suggest that they very carefully fiddled the numbers to look reasonable—but not carefully enough to fool well-educated statisticians!

13.6 Entanglement-coded messages. In *symmetric key systems* the sender A and receiver B use the same key to encrypt/decrypt a message. The "key" is normally a random number fed into an encryption algorithm, which determines its behaviour. A point of weakness is when A and B exchange the key to be used; if spy C intercepts the key, they will be able to eavesdrop, and A and B may never know this is happening. In *asymmetric key* systems, the eavesdropping problem is solved by using an algorithm that has separate encryption and decryption keys. The encryption key is public, but A keeps the decryption key completely private. However, this method is prone to "reverse engineering" unless the keys are extremely long.

In quantum key distribution A creates entangled pairs, sending one stream to themselves and the other to the distant B. The random sequence of up and down states constitutes the binary random number key; B simply has to invert their sequence to know what A has. But suppose C operates some kind of experiment in the middle, detecting the state of particles that pass by, in some non-destructive way?

The idea is that A and B reserve part of the sequence to test, setting up the same kind of experiment as used in Bell's inequality-style tests, with magnets rotated to three different angles. They choose the angles at random but after the test exchange information on the angles they used. The correlations between the states they found should follow the inequality in equation (13.6). If someone has been eavesdropping, i.e. making any other kind of measurement, this will always destroy these correlations, and equation (13.6) will not be obeyed.

Exercises from Chap. 14: Entropy, Complexity, and the Arrow of Time

14.1 Thermalisation and mixing timescales. The mass of a Nitrogen atom is 14.007 a.m.u, but a molecule is twice that. (a) The r.m.s. velocity is then $v = \sqrt{(3kT/m)} = 493\ \text{ms}^{-1}$. Thermalisation will take place roughly on the timescale it takes any one molecule to collide with another which is $\tau = \lambda/v \sim 10^{-10}$ s. (b) Mixing will happen on the timescale it takes for a typical molecules to diffuse from the middle of one half to the other, i.e. $L = 1.0\,\text{m}$. We can assume the two isotopes have more or less the same mass. Following Chap. 6, the time to do a random walk of net length L will be $t = L^2/2\lambda v = 4.7$ h. Much longer!

14.2 Thermalisation and mixing entropy changes. We have to be quite careful about our definitions! For (a), the thermodynamic entropy is undefined before the relaxation, but the statistical (Shannon) entropy increases; the Maxwell–Boltzmann distribution is that which maximises the multiplicity of the macrostate. For (b), the answer depends on whether the observer is capable of distinguishing the difference between the two isotopes. If they can distinginguish, the locational entropy increases—the initial state is a highly unlikely one. If they cannot distinguish, the locational entropy stays the same. Note however that both observers agree that the second law is obeyed—entropy will not spontaneously decrease.

14.3 Low pressure timescales. The mean free path should become 10^5 times longer, i.e 5.9 cm. The thermalisation timescale gets longer—$\tau \sim 0.1$ ms. The mixing timescale however gets much faster—$t \sim 17$ ms—now quite comparable to thermalisation.

14.4 One percent fluctuations. The population of one compartment undergoes a 1D random walk with step size ± 1. The distribution of populations for many sample paths will be centred on N, but with Gaussian spread $\sigma = \sqrt{n_{\text{steps}}}$. Half the paths will end up outside $\pm 0.67\sigma$. If the final population is $N \pm x$, then we are looking for $x/N = 0.01$. So $n_{\text{steps}} = (0.01N/0.67)^2$. For $N = 1000$, there is a 50% chance of a 1% fluctuation after 223 steps, i.e. 0.2 s. For $N = 10^6$ we need 10^8 steps, which takes 62 h. For $N = 10^9$ we need 7 thousand years. For $N = 10^{12}$ we need 7 billion years, half the age of the Universe. Considering that a gram of Nitrogen contains about 10^{22} molecules, you can see that uneven distributions in practice never happen spontaneously.

14.5 Probability asymmetry? The apparent asymmetry arises because we have created an artificial initial condition—all paths start from the same x value—and have then considered $t = 1, 2, 3...$ with multiplying possibilities. However, we could consider $t = -1$, and ask "what x-values could have led to the observed x value at $t = 0$?", and then for each of those, ask at $t = -2$, "what x-values could have led to to the $t = -1$ values", and so on. This would produce a set of paths diverging backwards in time. The concepts of randomness and probability are therefore time reversible, so this is not the origin of the arrow of time.

14.6 Communicating the arrow of time. Our understanding of gravity is via a field theory. A change in gravitional field is communicated by disturbances in space-time—this is not instantaneous, but travels at the speed of light. The cosmological arrow of time proposes a global entropy gradient, but (so far at least) there is no equivalent field theory, so it is not obvious how changes are communicated and points far apart in space see the same arrow of time. Definitely a challenge for the next generation of theorists!

Index

A. Lawrence, *Probability in Physics*, Undergraduate Lecture Notes in Physics,
https://doi.org/10.1007/978-3-030-04544-9

M

O

P

Printed in the United States
By Bookmasters